Classroom Manual for

Basic Automotive Service and Systems

Second Edition

Classroom Manual for
Basic Automotive Service and Systems
Second Edition

Jay Webster

California State University
Long Beach, California

Clifton E. Owen

Griffin Technical Institute
Griffin, Georgia

Jack Erjavec

Series Advisor

Professor Emeritus, Columbus State Community College
Columbus, Ohio

Africa • Australia • Canada • Denmark • Japan • Mexico • New Zealand • Philippines
Puerto Rico • Singapore • Spain • United Kingdom • United States

Delmar Staff:

Business Unit Director: Alar Elken
Executive Editor: Sandy Clark
Acquisitions Editor: Vernon R. Anthony
Development Editor: Catherine Wein
Editorial Assistant: Bridget Morrison
Executive Marketing Manager: Maura Theriault

Channel Manager: Mona Caron
Executive Production Manager: Mary Ellen Black
Project Editor: Christopher Chien
Production Coordinator: Karen Smith
Art/Design Coordinator: Cheri Plasse
Cover Image by David Kimball

For more information, contact Delmar, 3 Columbia Circle, PO Box 15015, Albany, NY 12212-0515; or find us on the World Wide Web at http://www.delmar.com

Asia
Thomson Learning
60 Albert Street, #15-01
Albert Complex
Singapore 189969
Tel: 65 336 6411
Fax: 65 336 7411

Australia/New Zealand
Nelson/Thomson Learning
102 Dodds Street
South Melbourne, Victoria 3205
Australia
Tel: 61 39 685 4111
Fax: 61 39 685 4199

Canada
Nelson/Thomson Learning
1120 Birchmount Road
Scarborough, Ontario
Canada M1K 5G4
Tel: 416-752-9100
Fax: 416-752-8102

International Headquarters
Thomson Learning
International Division
290 Harbor Drive, 2nd Floor
Stamford, CT 0692-7477
Tel: 203-969-8700
Fax: 203-969-8751

Japan
Thompson Learning
Palaceside Building 5F
1-1-1 Hitotsubashi, Chiyoda-ku
Tokyo 100 0002 Japan
Tel: 813 5218 6544
Fax: 813 5218 6551

Latin America
Thomson Learning
Seneca, 53
Colonia Polanco
11560 Mexico D. F. Mexico
Tel: 525-281-2906
Fax: 525-281-2656

South Africa
Thomson Learning
Zonnebloom Building
Constantine Square
526 Sixteenth Road
P.O. Box 2459
Halfway House, 1685
South Africa
Tel: 27 11 805 4819
Fax: 27 11 805 3648

Spain
Thomson Learning
Calle Magallanes, 25
28015-MADRID
ESPANA
Tel: 34 91 446 33 50
Fax: 34 91 445 62 18

UK/Europe/Middle East
Thomson Learning
Berkshire House
168-173 High Holborn
London
WC1V 7AA United Kingdom
Tel: 44 171 497 1422
Fax: 44 171 497 1426

Thomas Nelson & Sons LTD
Nelson House
Mayfield Road
Walton-on-Thames
KT 12 5PL United Kingdom
Tel: 44 1932 2522111
Fax: 44 1932 246574

Library of Congress Cataloging-in-Publication Data

Webster, Jay.
    Basic automotive service and systems / Jay Webster, Clifton Owen.
    — 2nd ed.
        p.   cm. — (Today's technician)
    Includes index.
    ISBN 0-8273-8544-7 (alk. paper)
    1. Automobiles—Maintenance and repair.   I. Owen, Clifton.
II. Title.   III. Series.
TL152.W38   1999
629.28'72—dc21
                                                                99-35324
                                                                CIP

CONTENTS

PREFACE

Thanks to the support the *Today's Technician* series has received from those who teach automotive technology, Delmar Publishers is able to live up to its promise to provide new editions every three years. We have listened to our critics and our fans and present this new, revised edition. By revising our series every three years, we can and will respond to changes in the industry, changes in the certification process, and to the ever-changing needs of those who teach automotive technology.

The *Today's Technician* series, by Delmar Publishers, features textbooks that cover all mechanical and electrical systems of automobiles and light trucks. Principal titles correspond with the eight major areas of ASE (National Institute for Automotive Service Excellence) certification. Additional titles include remedial skills and theories common to all of the certification areas and advanced or specialized subject areas that reflect the latest technological trends.

Each title is divided into two manuals: a Classroom Manual and a Shop Manual. Dividing the material into two manuals provides the reader with the information needed to begin a successful career as an automotive technician without interrupting the learning process by mixing cognitive and performance-based learning objectives.

Each Classroom Manual contains the principles of operation for each system and subsystem. It also discusses the design variations used by different manufacturers. The Classroom Manual is organized to build upon basic facts and theories. The primary objective of this manual is to allow the reader to gain an understanding of how each system and subsystem operates. This understanding is necessary to diagnose the complex automobile systems.

The understanding acquired by using the Classroom Manual is required for competence in the skill areas covered in the Shop Manual. All of the high priority skills, as identified by ASE, are explained in the Shop Manual. The Shop Manual also includes step-by-step instructions for diagnostic and repair procedures. Photo Sequences are used to illustrate many of the common service procedures. Other common procedures are listed and are accompanied with drawings and photographs that allow the reader to visualize and conceptualize the finest details of the procedure. The Shop Manual also contains the reasons for performing the procedures, as well as when that particular service is appropriate.

The two manuals are designed to be used together and are arranged in corresponding chapters. Not only are the chapters in the manuals linked together, the contents of the chapters are also linked. Both manuals contain clear and thoughtfully selected illustrations. Many of the illustrations are original drawings or photos prepared for inclusion in this series. This means that the art is a vital part of each manual.

The page layout is designed to include information that would otherwise break up the flow of information presented to the reader. The main body of the text includes all of the "need-to-know" information and illustrations. In the side margins are many of the special features of the series. Items such as definition of new terms, common trade jargon, tools list, and cross-referencing are placed in the margin, out of the normal flow of information so as not to interrupt the thought process of the reader.

Jack Erjavec

Highlights of this Edition—Classroom Manual

The Classroom Manual content and organization has been based on ASE testing areas. Other content updates include information on duties and responsibilities in an automotive repair shop, locating and applying for a position with a repair facility, safety inspection, repair orders, the use of computers in the repair shop, and electronic devices. Electronic devices and systems that are expected to be standard equipment in a few years are introduced.

The manual is arranged so that the basic theories and general information chapters are followed with a description and operation of the various systems of a typical car or light truck.

Highlights of this Edition—Shop Manual

The Shop Manual has been updated to correlate with the new content of the Classroom Manual. More information has been added on securing a job, technical and legal certification for the technician, entry level tasks and procedures, inspection and maintenance checks, supplement restraint systems and climate control operation. Some of the tasks described require close supervision to be performed correctly, but the entry-level technician can be expected to assist in the repair or perform the repair on their own within a few months of employment. Those tasks are presented as part of an inspection routine or part of a larger repair.

Job Sheets have been added to the end of each chapter. The Job Sheets provide a format for students to perform some of the tasks covered in the chapter. In addition to walking a student through a procedure, step-by-step, these Job Sheets challenge the student by asking why or how something should be done, thereby making the students think about what they are doing.

Classroom Manual

To stress the importance of safe work habits, the Classroom Manual dedicates one full chapter to safety. Included in this chapter are common safety practices, safety equipment, and safe handling of hazardous materials and wastes. This includes information on MSDS sheets and OSHA regulations. Other features of this manual include:

Cognitive Objectives

These objectives define the contents of the chapter and define what the student should have learned upon completion of the chapter. *Each topic is divided into small units to promote easier understanding and learning.*

Marginal Notes

New terms are pulled out and defined. Common trade jargon also appears in the margin and gives some of the common terms used for components. This allows the reader to speak and understand the language of the trade, especially when conversing with an experienced technician.

A Bit of History

This feature gives the student a sense of the evolution of the automobile. This feature not only contains nice-to-know information, but also should spark some interest in the subject matter.

References to the Shop Manual

Reference to the appropriate page in the Shop Manual is given whenever necessary. Although the chapters of the two manuals are synchronized, material covered in other chapters of the Shop Manual may be fundamental to the topic discussed in the Classroom Manual.

Cautions and Warnings

Throughout the text, cautions are given to alert the reader to potentially hazardous materials or unsafe conditions. Warnings are also given to advise the student of things that can go wrong if instructions are not followed or if a nonacceptable part or tool is used.

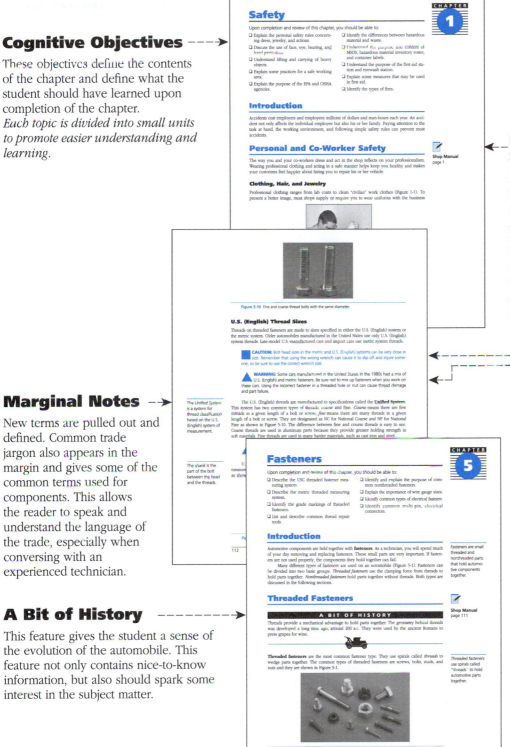

Terms to Know

A list of new terms appears next to the Summary. Definitions for these terms can be found in the Glossary at the end of the manual.

Review Questions

Short answer essays, fill-in-the-blanks, and multiple-choice type questions follow each chapter. These questions are designed to accurately assess the student's competence in the stated objectives at the beginning of the chapter.

Summaries

Each chapter concludes with summary statements that contain the important topics of the chapter. These are designed to help the reader review the contents.

Figure 5-43 The seal is trapped between the mating connectors when they are plugged together.

Terms to Know

Bolt
Butt connector
Dies
Fasteners
Helicoil
Keys
Nuts
Pins
Pitch gauge
Rivets
Screws
Shank
Snap rings
Society of Automotive Engineers
Solder
Splines
Stud
Taps
Tensile strength
Terminal connectors
Thread-restoring files
Threaded fasteners
Thread chasers
Unified System
Washers
Weather-tight

Summary

❑ Threaded fasteners use threads to hold automotive parts together. Common threaded fasteners include bolts, screws, studs, nuts, and washers.

❑ Threads are measured and classified according to the U.S. (English) and metric systems. U.S. (English) threads are made to standards of the Unified System.

❑ Unified System threads are classified according to the bolt shank diameter, length, and number of threads per inch. Threads in this system can be classified as fine or coarse.

❑ Metric system threads are classified according to shank diameter, length, and pitch.

❑ Both metric and U.S. (English) fasteners have grade markings to show fastener strength. U.S. (English) fasteners use marks on the bolt head to indicate grades. Metric bolts use property numbers to indicate grades.

❑ A pitch gauge can be used to determine the size of a fastener.

❑ Fasteners must be torqued to ensure they are not over- or undertightened. Overtightened fasteners loose their strength.

❑ Damaged threads may be repaired with a tap, die, or by installing a helicoil.

❑ Many automotive parts are held together with nonthreaded fasteners. Common nonthreaded fasteners include keys, snap rings, rivets, splines, and pins.

❑ Electrical system repair involves the use of automotive wire and electrical connectors. Wire is sized according to American Wire Gauge sizes. The smaller the cross section of the wire core, the larger the AWG number. Replacement wires must be the same gauge and should be the same color.

❑ Electrical terminal connectors fit on the end of the wire. Butt connectors join two wires together. Connectors must be the correct shape and wire gauge size.

❑ Weather-tight connectors are used to protect the interior of the connector from moisture.

Review Questions

Short Answer Essays

1. Describe the difference between a bolt and a screw.
2. Explain the difference between a bolt and a stud.
3. Describe how the length of a bolt is measured.
4. Explain how to tell the strength of a U.S. (English) bolt.

Review Questions

Short Answer Essays

1. Explain the measurement markings on a USC micrometer.
2. List and describe the different components on a micrometer.
3. Explain how stepped or go/no-go feeler gauges are machined.
4. Explain how to convert 3 inches to millimeters.
5. Discuss the components of a dial caliper.
6. Explain how pressure and vacuum are used to move air into the engine's cylinders.
7. List and discuss the differences between a USC and a metric dial caliper.
8. Describe the standard for measuring vacuum.
9. Explain how liquid can be used to transfer force.
10. Explain the metric system of measurement.

Fill-in-the-Blanks

1. The calipers on a dial caliper are moved by a(n) _____ _____.
2. Some micrometers have an additional scale called a(n) _____ _____.
3. One rotation of the spindle on a USC micrometer moves the spindle _____ _____.
4. A 3- to 4-inch micrometer can measure _____ inch.
5. A typical dial caliper will measure up to _____ inch(es) in _____-inch segments.
6. A stepped feeler gauge may be known as a(n) _____ gauge.
7. Large bores may be measured with a micrometer and_____ or a(n) _____.
8. A dial indicator may have a(n) _____ or a(n) _____ scale.
9. A micrometer has a(n) _____ to prevent an over torque of the spindle.
10. The vernier scale is used to read _____ of an inch.

Shop Manual

To stress the importance of safe work habits, the Shop Manual also dedicates one full chapter to safety. Other important features of this manual include:

Performance Objectives

These objectives define the contents of the chapter and what the student should have learned upon completion of the chapter.

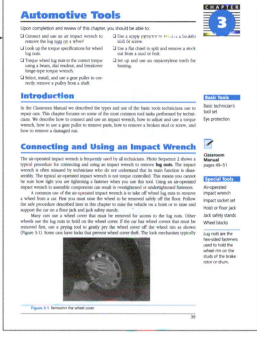

Photo Sequences

Many procedures are illustrated in detailed Photo Sequences. These photographs show the students what to expect when they perform particular procedures. They also familiarize students with a system or type of equipment that the school may not have.

Tools Lists

Lists of tools needed to perform tasks are included in each chapter. Whenever a special tool is required to complete a task, it is listed in the margin next to the procedure.

Marginal Notes

New terms are pulled out and defined. Common trade jargon also appears in the margins and gives some of the common terms used for components. This allows the reader to speak and understand the language of the trade, especially when conversing with an experienced technician.

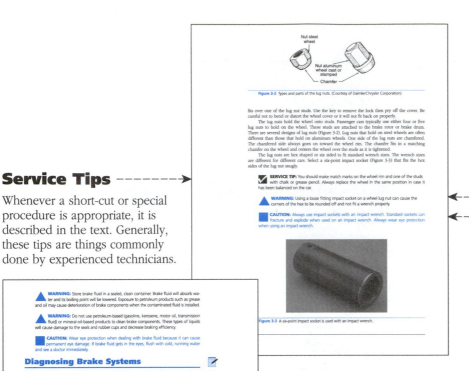

Figure 3-2 Types and parts of the lug nuts. (Courtesy of DaimlerChrysler Corporation)

fits over one of the lug nut studs. Use the key to remove the lock then pry off the cover. Be careful not to bend or distort the wheel cover or it will not fit back on properly.

The lug nuts hold the wheel onto studs. Passenger cars typically use either four or five lug nuts to hold on the wheel. These studs are attached to the brake rotor or brake drum. There are several designs of lug nuts (Figure 3-2). Lug nuts that hold on steel wheels are often different than those that hold on aluminum wheels. One side of the lug nuts are chamfered. The chamfered side always goes on toward the wheel rim. The chamfer fits in a matching chamfer on the wheel and centers the wheel over the studs as it is tightened.

The lug nuts are hex shaped or six sided to fit standard wrench sizes. The wrench sizes are different for different cars. Select a six-point impact socket (Figure 3-3) that fits the hex sides of the lug nut snugly.

SERVICE TIP: You should make match marks on the wheel rim and one of the studs with chalk or grease pencil. Always replace the wheel in the same position in case it has been balanced on the car.

WARNING: Using a loose fitting impact socket on a wheel lug nut can cause the corners of the hex to be rounded off and not fit a wrench properly.

CAUTION: Always use impact sockets with an impact wrench. Standard sockets can fracture and explode when used on an impact wrench. Always wear eye protection when using an impact wrench.

Figure 3-3 A six-point impact socket is used with an impact wrench.

Service Tips

Whenever a short-cut or special procedure is appropriate, it is described in the text. Generally, these tips are things commonly done by experienced technicians.

Warnings and Cautions

Throughout the text, cautions are given to alert the reader to potentially hazardous materials or unsafe conditions. Warnings are also given to advise the student of things that can go wrong if instructions are not followed or if a nonacceptable part or tool is used.

References to the Classroom Manual

Reference to the appropriate page in the Classroom Manual is given whenever necessary. Although the chapters of the two manuals are synchronized, material covered in other chapters of the Classroom Manual may be fundamental to the topic discussed in the Shop Manual.

WARNING: Store brake fluid in a sealed, clean container. Brake fluid will absorb water and its boiling point will be lowered. Exposure to petroleum products such as grease and oil may cause deterioration of brake components when the contaminated fluid is installed.

WARNING: Do not use petroleum-based (gasoline, kerosene, motor oil, transmission fluid) or mineral-oil-based products to clean brake components. These types of liquids will cause damage to the seals and rubber cups and decrease braking efficiency.

CAUTION: Wear eye protection when dealing with brake fluid because it can cause permanent eye damage. If brake fluid gets in the eyes, flush with cold, running water and see a doctor immediately.

Diagnosing Brake Systems

Classroom Manual pages 315–323

Brake system failures can be classed as either *hydraulic* or *mechanical* failures. Both will create symptoms that can be readily connected to a component or assembly. Mechanical failures will be discussed by common symptoms and their probable causes.

There are two conditions that constitute hydraulic failure: *external leaks* and *internal leaks*. Although mechanical conditions will create the leaks, they will be addressed as hydraulic problems to clarify the operational symptoms. An external leak is defined as "brake fluid exiting the system completely." An internal leak is defined as "a fluid leak within the system that does not exit the system."

External Leaks

A driver may complain that when the brakes were applied, the pedal dropped drastically and it took much more distance to stop the vehicle. The most common cause of this problem is an external leak somewhere in the system. A complete leak where the fluid exits the system results in a complete loss of pressure. The pedal will drop much lower than normal or go to the floorboard immediately upon brake application. If a single-piston master cylinder is being used, the pedal will hit the floorboard and cause a complete loss of all brakes. Pumping the pedal in an effort to regain control only compounds the problem. On a dual-piston master cylinder, the pedal will drop much lower than normal, but the driver will have brakes on one of the split systems and can stop the car. Again, pumping the pedal will not improve braking. An external leak of this type is noticeable when performing a visual inspection. The brake fluid level will also drop in the reservoir.

If the master cylinder is leaking externally, brake fluid will be leaking between the rear of the master cylinder and power booster or firewall (Figure 11-3). Usually, an external master

Figure 11-3 Brake fluid between the power booster and master of the master cylinder. (Reprinted with permission)

Customer Care

This feature highlights those little things a technician can do or say to enhance customer relations.

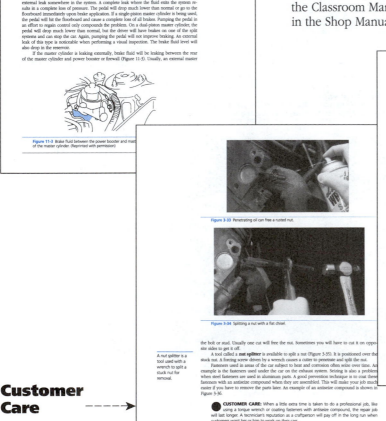

Figure 3-33 Penetrating oil can free a rusted nut.

Figure 3-34 Splitting a nut with a flat chisel.

the bolt or stud. Usually one cut will free the nut. Sometimes you will have to cut it on opposite sides to get it off.

A tool called a **nut splitter** is available to split a nut (Figure 3-35). It is positioned over the stuck nut. A forcing screw driven by a wrench causes a cutter to penetrate and split the nut.

Fasteners used in areas of the car subject to heat and corrosion often seize over time. An example is the fasteners used under the car on the exhaust system. Seizing is also a problem when steel fasteners are used in aluminum parts. A good prevention technique is to coat these fasteners with an antiseize compound when they are assembled. This will make your job much easier if you have to remove the parts later. An example of an antiseize compound is shown in Figure 3-36.

A *nut splitter* is a tool used with a wrench to split a stuck nut for removal.

CUSTOMER CARE: When a little extra time is taken to do a professional job, like using a torque wrench or coating fasteners with antiseize compound, the repair job will last longer. A technician's reputation as a craftsperson will pay off in the long run when customers want her or him to work on their cars.

Job Sheet 13

Name _____ Date _____

Removing, Inspecting, and Replacing Spark Plugs and Wires

Upon completion of this job sheet, you should be able to replace the spark plugs and wires.

Tools Needed

Hand tool set
Service manual

Procedures

1. Locate and record the specifications for the engine.
   Engine ID (type, size, valve)_____
   Spark plug type _____
   Spark plug torque_____
   Spark plug gap_____
   Spark plug wire type _____
2. Determine the best method of reaching the spark plugs.
3. Remove the wire from the first plug to be removed.
4. Remove the spark plug and inspect it. Record all findings.
   Cyl #1 _____ Cyl #2 _____
   Cyl #3 _____ Cyl #4 _____
   Cyl #5 _____ Cyl #6 _____
   Cyl #7 _____ Cyl #8 _____
5. Set the air gap on the new plug and install into the cylinder head.
   What type of tool was used to measure the gap?_____
6. Select the new spark plug wire by matching it to the old one.
7. Route the new wire alongside the old wire to the distributor cap or coil pack. Replace the old wire at the cap or coil terminal.
8. Connect the wire to the spark plug.
9. Start the engine to check the installation of this spark plug.
10. Repeat steps 3 through 10 until all plugs and wires are replaced.
11. Recheck the placement of the wires and their connections at the plug and distributor/coil pack.

215

Job Sheets

Located at the end of each chapter, the Job Sheets provide a format for students to perform procedures covered in the chapter. A reference to the ASE Task addressed by the procedure is referenced on the Job Sheet.

Case Studies

Case Studies concentrate on the ability to properly diagnose the systems. Each chapter ends with a Case Study in which a vehicle has a problem, and the logic used by a technician to solve the problem is explained.

ASE Style Review Questions

Each chapter contains ASE Style Review Questions that reflect the performance objectives listed at the beginning of the chapter. These questions can be used to review the chapter as well as to prepare for the ASE certification exam.

Diagnostic Charts

Chapters include detailed diagnostic charts linked with the appropriate ASE task. These charts list common problems and most probable causes. They also list a page reference in the Classroom Manual for better understanding of the system's operation and a page reference in the Shop Manual for details on the procedure necessary for correcting the problem.

Terms to Know

Terms in this list can be found in the Glossary at the end of the manual.

CASE STUDY

Using the wrong type of soldering flux can have disastrous results. A friend related a problem several years ago. He was doing a ground-up restoration on an old 356 Porsche. All the old wiring had been pulled out of the car in order to put in a complete new wiring harness. After pricing commercially available wiring harnesses, he decided to make his own. He bought and cut hundreds of feet of new wire. He wanted to do a good job, so he soldered each terminal and connector in his new loom.

The newly restored car had been on the road a couple of years when he started having electrical problems. Light switches, turn signals, and accessories stopped working for no apparent reason. He began to notice that a corrosion was building up on his electrical connections. Further investigation showed that the corrosion was caused by his use of an acid core solder. He was quite unhappy when he found out that all of the hundreds of connections had to be redone with the new terminals and the correct type of rosin core solder.

Terms to Know

Antiseize	Soldering iron	Tap wrench
Blind hole	Splice	Taper tap
Bottoming tap	Stripped threads	Threadlocking compound
Crimping	Stripping	Thread sealing compound
Flux	Stripping and crimping pliers	Through hole
Soldering gun	Stud remover	Tinning

ASE Style Review Questions

1. The use of a pitch gauge is being discussed. *Technician A* says the pitch gauge can be used to identify English (U.S.) threads. *Technician B* says the pitch gauge can be used to identify metric threads. Who is correct?
   A. A only
   B. B only
   C. Both A and B
   D. Neither A nor B

2. The use of a tap to repair internal threads is being discussed. *Technician A* says the tap must have the correct thread size on the shank. *Technician B* says of thread sizes. Who is
   A. A only
   B. B only
   C. Both A and B
   D. Neither A nor B

3. The use of a tap to repair internal threads is being discussed. *Technician A* says a taper tap is used for a through hole. *Technician B* says a taper tap is used for a blind hole. Who is correct?
   A. A only
   B. B only
   C. Both A and B
   D. Neither A nor B

4. The use of a tap to repair internal threads is being discussed. *Technician A* says a taper tap has numerous incomplete threads at its bottom.

Table 8-5 ASE Task

Inspect, replace, and adjust drive belts, tensioners, and pulleys.

Problem Area	Symptoms	Possible Causes	Classroom Manual	Shop Manual
Performance	Belt squealing.	1. Belt loose.	181	201–205
	Chirping.	2. Belt worn.	—	201–205
	Poor	3. Tensioner or pulley worn.	—	201–205
	accessory	4. Accessory damaged.	—	201–205
	performance.			
SAFETY				

Wear safety glasses.

Table 8-6 ASE Task

Inspect, test, service, repair or replace ignition system secondary circuit wiring and components.

Problem Area	Symptoms	Possible Causes	Classroom Manual	Shop Manual
Performance	Engine hard to start.	1. Worn spark plugs.	214	196–197
		2. Corroded, damaged spark plug wires.	214	196–197
	Poor idle.	3. Routine service.	214	195–198
	Time/mileage.			
SAFETY				

Wear safety glasses.
Allow engine to cool.

Table 8-7 ASE Task

Inspect, service, and replace battery, battery cables, clamps, and hold-down devices. Perform battery capacity load, high-rate discharge test; determine needed repairs.

Problem Area	Symptoms	Possible Causes	Classroom Manual	Shop Manual
Engine	Will not start.	1. Corroded cables, terminals.	181	191
	Battery will not hold	2. Battery shorted, open.	181	191
	charge.	3. Low electrolyte.	181	192
		4. Inoperative charging system.	181	—
SAFETY				

Wear safety glasses.
Allow engine to cool.
Disconnect negative cable first and reconnect last.

209

Reviewers

Special thanks to the following instructors for reviewing this material:

Charles Capsel, Antelope Valley College

Rick Curlee, Hill College Technical

Donald R. Deal, Red River AVTS

Dan Encinas, East Los Angeles College

Thomas J. Fitch, Monroe Community College

Earl J. Friedell, Jr., DeKalb Technical Institute

Jon Gerdy, New Castle School of Trades

James Helmle, Milwaukee Area Technical College

Bob Klauer, Metro Tech

Norris Martin, Texas State Technical College – Waco

Don Moseley, Monterey Peninsula College

Alan Penuela, Ventura College

Michael Stiles, Indian River Community College

Contributing Companies

I would also like to thank these companies who provided technical information and art for this edition:

A & E Manufacturing Company

Actron Manufacturing Company

American Honda Motor Co., Inc

Brake Parts, Inc.

Breton Publishers

C. Thomas Olivo Associates

Central Tools, Inc.

Cooper Automotive/Champion Spark Plug Company

Chicago Rawhide

CRC Industries, Inc.

DaimlerChrysler Corporation

Danaher Tools

Davis

Deere & Company

Delco-Remy Co.

DuPont Automotive Finishes

EIS Brake Parts

Federal-Mogul Corporation

Fel-Pro, Inc.

Ford Motor Company

General Motors Corporation, Service Operations

Gray Automotive

Griffin Technical Institute

Goodson Shop Supplies

Hunter Engineering Company

Klein Tools

The L.S. Starrett Co.

Lisle Tools Corp.

MATCO Tool Company

Mazda Motor of America, Inc.

Mitsubishi Motor Sales of America, Inc.

Moog Automotive, Inc.

Mitchell International

© National Institute for Automotive Service Excellence (ASE)

National Safety Council

Nissan North America

National Lubricating Grease Institute

Owatonna Tool Company

POP Fasteners

Proto Tools

Sealed Power Corp.

Sears Industrial Tools

Siebe North, Inc.

Society of Automotive Engineers, Inc.

Snap-on Tools Company

Spalding Lincoln-Mercury-Mazda-Dodge

©1998 Stanley Tools, a Product Group of The Stanley Works, New Britain, CT

The Timken Co.

U.S. Navy

Volvo Car Corporation

Western Emergency Equipment

York Technical College

Safety

CHAPTER 1

Upon completion and review of this chapter, you should be able to:

❏ Explain the personal safety rules concerning dress, jewelry, and actions.

❏ Discuss the use of face, eye, hearing, and hand protection.

❏ Understand lifting and carrying of heavy objects.

❏ Explain some practices for a safe working area.

❏ Explain the purpose of the EPA and OSHA agencies.

❏ Identify the differences between hazardous material and waste.

❏ Understand the purpose and content of MSDS, hazardous material inventory roster, and container labels.

❏ Understand the purpose of the first-aid station and eyewash station.

❏ Explain some measures that may be used in first aid.

❏ Identify the types of fires.

Introduction

Accidents cost employers and employees millions of dollars and man-hours each year. An accident not only affects the individual employee but also his or her family. Paying attention to the task at hand, the working environment, and following simple safety rules can prevent most accidents.

Personal and Co-Worker Safety

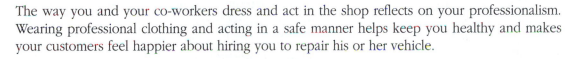

Shop Manual
page 1

The way you and your co-workers dress and act in the shop reflects on your professionalism. Wearing professional clothing and acting in a safe manner helps keep you healthy and makes your customers feel happier about hiring you to repair his or her vehicle.

Clothing, Hair, and Jewelry

Professional clothing ranges from lab coats to clean "civilian" work clothes (Figure 1-1). To present a better image, most shops supply or require you to wear uniforms with the business

Figure 1-1 Work clothes should fit neatly without binding or looseness.

logo. When you go to lunch, the logo and your clean uniform advertise your business and may cause a remembrance in the mind of some future customer at the right time. Uniforms are fitted to make sure no loose sleeves or shirttails can become tangled. If uniforms are not required, follow the same procedures with your civilian work clothes. Extra looseness is not beneficial in the automotive shop environment.

Long or short hair is, in most cases, not an issue for employment. Long hair in a shop, however, can be dangerous. If you have hair that is long enough to get tangled in rotating parts, it must be pinned or fastened securely out of the way. Never assume it is short enough. Under some circumstances, hair that is slightly more than shoulder length can cause problems.

Jewelry is one item that technicians do not like to remove, particularly wedding rings and religious images. While your spouse and a Divine Entity may look out for you, it is up to you to take care of certain things for yourself. One is reducing the chance of being seriously injured at work. Do not wear rings when working in an automotive shop. Rings tend to conduct electricity and react in strange ways when exposed to chemicals. Typical of personal religious symbols is a long necklace with the jewelry hanging inside the shirt or blouse. Long necklaces should be removed before beginning work. Ask your supervisor if there are any storage areas you may use for storage of valuable jewelry.

Eye and Face Protection

You can go to almost any shop and see technicians working without eye or face protection. Most of them feel the protection is uncomfortable or cuts visibility. Most of them also have not seen serious eye or face injuries first hand. The wearing of eye protection during work is the safest method for protecting your sight (Figure 1-2). Most eye injuries in the automotive shop come from a piece of rust or metal from a routine job such as exhaust work. Face injuries often happen during cleaning or grinding parts.

While it may feel uncomfortable when first wearing eye protection, you will become accustomed to it with daily wear and eventually feel naked without it. Remember your first ring or your first watch? For a few days, it worried you to wear it. Now, forgetting that ring or watch throws the whole day off. The same applies to eye and face protection. The vision problem is corrected in two ways. First, buy a set of safety glasses that are worth the money and, second, wear them every day. Your tools are your best investment in your career. Eye and face protection devices are part of your tool set.

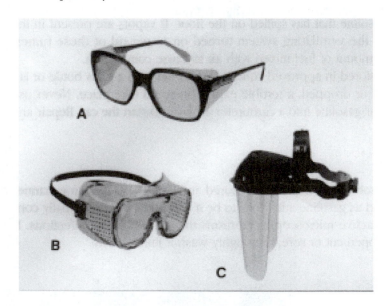

Figure 1-2 Eye and face protection are requirements for safety. (Courtesy of Goodson Shop Supplies)

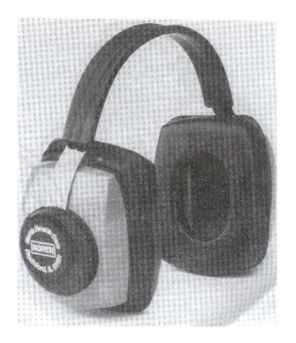

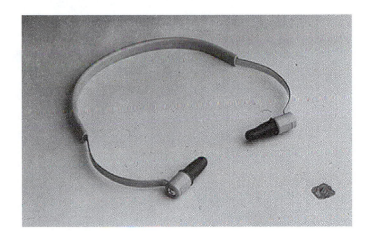

Figure 1-3 Ears can be damaged by sharp, loud noises and dusty conditions. (Courtesy of Siebe North, Inc.)

CAUTION: Ordinary prescription glasses are not safety glasses even though they may offer some impact protection. Order safety prescription glasses to include the frame. You are are not protecting yourself if the glasses do not meet occupational safety glasses requirements.

Hearing and Hand Protection

With some of the loud music being played on vehicle audio equipment, it is hard to imagine that an automotive shop can cause ear damage. A typical repair business does not have a great deal of *continuous noise* such as a body shop, but there are instances when hearing needs to be protected. Earmuffs or plugs are excellent for temporary protection on the job (Figure 1-3).

Hands are subject to cuts and abrasions by sharp tools and burning by chemicals. Be careful how you work your hands into those crevices on the vehicle and use good hand tools. When cleaning parts, wear chemical protection gloves to prevent burns. One thing about hearing and hand protection, most technicians use these types of protection devices without prompting. The pain and discomfort of exposure is immediate, as opposed to eye and face protection where the danger is not readily apparent. Lack of eye, face, hearing, or hand protection may put your career on hold for some time.

Lifting and Carrying

Shop Manual
page 1

Back pain does not normally result when a person tries to lift something above his or her capabilities. It usually happens when your body is in a position that will not accept the weight being moved. Before lifting any object, always position your body so the additional weight is placed over your legs. If lifting from the floor, bend at the knees and attempt to keep your back straight (Figure 1-4). This will allow your legs to perform the lifting motion. The reverse applies if an object is to be placed on the floor or a lower surface. Secure the object and lower it to the floor by bending your knees. This may be awkward and could cause a loss of control with heavier objects. Before moving heavy objects, ask another person to help distribute the weight and maintain control. If the object is dropped or moves off balance, the best method is to step back and allow the object to drop.

Also, if a change of direction is needed, such as moving an object from one table to another, do not twist your body. Turn your entire body with the object. Keeping your feet and legs in place and twisting at your waist places the weight of the object at an angle to your muscles. Your back and shoulder muscles would not be able to fully support the weight and may be over exerted.

Carrying heavier objects are dangerous to you, the object, and personnel working around you. If the object cannot be carried comfortably by you, ask a second person to help. Do not wait until the object becomes unstable.

> **CAUTION:** Trying to stop a falling object can result in injury to you without regaining control of the object. Step back and allow the object to drop. If possible, you may try to control it after it hits the floor, but do not let yourself be placed in jeopardy.

Co-Worker Safety

Another safe practice is to watch out for your co-workers. You need to have an idea of what is happening around you. The person in the next bay may be busy moving something, not knowing you are approaching from the blind side. The same rule applies in your case. In addition, your co-workers are like everyone else. Do them a favor and remind them of eye protection and other safety violations they may have missed.

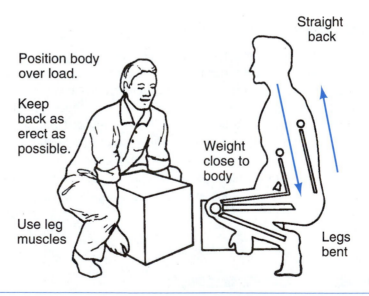

Figure 1-4 With the back kept straight, the load will be placed on the legs.

Maintaining a Safe Work Environment

Shop Manual
pages 1–5

There is a great deal of equipment and tools in the typical shop. They will be covered in detail in Chapter 3, "Automotive Tools and Equipment." However, the safe operation of equipment is not the only item to consider when entering the shop.

When a vehicle is parked inside the shop's premises, make sure it is shifted into PARK or a forward drive gear with manual transmissions. The vehicle should have the wheels blocked, particularly when being lifted by a floor jack. The blocks should be placed at the end of the vehicle not being lifted. Before starting the engine, connect the shop's exhaust system to the vehicle's exhaust system (Figure 1-5).

The shop's work and walking areas must be kept clean of oils, grease, rags, tools, or any other material that may cause an accident. Tools should be in use or in a secure place like the toolbox. Spills of any sort need to be removed and the area cleaned of all residues. Dirty rags are like tools. They are either being used or being placed in the proper storage container. **Mechanic creepers** have a tendency to be left on their wheels (Figure 1-6). The creeper should stand on end in a safe place unless someone is actually using it.

A final item to watch for when entering or working around the shop is the raised vehicle. The height of the lift, the vehicle's body, or the frame edge may be at the correct height to connect with an eye or head. Do not allow yourself to get distracted enough to walk into the edge of a vehicle or the lift. The best method to avoid accidents in the shop is to pay attention to what you and others are doing.

A *mechanic creeper* is a low, wheeled platform on which a technician can lie and roll under the vehicle to perform repairs.

Figure 1-5 Carbon monoxide is a colorless, odorless, deadly gas and is present in most vehicle exhaust gases.

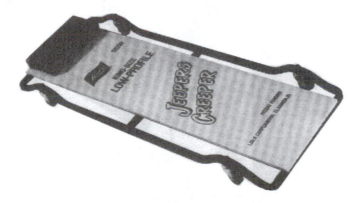

Figure 1-6 Mechanic creepers are handy tools, but they can be dangerous if stored improperly. (Courtesy of Sears Industrial Tools)

Shop Manual
page 7

The *Environmental Protection Agency* and the *Occupational Safety and Health Administration* were created by Congress to help control environmental and health hazards to the public.

EPA and OSHA Overview

The **Environmental Protection Agency (EPA)** and the **Occupational Safety and Health Administration (OSHA)** are two federal departments formed to set regulations for environment protection and worker safety. They have separate functions, but some of their authority overlaps.

EPA

The EPA was formed to set and enforce regulations covering waste disposal, disposal equipment, waste produced by manufacturer and service businesses, and landfills for household and industrial garbage. The harmful emissions of gasoline and diesel engines are tested by the various emission tests mandated by the EPA. While the control of those emissions is a daily part of an automotive technician's work, the waste produced by the shop is of main interest in this chapter.

OSHA

Worker safety is the prime goal of OSHA. The agency has the authority to inspect almost any business at almost any time. Its function is to check the work area for unsafe conditions. The items checked range from safety training and documentation to how hazardous materials are stored. Of primary interest to an OSHA inspector is fire prevention, exit routes, and the general condition of the shop. Cleanliness is not a prime concern of OSHA, but if the shop is dirty and unorganized to the point of being unsafe, it will probably receive a warning citation and the promise of a return visit.

The initial introduction of the EPA and OSHA caused many conflicts between businesses and government. The main reason was the cost of meeting the new standards. During the years since their inception, both business and government have learned to work together. The final results are not available, but businesses have found that cleaner, safer workplaces tend to increase production. Workers, businesses, and citizens have profited from a safe work environment and a cleaner atmosphere.

Shop Manual
pages 7–8

Hazardous waste is the dangerous byproduct of manufacturing and maintenance. The EPA regulates its storage and disposal.

Hazardous Waste and Material Control and Recycling

Oil, antifreeze, solvents, and almost any other shop waste is classed as **hazardous waste** and must be controlled. However, the EPA does not deem an item as *waste* until the generator (business) determines when the item will be disposed of or recycled, usually when a certain amount has been stored. Used oil and antifreeze must be stored in containers marked "used" (Figure 1-7). When

Figure 1-7 Containers of used materials must be labeled according to the contents.

the time comes to dispose of or recycle, it then falls under the classification of EPA-regulated waste and must be handled accordingly. While stored as "used," the container must meet EPA guidance. A leak in the container means the entire contents is waste. The cost of a cleanup can run from thousands to millions of dollars. Batteries, oily rags, tires, A/C refrigerant, and some vehicle parts are also hazardous waste.

Disposal of Waste

Each type of waste must be properly disposed of according to the chemical make-up of the material in it. Clean, "used" oil and antifreeze can be treated and reused or the oil can be treated and burned in the furnace of electric generating utilities or other manufacturing processes. Other waste requires special handling and possible long-term storage at a secure site (Figure 1-8). It must be remembered that the **waste generator** is responsible for the waste; the transportation or disposal company may be charged with violation of the law or its legal contract, but any cleanup is at the cost of the generator.

Hazard Materials

Hazardous materials are covered to some extent by the EPA. These materials are items used in the shop daily or stored until disposal. The EPA is primarily concerned with the storage and accidental release of hazardous materials waiting for disposal. OSHA sets forth the safety features of the business, one of which is the proper storage of hazardous materials stored for use.

Recycling Waste

Recyclable materials are being used more each model year. **Recycling** liquids and batteries has been done for years, but some of the other materials used to build cars and trucks have typically been sent to salvage yards or landfills. Tires have filled backyards and landfills for years. Burning them pollutes the air and, once set afire, tires burn until the material is completely gone. There is almost no way to put out tire fires. But new recycling technology has made it possible to recycle most vehicle components. Some vehicle manufacturers claim that

A hazardous *waste generator* is the business (or person) that created the waste. Vehicle repair shops collect used engine oil and, at the time of disposal, the shop becomes the generator. Until the oil is sent for disposal, it is classified as *hazardous material*, as is the new oil stored on the parts shelf.

Recycling is treating or processing materials for reuse. Treating or processing may include complete meltdown and reforming of the material or using serviceable used components to replace unserviceable components.

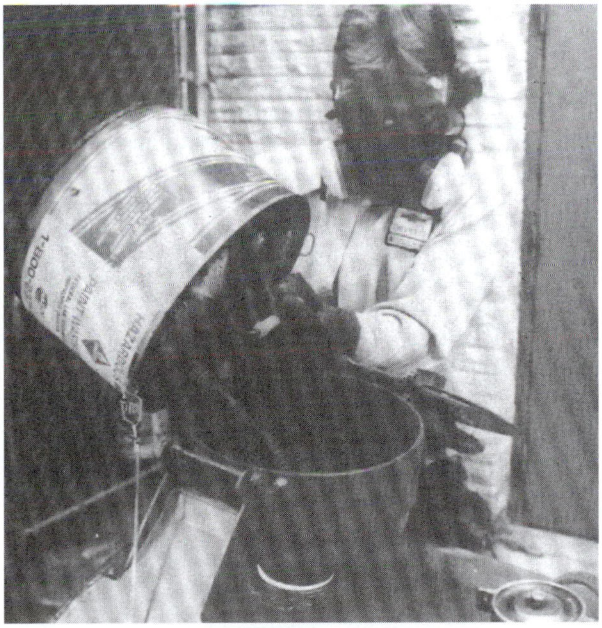

Figure 1-8 Some hazardous materials require special protection and processing during handling. (Courtesy of DuPont Automotive Finishes)

over 80 percent of their vehicles can be recycled into new components. This helps lower the cost of the vehicle and reduces the amount of waste. By definition, a major automotive recycling business is the local salvage yard.

Shop Manual
pages xx

Right-to-know laws state that an employer must inform his or her employees of hazardous waste, materials, or other dangers in the automotive shop. This information must also include safety measures and first-aid treatment.

Material Data Safety Sheets (MSDS) list names, chemical properties, safety, cleanup procedures, reactivity, and other information that might be important if an incident occurred in the shop.

Right-to-Know

Each employee is entitled to know the hazardous materials or waste they may face each workday. The **right-to-know laws** state that the employer must inform the employees of the dangers. This includes first-aid treatment and safety measures that must be followed. There are three documents that the employee must be aware of and given the opportunity to examine.

Material Safety Data Sheets

The **Material Safety Data Sheet (MSDS)** lists the name, chemical and physical properties of the agent, safety clothing and equipment, its effects on humans, reactivity if exposed to other materials, and cleanup procedures (Figure 1-9). The MSDS must be filed in a location accessible to any employee at any time during the workday. The MSDS must also be available for use by the employee's doctor.

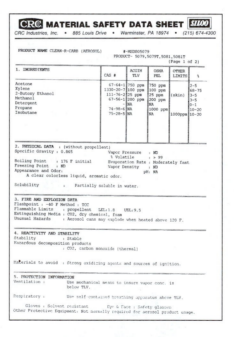

Figure 1-9 MSDS must be available to all employees and employees doctors. (Courtesy of CRC Industries, Inc.)

Hazardous Material Inventory Rosters

A second document, the **hazardous material inventory roster**, lists all hazardous materials stored *in* or *on* the business premises (Figure 1-10) but in an area away from the "dangerous" side of the business. The inventory roster is used by firefighters and other emergency personnel to determine the danger they may face and the actions needed to control a situation. While not immediately available to every employee, each employee may request to review the roster.

Container Labels

The third protective document can be found on almost any product from cigarettes to nuclear waste. It is the container label. The *container label* contains much of the same information included on the MSDS plus first-aid treatment and provides on-site information when the MSDS is not readily available. However, there may be some problems with labels. They may be torn or unreadable because of grease, oil, dirt, or printing in smaller letters. Sometimes the agent in a small container comes from a larger one. For instance, paint thinner may be shipped and

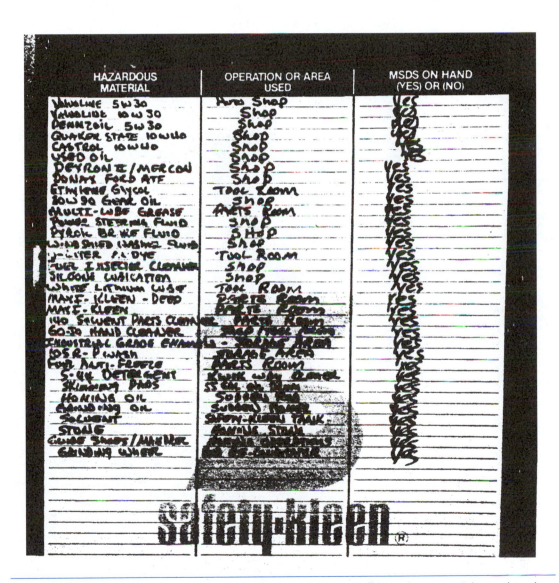

Figure 1-10 The hazardous material inventory roster must list all hazardous material stored on the premises.

stored in fifty-gallon drums. The thinner is placed in smaller, quart or gallon containers for easier use. The law states that the smaller, useable container must be labeled the same as the storage container. Failure in labeling the small container could cause an accident, fines, or both. Any container of hazardous material or waste must be labeled according to the contents. It should be noted that the law does not apply to tools like the paint reservoir for a paint gun that is in use.

Shop Manual
pages 5–7

First aid is defined as any medical assistance provided for an injured person to prevent further injuries. It also stabilizes the patient until medical personnel become available.

First Aid

Administrating **first aid** may be essential to life and limb. Each employee needs to know the location of the aid station, eyewash station, and common methods of applying first aid to others. Upon your arrival in a new shop, locate the first-aid station and eyewash. A large firm may have several sites for each.

The typical first-aid station is designed strictly for initial treatment. The station should have different types of bandages, burn ointment, and cream for the external treatment of cuts (Figure 1-11). This is an initial treatment station. The intent is to protect the injured and treat the wound until professional medical personnel are available.

The eyewash station is also used for immediate treatment. It may be mounted over a sink and designed to direct a large volume of low-pressure water into the eyes. Some systems consist of an adapter on a sink faucet while others are stand-alone units that may be able to sterilize the water (Figure 1-12).

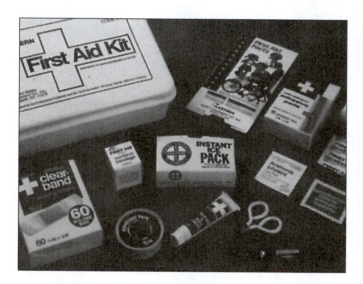

Figure 1-11 A typical first-aid kit will contain different types of bandages and ointments needed for minor first-aid treatment.

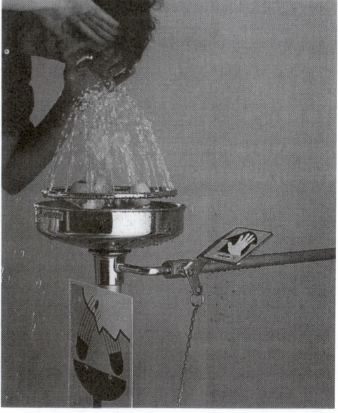

Figure 1-12 The eyewash station must be capable of delivering high-volume, low-pressure, clean water to flush the eyes. (Courtesy of Western Emergency Equipment)

Applying First Aid

The condition of the injured person determines the steps and procedures for first aid. Some initial steps are listed here.

Shock is one of the greatest dangers to an injury victim. If symptoms of shock appear, the injured person is conscious, and the situation allows it, calm the person and get him or her to lie on their back, using a blanket or similar object for warmth. Elevate the feet above the level of the heart. This tends to push the blood toward the brain. Apply a **pressure dressing** (bandage) over a bleeding wound and apply pressure. This will help reduce the bleeding.

Unconscious persons, or those with head or back injuries, should not be moved unless there is a greater danger present, such as a fire. Bleeding limbs should be pressure bandaged and slightly elevated above the level of the heart to reduce blood pressure to the wound and help control bleeding. When seeking help, avoid leaving an injured person alone if possible. Yell out to employees or customers in the next room or to pedestrians on the street. Use a cellular phone if available. Seek help from *anyone* and notify your supervisor as soon as possible.

Most businesses offer some type of first-aid training. If they do not, suggest such training to your supervisor. The Red Cross offers first-aid classes in almost every community. Attendance at one class may mean the difference between life and death for you, a loved one, or a co-worker.

Fire Prevention and Control

Fire in an automotive shop can be extremely dangerous. There are numerous places for a fire to start. Improperly stored, oily, or greasy rags can **spontaneously combust**. That means they will generate enough heat to start a flame. Once started, the fire can spread quickly if not correctly and immediately extinguished.

The different types of fires are shown in Figure 1-13. Note the materials in types A and C. While an A-type fire would be unusual in an automotive repair shop, it could still be a possible danger in or around the shop's storage area. A Class C fire could occur on a vehicle's electrical motor because of a short or ground, but a major electrical fire is uncommon in this business.

The most common fire in an automotive shop is a Class B type. This fire is fed by flammable liquids such as fuel, solvent, or other flammable liquid. Following standard safety and firefighting procedures can control almost all types of fires.

WARNING: Before turning on the fuel system after repairs, double-check the line fittings (Figure 1-14). Fuel leaks from line fittings on or near the engine tend to puddle on the engine or drip onto hot metal components or electrical systems. Burning fuel within the engine compartment can cause extensive damage or injury.

The technician must use caution when dealing with flammable liquids around the engine and working on electrical systems. Fuel lines must be tightened correctly and firmly. If fuel pressure has to be released on a fuel injection system, use the manufacturer's procedures (Figure 1-15). A catch basin can be used on low-pressure fuel systems. Once the lines are disconnected, block them to prevent fuel from escaping from the tank or fuel delivery system.

Shock is caused by a great disturbance to the mind or body and is characterized by a sudden drop in blood pressure, sweaty hands, faintness, and possible loss of consciousness.

A *pressure dressing* is a sterile bandage placed on an injury followed by pressure applied against the bandage to slow or stop bleeding.

Shop Manual pages 9–10

Spontaneous combustion is a fire produced by the heat of internal chemical action.

	Class of Fire	Typical Fuel Involved	Type of Extinguisher
Class **A** Fires (green)	**For Ordinary Combustibles** Put out a class A fire by lowering its temperature or by coating the burning combustibles.	Wood Paper Cloth Rubber Plastics Rubbish Upholstery	Water*[1] Foam* Multipurpose dry chemical[4]
Class **B** Fires (red)	**For Flammable Liquids** Put out a class B fire by smothering it. Use an extinguisher that gives a blanketing, flame-interrupting effect; cover whole flaming liquid surface.	Gasoline Oil Grease Paint Lighter fluid	Foam* Carbon dixoide[5] Halogenated agent[6] Standard dry chemical[2] Purple K dry chemical[3] Multipurpose dry chemical[4]
Class **C** Fires (blue)	**For Electrical Equipment** Put out a class C fire by shutting off power as quickly as possible and by always using a nonconducting extinguishing agent to prevent electric shock.	Motors Appliances Wiring Fuse boxes Switchboards	Carbon dioxide[5] Halogenated agent[6] Standard dry chemical[2] Purple K dry chemical[3] Multipurpose dry chemical[4]
Class **D** Fires (yellow)	**For Combustible Metals** Put out a class D fire of metal chips, turnings, or shavings by smothering or coating with a specially designed extinguishing agent.	Aluminum Magnesium Potassium Sodium Titanium Zirconium	Dry powder extinguishers and agents only

*Cartridge-operated water, foam, and soda-acid types of extinguishers are no longer manufactured. These extinguishers should be removed from service when they become due for their next hydrostatic pressure test.

Notes:

(1) Freezes in low temperatures unless treated with antifreeze solution, usually weighs over 20 pounds, and is heavier than any other extinguisher mentioned.

(2) Also called ordinary or regular dry chemical (sodium bicarbonate).

(3) Has the greatest initial fire-stopping power of the extinguishers mentioned for class B fires. Be sure to clean residue immediately after using the extinguisher so sprayed surfaces will not be damaged (potassium bicarbonate).

(4) The only extinguishers that fight A, B, and C classes of fires. However, they should not be used on fires in liquefied fat or oil of appreciable depth. Be sure to clean residue immediately after using the extinguisher so sprayed surfaces will not be damaged (ammonium phosphates).

(5) Use with caution in unventilated, confined spaces.

(6) May cause injury to the operator if the extinguishing agent (a gas) or the gases produced when the agent is applied to a fire is inhaled.

Figure 1-13 Classes B and C type fires present the greatest fire concern in an automotive shop.

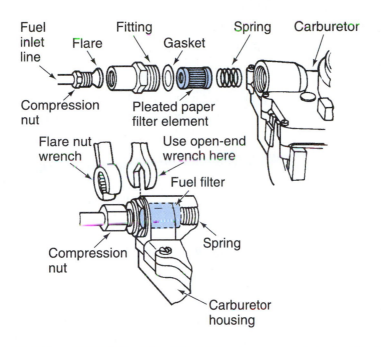

Figure 1-14 Fuel connections similar to the ones shown can be found in carburetor and fuel injection engines.

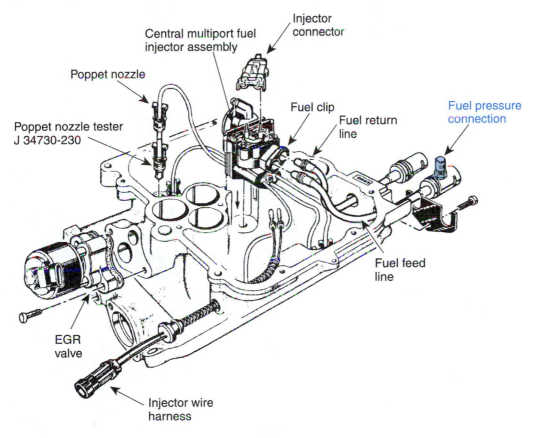

Figure 1-15 Properly relieving the fuel pressure gauge through the gauge connection is the safest method to reduce fuel injection fuel spills. (Courtesy of Chevrolet Motor Division, General Motors Corporation)

▲ **WARNING:** Double check the cable and **polarity** of the battery post before making battery connections. Batteries can explode with dangerous force. The acid and explosion can cause damage to personnel and equipment.

Batteries can be a source of fire or explosion if not handled properly. Before connecting a battery, the technician must be certain there is no load on the electrical system that may cause a spark. A battery that is being load tested or recharged must have the cables connected correctly by polarity.

■ **CAUTION:** Batteries emit hydrogen gas during discharging. The gas may be present during repairs. Hydrogen gas is extremely *volatile* and can be easily ignited by a small spark.

When you first arrive at a new shop, walk around to find the fire extinguishers and note each type. One common type is a Class B for use against Class B fires. It will be marked with a square and a B (Figure 1-16). Most shops use multipurpose fire extinguishers. This type will have markings for Class A, B, and C fires. Class B or multipurpose fire extinguishers are used against Class B fires. The extinguisher is aimed at the base of the fire, which cuts off the oxygen. Water should never be used against a flaming liquid fire. It will only spread the fire's fuel over a larger area.

While prevention is always the best method for fire control, there may come a time when you must fight a fire. Do not keep the secret to yourself. Let everyone know. While fighting a fire, do not place yourself in additional danger. You need to keep track of the actions around you. Fires have a tendency *not* to follow rules. They can quickly cut off your escape route. Many times we allow ourselves to attempt something that is outside our capabilities. While that may be part of learning, fighting a fire beyond our control will probably result in additional injuries or damage. If the fire appears to be too large for you, then it is time to leave and let the professionals handle it. A five-pound or ten-pound fire extinguisher will only go so far. Do not place yourself or others in danger.

Figure 1-16 The multipurpose fire extinguisher is the type most suitable for automotive shops.

Summary

❑ Work clothing should not be loose enough to catch on equipment.

❑ Hair should be short or tied back out of the way.

❑ Jewelry should always be removed before beginning work.

❑ Moving an object can cause injury to you or damage to the object if proper lifting procedures are not followed.

❑ The EPA regulates the storage, disposal, and recycling of hazardous waste.

❑ OSHA inspects businesses for safety hazards to the employees.

❑ First aid is the initial medical treatment for an injury.

❑ Eyewash stations are set up to flush eyes that have been exposed directly to hazardous materials.

❑ Fire prevention is the best way to stop fires because it removes the fire danger.

❑ The most common type of fire in an automotive shop is Class B.

Terms to Know

Environmental Protection Agency

Hazardous materials

Hazardous waste

Mechanic creepers

Occupational Safety and Health Administration

Polarity

Pressure dressing

Preventive maintenance

Right-to-know laws

Volatile

Waste generator

Review Questions

Short Answer Essays

1. Explain the four types of fires and the materials involved in each.

2. Explain your actions if a fire breaks out in your work area.

3. A co-worker has a serious bleeding injury on the lower arm. Explain some actions for first aid.

4. List the three documents an employee must be made aware of under the right-to-know laws.

5. List the five major areas of an MSDS.

6. Explain the differences between an MSDS and a container label.

7. Explain why the hazardous material inventory roster must be available to emergency workers.

8. Define the difference between *hazardous material* and *hazardous waste*.

9. Briefly explain the purpose of the EPA and OSHA.

10. Explain why water should *not* be used against flammable liquid fires.

Fill-in-the-Blanks

1. Protection of the face during grinding is accomplished by wearing _____ _____ .

2. A possible location for the eyewash station is the _____ _____ or a(n) _____ _____ unit.

3. The fire extinguisher nozzle should be aimed at the _____ of the fire.

4. Class B fires are extinguished by cutting the flow of _____ .

5. Hazardous material comes under EPA regulations when it is prepared to be _____ or _____.

6. A leak in a used-liquid container means all of the material is considered to be _____.

7. A mechanic creeper should always be _____ when not in use.

8. Jewelry should be _____ while in the shop.

9. A container label has almost the same information as the _____ _____ _____ _____.

10. All of the hazardous material on the business' premises are listed on the_____ _____ _____ _____.

ASE Style Review Questions

1. Fire fighting is being discussed. *Technician A* says to keep track of the situation around you.
   Technician B says one of the first actions to be taken is to alert others of the fire. Who is correct?
   A. A only
   B. B only
   C. Both A and B
   D. Neither A nor B

2. Hazardous waste is being discussed. *Technician A* says the person or company performing the disposal is the one responsible for the waste.
   Technician B says good business ethics include the proper handling of waste. Who is correct?
   A. A only
   B. B only
   C. Both A and B
   D. Neither A nor B

3. *Technician A* says face protection is required when cleaning parts in a solvent tank.
   Technician B says eye protection should be worn anywhere in the shop. Who is correct?
   A. A only
   B. B only
   C. Both A and B
   D. Neither A nor B

4. The selection of fire extinguishers is being discussed. *Technician A* says an extinguisher marked with a circle and a B should be used against a flammable liquid fire.
   Technician B says cut the oxygen from a Class B fire with water. Who is correct?
   A. A only
   B. B only
   C. Both A and B
   D. Neither A nor B

5. Fires are being discussed. *Technician A* says a type A fire is the result of burning fuel.
   Technician B says a type B fire will usually be located around electrical circuits. Who is correct?
   A. A only
   B. B only
   C. Both A and B
   D. Neither A nor B

6. Personal safety is being discussed. *Technician A* says tucking long hair inside a shirt or blouse is sufficient for safety.
   Technician B says jewelry allowed to hang inside the blouse or shirt satisfies safety rules. Who is correct?
   A. A only
   B. B only
   C. Both A and B
   D. Neither A nor B

7. Work practices are being discussed. *Technician A* says a mechanic creeper should stand on end unless some one is lying on it.
   Technician B says looking out for co-workers increases your safety. Who is correct?
   A. A only
   B. B only
   C. Both A and B
   D. Neither A nor B

8. A fire has started near a hazardous material storage area. *Technician A* says to alert others and attempt to control the fire.
   Technician B says the MSDS should be made available to emergency crews. Who is correct?
   A. A only
   B. B only
   C. Both A and B
   D. Neither A nor B

9. An employee has been injured at a distance from the main area. *Technician A* says to contact help first.
   Technician B says to assess the situation, apply first aid if possible, and make arrangements to alert others. Who is correct?
   A. A only
   B. B only
   C. Both A and B
   D. Neither A nor B

10. Safe working areas are being discussed. *Technician A* says the actions of an unsafe co-worker may affect other employees.
    Technician B says unsafe working conditions lead to injuries and loss of profit for the employer. Who is correct?
    A. A only
    B. B only
    C. Both A and B
    D. Neither A nor B

The Automotive Business

Upon completion and review of this chapter, you should be able to:

❏ Discuss the different types of businesses supporting the automotive industry.

❏ Describe the main differences between dealerships, independents, franchises, and service stations.

❏ Discuss how departments can increase business.

❏ Discuss the general duties and responsibilities of the service manager.

❏ Discuss the legal and ethical responsibilities of a business and its employees.

❏ Discuss the role of accrediting agencies within the automotive industry.

Introduction

Employees must work effectively within the organization so they can show a profitable return for their investment within the company. This includes knowing the company's organization, general duties of each department, working with the various departments, and protecting the **investment** through safe, profitable, and ethical work practices.

Investment is the amount of time, money, and other resources used to establish a business.

Shop Manual
page 17

The Business

Automotive service centers used to be fairly small operations. Usually there was a person who owned and managed the business while performing repairs and customer service at the same time. Even manufacturer repair service centers were small-scale compared to today's multimillion-

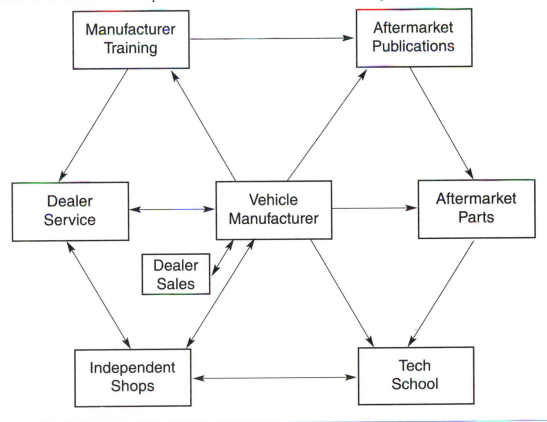

Figure 2-1 Relationship of automotive business.

dollar shops. Small operations still exist today, but the majority of automotive repair businesses must meet the demand for highly trained technicians, expensive diagnostic equipment, customer waiting rooms, extensive parts departments, and the facilities and staff to support them. The various levels of business interlock throughout the industry (Figure 2-1).

A modern-day automotive repair business includes every type of employee position from custodian to general manager. There are accountants, clerks, sales people, service writers, technicians, department and service managers, vehicle preps, and a general manager to oversee the entire operation. In any case, the person who performs the work that directly impacts profit-making is the technician.

Dealership Operations

Normally, **dealerships** are automotive businesses operating under independent ownership in conjunction with a vehicle manufacturer. Operations of this type usually sell and service only certain makes of vehicles. The owners conduct day-to-day business, either directly or through a **general manager.** The business must conform to certain standards set forth by the vehicle manufacturer (Figure 2-2). These standards may include policies on new car sales, accounting and financing procedures, service operations, and the layout of the facilities. In some instances, the manufacturer may withdraw the **franchise** license if standards and profit margins are not met.

Sales and service departments working together can create a customer relationship that draws repeat business and favorable word-of-mouth advertising. The technician is required to have extensive knowledge of the makes and models sold at the dealership, and is required to perform warranty repairs.

Shop Manual
page 17

Dealerships are usually much larger than the typical shop. They may sell and service several different makes and models and mix domestic and import brands.

A *general manager* is responsible for the daily operation of a business and reports to the owners.

Franchises are businesses locally-owned and operated under the policies of a larger company. McDonald's is a franchise operation.

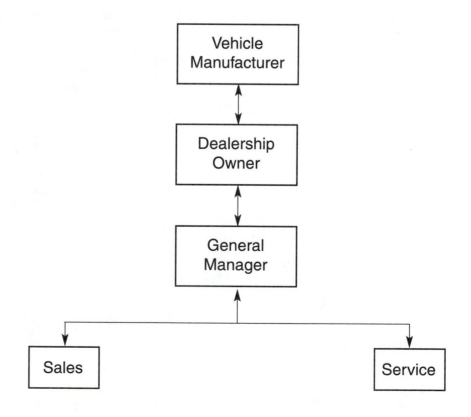

Figure 2-2 Relationship between dealership and manufacturer.

Figure 2-3 Independent shops repair almost any type and make of vehicle.

Independent Repair Facilities

Independent repair shops usually concentrate on vehicle repairs only (Figure 2-3). They may stock some of the most common repair parts, but the majority of needed parts come from after-market **vendors** or the local dealership. Independents usually take in any vehicle that needs repair. This places a heavy technical load on the technicians. They must know how to repair almost any vehicle and all automotive systems. Independent shop owners usually have to pay for their technicians' update training in addition to salaries and travel expenses. Most independents make use of classes offered by local parts vendors or the local technical school.

Other Automotive Repair Businesses

Franchise automotive businesses are independently owned and operated. They are licensed under the business umbrella of a national or regional chain (Figure 2-4). The owner must follow the policies of the company issuing the franchise license. Precision Tune and Quik Lube are two examples. These types of businesss are set up to perform only certain types of repairs

Shop Manual
page 17

Vendors are businesses that sell parts or service to other businesses and the public. NAPA is an automotive vendor.

Figure 2-4 Franchise businesses usually service one system of the vehicle, but may perform related system repairs.

or maintenance. Businesses like Precision Tune make engine performance repairs or maintenance while Quik Lube types usually perform oil and lubrication work. Since their inception, each business has changed its operations to include work not directly related to its original base business.

Local service stations are another part of the repair business. Usually, they handle brakes, tires, minor tune-ups, and other light repairs. Other local businesses repair only certain systems of the automobiles. Generally, they specialize in brakes, tires, exhaust, and related repairs.

Each of these automotive businesses make up the automotive repair facilities in this country. To support them, there are numerous parts makers and suppliers, hazardous waste disposal companies, and many other local, national, and international firms that contribute to keeping a vehicle on the road.

Management and Service

Shop Manual
pages 17–19

Managers are responsible for the daily operation of a segment of a business and report to the general manager.

Without a concerted effort by all members of an organization, a customer may buy a vehicle or part from a one business then go elsewhere to have it installed or maintained. This reduces profit and basically loses a customer.

The **managers** of each department must work together to ensure the customer receives first-class service from any employee they may contact. Some dealerships, notably Saturn, introduce their new-car customers to the service department within a short time of purchase. The Saturn dealers have monthly gatherings where their newest car buyers meet a service technician. The customer's perception of receiving individual, personalized treatment boosts the prestige of the company and its employees.

Independents, on the other hand, are usually smaller and locally-owned operations. Most of the employees and the owner are from the local area and know their customers first hand. Many times, independents will work with local dealerships to handle work that is outside the independent's capabilities. Both operations profit from this arrangement in monetary terms and customer trust.

The *service manager* controls the operation of the service department, including parts, technicians, support staff, and the training all department employees.

Service Departments

Usually, a **service manager**, who is responsible for the **service writer**, technicians, the department support staff, and parts, heads the service department. He or she works with vehicle sales in a dealership to bring new-car customers back for service when needed (Figure 2-5).

The *service writer* meets the customer, writes the repair order, and assigns a technician based on company policy.

GENERAL MANAGERIAL DUTIES

General Manager	Sales Manager	Service Manager
Reports to owner	Reports to general manager	Reports to general manager
Delegates authority	Supervises sales staff	Supervises service staff
Supervises department managers		
General budget reports	Sales budget reports	Service budget reports
Hires, fires, trains managers and general staff personnel	Hires, fires, trains sales and staff personnel	Hires, fires, trains parts personnel, staff, and technicians
Supervises interactions between departments	Coordinates with general manager and service to gain and keep customers for the business	Coordinates with general manager and sales to gain and keep customers for the business

Figure 2-5 Managers share some common duties so the various departments can work together to build and maintain the business.

The parts department has a manager who reports to the service manager. In an independent shop, the number of people actually involved in these types of duties and responsibilities is often much less than that of a dealership.

Legal Responsibilities

Not only are the technicians expected to be technically proficient, they must understand the legal side of automotive repair. The customer is paying for professional service and has a right to expect it. Poor repair work, damage to the vehicle, and poor attitude toward the customer may lead to legal proceedings against the technician and the shop.

In addition to legal responsibilities to the customer, federal, state, and local laws and regulations require business owners and employees to help protect the community and environment from harm. Automotive technicians share this legal responsibility. Proper handling of hazardous materials and waste is of prime concern in the automotive industry. Not only can a poorly tuned engine emit hazardous materials and gases, but the byproducts of repair work also may damage the environment. The business owner is ultimately responsible for the actions of the firm's employees, but the technician plays a large part in fulfilling the legal responsibilities of the company.

Shop Manual
page 19

Business Ethics

Most people hold the automotive industry in high esteem. Two segments of the industry, sales and service, are the exceptions. Much of this distrust stems from the high cost of vehicles and their repairs. Further supporting this distrust is the fact that most people do not understand the workings of an automobile, and they have no reference with which to compare the costs of buying and maintaining a car to that of other large purchases. Educating customers is one way to resolve the problem. Very few automobile owners understand that their vehicle's computer is almost as powerful as the personal computer they have at home or that the technicians are constantly attending some type of training. A business will survive much better in today's highly competitive automotive arena if it can educate its customers to some extent. A customer may pay well over $2,500 for a personal computer (PC) and balk at paying $600 for a vehicle computer. A technician or service writer can win over the customer by showing the capabilities of that $600 computer versus the PC.

Automotive businesses expect to be treated fairly by their suppliers and must treat their own customers in the same manner. By treating each customer as a valued individual, charging a fair price for service, and admitting and correcting mistakes, the shop and all its employees can expect to be in business for a long time.

Business ethics are hard to define. While one person's idea may be different from another's, the managers set the rules. Treating customers, managers, and employees fairly should be the basic rule.

National Institute for Automotive Service Excellence (ASE)

Shop Manual
page 21

ASE tests technicians for their technical knowledge. Technicians are authorized to wear the shoulder patch only if they have successfully completed a written test and have two years working experience. They must re-certify every five years to maintain their standing. The Master Automotive Technician patch is given only if a technician passes all eight automotive tests and meets the experience requirement (Figure 2-6). Certifications are also offered in heavy truck, alternative fuels, parts, machining, and school buses. The automotive tests given in May and November each year are engine rebuilding, engine performance, brakes, steering and suspension, heating and air conditioning, electrical, automatic transmission, and manual drivelines.

ASE is the logo for the National Institute for Automotive Service Excellent.

Figure 2-6 Master technicians must have two years of experience and pass all eight exams.

The shop can be certified if it meets ASE standards for shop operation and employee technical certification. The shop can then display the blue ASE sign and use the ASE symbol in its advertising. The ASE sign and shoulder patch can be a valid advertising tool. ASE also publicizes the technician's role within the automotive business.

National Automotive Technicians' Education Foundation (NATEF)

Shop Manual
pages 23–25

The *NATEF* key is the logo for a certification administrated by the National Automotive Technicians' Education Foundation. It is a representation of a car key with the certified program's title.

NATEF, a division of ASE, certifies automotive training programs. The programs may be public and private high schools or post-secondary schools. Students who graduate from a NATEF certified program have an advantage over other persons applying for an automotive technician position.

To meet NATEF standards for certificates, the training program must be capable of teaching standard tasks published by NATEF. The program must have all of the tools required by NATEF and have ASE-certified instructors in each area requesting certification. The program is certified for five years with an update at the halfway point. Once certified, the school can use the NATEF key in its advertising and catalog (Figure 2-7).

AUTOMOTIVE TECHNOLOGY

Figure 2-7 The right to use the NATEF key in advertising is awarded after successfully completing the NATEF training program procedures.

Automatic Transmission Rebuilders Association (ATRA)

Shop Manual
page 17

ATRA is an association of transmission rebuilders. It offers a certification to any technician who desires the training, but it is directed toward technicians who only do automatic transmission repairs. The test is given annually at selected sites and is brand and model specific in its questions. The test is designed for technicians who have experience in transmission diagnosing and repairs.

ATRA and *MACS* are two examples of automotive certification organizations. There are others, but the most nationally recognized are ASE and NATEF.

Mobile Air Conditioning Society (MACS)

Shop Manual
page 17

MACS is similar to ATRA in its testing. It is designed for automotive air conditioning technicians, but, like ASE, is offered to any technician. An EPA requirement for all air conditioning technicians is to be certified by an approved agency. This test will be discussed further in Chapter 12, "Auxiliary Systems and Climate Control."

Summary

❑ Management is responsible for all business operations.
❑ All employees share the legal responsibilities of the business.
❑ Education of the customer is another way to gain customer trust.
❑ ASE provides a nationwide certification program for technicians.
❑ NATEF certifies automotive training programs.
❑ ATRA provides a means for automatic transmission technicians to gain specific skill recognition.
❑ MACS is designed for air conditioning technicians.

Terms to Know
ASE
ATRA
Business ethics
Dealership
Franchise
General manager
Investment
MACS
NATEF
NATEF key
Service manager
Service writer
Supplier
Vendors

Review Questions

Short Answer Essays

1. Describe the role of the independent shop within the automotive repair industry.

2. Explain how sales and service work together to gain customers.

3. Explain how certifying agencies can assist in promoting customer trust and validating technician technical knowledge.

4. Describe the purpose of NATEF.

5. Discuss how customer education can help an automotive repair business.

6. List the eight automotive areas tested semi-annually by ASE.

7. List the steps needed for a technician to be certified as a master automotive technician.

8. Discuss *your* definition of business ethics.

9. Explain why employees share in the legal responsibilities of the employer.

10. List the level of management typically found in an automotive shop.

Fill-in-the-Blanks

1. The _____ department manager reports directly to the service manager.

2. The service writer completes and _____ the repair order per company policy.

3. NATEF conducts certification of _____ _____.

4. The logo for NATEF is a(n) _____ with the _____ name on it.

5. ASE certified technicians wear a(n) _____ ASE patch.

6. Vehicle manufacturer training programs can be certified by _____.

7. Air conditioning technicians can be certified through _____.

8. ATRA offers certification to _____ _____ technicians.

9. Air conditioning technicians must be _____ certified by either MACS or ASE.

10. A(n) _____ is also known as a supplier.

ASE Style Review Questions

1. Business operations are being discussed. *Technician A* says the department managers may report directly to the owner.
   Technician B says the parts manager works directly with the service manager. Who is correct?
   A. A only
   B. B only
   C. Both A and B
   D. Neither A nor B

2. The general manager's duties are being discussed. *Technician A* says this manager is responsible for the parts and service departments.
   Technician B says the owner reports to the general manager. Who is correct?
   A. A only
   B. B only
   C. Both A and B
   D. Neither A nor B

3. Legal responsibilities of the business are being discussed. *Technician A* says each employee shares in legal responsibilities.
   Technician B says the owner bears ultimate responsibly. Who is correct?
   A. A only
   B. B only
   C. Both A and B
   D. Neither A nor B

4. Program certification is being discussed. *Technician A* says ATRA certifies air conditioning programs.
   Technician B says MACS certifies air conditioning technicians. Who is correct?
   A. A only
   B. B only
   C. Both A and B
   D. Neither A nor B

5. *Technician A* says that to achieve master automotive certification, nine tests must be successfully completed, including the Advanced Engine Performance Test.
Technician B says only eight tests must be taken and only two years of experience are needed for master certification. Who is correct?
   A. A only
   B. B only
   C. Both A and B
   D. Neither A nor B

6. *Technician A* says business ethics means treating all customers honestly and fair.
Technician B says business ethics means treating vendors correctly. Who is correct?
   A. A only
   B. B only
   C. Both A and B
   D. Neither A nor B

7. The service manager's duties and responsibilities are being discussed. *Technician A* says the training of a technician is his or her responsibility.
Technician B says he or she usually has authority over the parts manager. Who is correct?
   A. A only
   B. B only
   C. Both A and B
   D. Neither A nor B

8. Independent repair shops are being discussed. *Technician A* says this type of shop is needed to help reduce customer costs.
Technician B says independents may send work to the local dealership. Who is correct?
   A. A only
   B. B only
   C. Both A and B
   D. Neither A nor B

9. Technician certification and training is being discussed. *Technician A* says NATEF ensures certain tasks are taught.
Technician B says NATEF ensures that the technician is certified. Who is correct?
   A. A only
   B. B only
   C. Both A and B
   D. Neither A nor B

10. Air conditioning certification is being discussed. *Technician A* says an air conditioning technician can be EPA certified by ASE only.
Technician B says ATRA deals with air conditioning technicians. Who is correct?
   A. A only
   B. B only
   C. Both A and B
   D. Neither A nor B

Automotive Tools and Equipment

Upon completion and review of this chapter, you should be able to:

❑ Identify and describe the safe use of the common wrenches used in the automotive shop.

❑ Identify and describe the correct use of the power wrenches used in the automotive shop.

❑ Identify and explain the safe use of the metalworking tools commonly used in the automotive shop.

❑ Identify and explain the safe use of the common threading tools used in the automotive shop.

❑ Describe the basic parts of an oxyacetylene welding and cutting outfit.

❑ Describe the major types of lifts.

❑ Describe the purpose of the computerized test equipment.

❑ Describe electrical test equipment.

Introduction

Knowing how to use and care for tools properly is one of the most important parts of being an automotive technician. There is a correct tool for each and every automotive repair job. It is also important to use each tool in the correct way. Technicians use a number of these tools on every repair job. This chapter discusses the most common tools. Remember, a successful technician must know the name of each tool, what it does, and what tool works best for each automotive repair job.

Most of the tools used by a technician are called hand tools. They get their name from the fact that they are operated by the technician's own muscle power. Power tools, on the other hand, get their power from electricity, compressed air, or hydraulics.

Wrenches

Many automotive parts and components are held together or fastened with bolts and nuts. Wrenches, which come in many different sizes, are designed to tighten or loosen these bolts and nuts. Bolts and nuts also come in different sizes.

The size of a wrench is determined by the size of the nut or bolt head it fits on. The wrench shown in Figure 3-1 has the number 14 stamped on it. This means the opening of the wrench measures 14 millimeters across the opening and it will fit on a bolt head or nut that measures 14 millimeters across.

Wrench sizes are given either in metric or U.S. (English) system units (see Chapter 6, "Mechanical Measurements and Measuring Devices." Metric wrench sizes are given in millimeters, for example, 7, 8, 9, 10, 11, 12, and 13 millimeters. Sizes for U.S. (English) system wrenches are given in fractions of an inch, for example, $5/16$, $3/8$, $7/16$, $1/2$, and $9/16$ inches. Because both U.S. (English) and metric system bolts and nuts are common, the technician will need both metric and U.S. (English) wrench sets.

Open-End Wrenches

An **open-end wrench** is one common type of wrench (Figure 3-2). These wrenches have an opening at the end that is placed on the bolt or nut. The opening is usually at an angle to the handle. Often one of the gripping surfaces or jaws is thicker than the other. This makes turning in a tight space easier. A hexagonal nut can be turned continuously just by turning the

Shop Manual
page 39

An *open-end wrench* has an opening at the end that can be place on the bolt or nut.

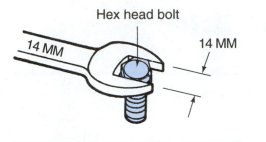

Hex head bolt

14 MM   14 MM

Figure 3-1 The size stamped on a wrench shows what bolt head or nut it fits.

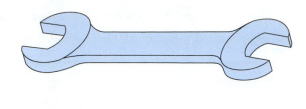

Figure 3-2 An open-end wrench.

wrench over between each swing (Figure 3-3). Open-end wrenches are made in many different sizes and shapes. Most open-end wrenches have two open ends of different sizes.

CAUTION: The technician must be careful to use the right size wrench with any given nut or bolt. If a wrench opening is larger than the nut, the corners of the nut or bolt head can be rounded off. The technician should always be careful or bruised knuckles or broken bones could result.

When using open-end wrenches, always try to pull the load as shown in Figure 3-4. When a wrench is pulled, it is usually away from some obstruction. When a wrench is pushed, it is toward an obstruction. A hand injury can result if the wrench slips off the nut or bolt head. If an open-end wrench must be pushed toward an obstruction, do not wrap fingers around the handle. Use the heel of the hand as shown in Figure 3-5. Be prepared with good footing and a clean floor in case something breaks.

Box-End Wrenches

A box-end wrench is designed to fit all the way around a bolt or nut.

Box-end wrenches like the one shown in Figure 3-6 are designed to fit around a bolt or nut. They usually have twelve points or corners to grip the nut or bolt head. They allow the technician to apply a lot of force with less chance of the wrench slipping off the nut or bolt. Like other wrenches, they come in many sizes. The angle of the head is offset to the handle to give turning room in a tight space. Two common offsets are 15 and 45 degrees as shown in Figure 3-7.

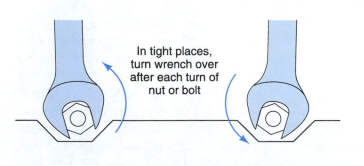

In tight places, turn wrench over after each turn of nut or bolt

Figure 3-3 An open-end wrench is flipped over to turn nuts in tight places.

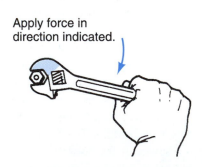

Apply force in direction indicated.

Figure 3-4 Pull on an open-end wrench when tightening.

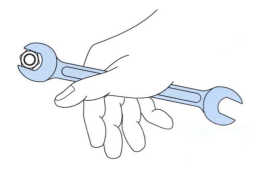

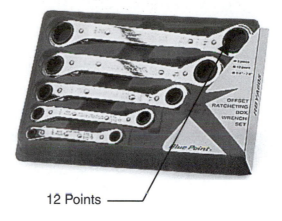

12 Points

Figure 3-5 Correct way to push on an open-end wrench.

Figure 3-6 A box-end wrench with twelve points. (Courtesy of Snap-on Tools Company)

Combination Wrenches

Combination wrenches like the one shown in Figure 3-8 are very useful. They are called combination wrenches because they have one box-end and one open-end. The open-end side is used where space is limited. The box-end side is used for final tightening or to begin loosening. They are usually the same size at both ends.

A combination wrench has one box end and one open end.

Socket Wrenches

Socket wrenches, or sockets, have the wrench part and the handle part made in two different pieces. The wrench part (Figure 3-9) fits all the way around the bolt head or nut with little danger of slipping off. The wrench part can be removed from the handle. Sockets of many different sizes can be used with one handle. Sockets usually come in sets and are made in all the English and metric sizes.

Socket wrenches and attachments are often grouped together and called "sockets." They fit all the way around a bolt or nut and can be detached from a handle.

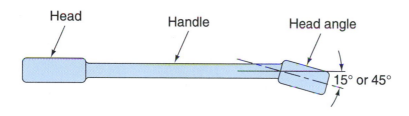

Figure 3-7 The box-end of the wrench may be offset to the handle 15 or 45 degrees.

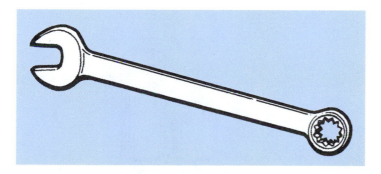

Figure 3-8 A combination wrench. (Courtesy of U.S. Navy)

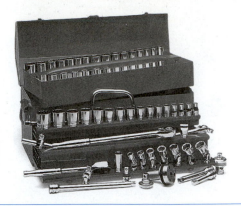

Figure 3-9 A socket wrench or socket. (Courtesy of Snap-on Tools Company)

Sockets are attached to a square drive lug on the handle by a square hole at one end of the socket (Figure 3-10). These drive holes and lugs are also made in different sizes. For small bolts and nuts, such as automotive trim parts, socket sets with a $^1/_4$-inch square drive are useful. For general purpose work, a $^3/_8$-inch drive set is popular. Heavier work requires a $^1/_2$-inch drive socket set. Even larger sizes, $^3/_4$ inch and 1 inch, are made for driving very large socket wrenches.

Socket wrenches come in two basic lengths: the standard or common length, and long or deep sockets. The latter is longer so that it fits over a long bolt or stud. A standard and deep socket are shown in Figure 3-11. Deep sockets come in all the same common wrench sizes as the standard size sockets.

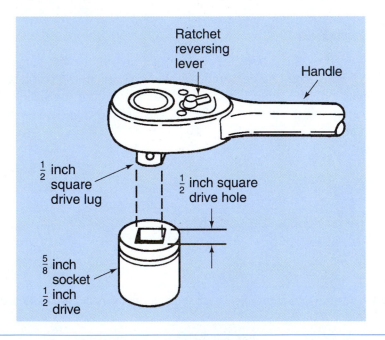

Figure 3-10 The square drive lug on the ratchet handle fits the square hold in the socket.

Figure 3-11 A deep and standard length socket.

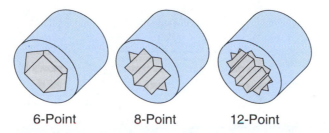

6-Point   8-Point   12-Point

Figure 3-12 Sockets are made with different numbers of points.

Sockets are also available with different numbers of corners, or points, inside the socket to grip the bolt or nut. The three common numbers of points are six, eight, and twelve (Figure 3-12). The twelve-point socket is easiest to slip over a bolt or nut because of its many corners. The six-point socket is the hardest to slip over the bolt or nut. The fewer the points or corners, the stronger the socket. The six-point socket is stronger than the twelve-point version and less likely to slip and round the corners of a fastener.

Socket Drivers, Handles, and Attachments

Socket wrenches require **handles** or **drivers** in order to be used. A large number of handles and **attachments** are available to drive socket wrenches. The most commonly used type of driver is called a *ratchet handle* (Figure 3-13). It has a square drive that fits into the square hole in the socket wrench. The socket is then placed over a bolt or nut. The bolt or nut is tightened or loosened by rotating the socket handle. A freewheeling or ratchet mechanism inside the ratchet handle allows it to drive the nut in one direction and to move freely in the other direction without driving the nut. This permits fast work in a small space because the socket does not have to be removed from the nut each time it is turned. A lever on the ratchet handle allows the mechanic to choose which direction the ratchet will drive and which direction it will turn free (Figure 3-14).

Socket drivers, handles, and attachments fit into the square hole in the socket wrench and are used to drive socket wrenches.

Figure 3-13 A ratchet handle for sockets. (Courtesy of Snap-on Tools Company)

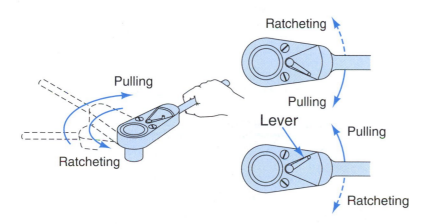

Figure 3-14 Switching the lever allows the ratchet to pull in either direction.

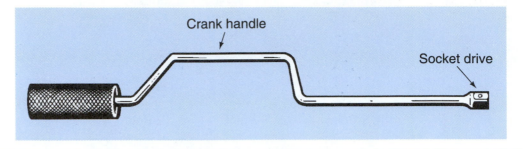

Figure 3-15 A speed handle. (Courtesy of U.S. Navy)

The *speed handle* (Figure 3-15) is another popular socket driver. The socket wrench is installed on the end of the speed handle. The technician pushes on the end of the handle to hold the socket firmly on the nut or bolt. At the same time, the technician turns the crank-shaped handle in a direction to tighten or loosen. The combination of a swivel handle and crank allows very quick driving of a socket. The speed handle is used when a large number of bolts or nuts must be removed or replaced. Technicians often use this handle when removing an automatic transmission fluid pan. A *breaker bar* (Figure 3-16) is another common socket driving tool. The breaker bar is used to break loose bolts or nuts that are very tight. The long handle allows the technician to get a lot of force into the turning. Its drive end has a hinge that will permit driving at different angles.

A *sliding T-handle* is also used to drive sockets. This tool has a handle shaped like the letter T. The handle slides through the driver part. The handle can be used on center or it can be moved outward to provide a longer lever for more torque. A sliding T-handle driver is shown in Figure 3-17.

Socket wrenches are often attached to the drivers through *extensions* (Figure 3-18). One end of the extension is connected to a handle or driver and the other to a socket wrench. The extension allows the socket to be used in an area where an ordinary handle would not have enough room to turn. Extensions come in a variety of lengths for the different jobs they have to do. The common lengths are 3, 6, 12, 18, 24, and 36 inches. Flexible extensions are also available that allow the technician to bend them around obstructions.

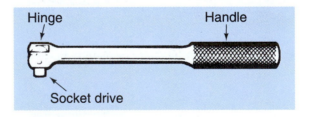

Figure 3-16 A breaker bar. (Courtesy of U.S. Navy)

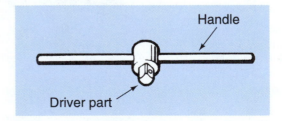

Figure 3-17 A sliding T-handle. (Courtesy of U.S. Navy)

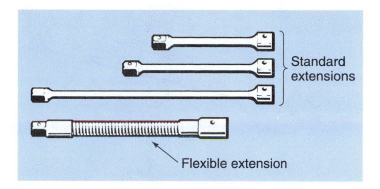

Standard
extensions

Flexible extension

Figure 3-18 Extensions are available in different lengths. (Courtesy of U.S. Navy)

Figure 3-19 A universal extension. (Courtesy of Proto Tools)

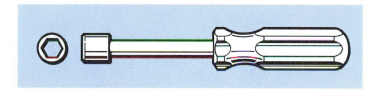

Figure 3-20 A nut driver. (Courtesy of U.S. Navy)

A special type of extension, or *universal*, is available when the extension must operate at a sharp angle (Figure 3-19). This is desirable when parts get in the way of driving a socket. It is made in two parts, which allow it to bend at near 90 degrees.

A socket, extension, and handle can be combined into one tool called a *nut driver* (Figure 3-20). The handle is often made like a screwdriver handle. Nut drivers are usually made in small sizes for very small nut and bolt heads.

Adjustable Wrenches

The wrench shown in Figure 3-21 is called an **adjustable wrench** because it adjusts to fit bolts and nuts of different sizes. There are adjustable wrenches from about 4 inches to about 20 inches long. The longer the wrench, the larger the opening will adjust. For example, a 6-inch adjustable opens $3/4$ inch wide and the 12-inch opens $1 5/16$ inches.

An *adjustable wrench* adjusts to fit bolts of different sizes.

Many people call the adjustable wrench a "Crescent wrench." This is due to the fact that the Crescent Tool Company made this type of wrench for many years. However, the correct name is adjustable wrench.

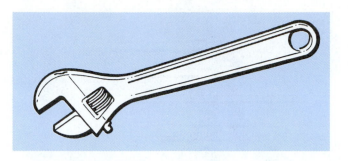

Figure 3-21 An adjustable wrench. (Courtesy of U.S. Navy)

CAUTION: The adjustable wrench is not as strong as a box-end or open-end wrench. The adjustable jaw can break if overloaded. The technician must be careful to adjust the jaws to fit snugly against the flats of the nut or bolt head. When tightening, the wrench must be placed on the bolt or nut so that stress falls on the stationary jaw. If the wrench is used incorrectly, the adjustable jaw can break and cause injury. The right and the wrong way to use an adjustable end wrench are shown in Figure 3-22.

Allen Wrenches

An *Allen wrench* is used to tighten or loosen Allen head screws.

Some automotive components are fastened with hollow head Allen screws. These screws require special **Allen wrenches** (Figure 3-23). The hexagonal, or six-sided Allen wrench fits in the hexagonal head of the Allen head screw. The fit is tight and prevents the wrench from slipping out of the screw head. Allen wrenches are available in sets. They are sized according to the size of the Allen screw in which they fit. They are made in U.S. (English) system sizes such as $3/32$ and $1/8$. They are also made in metric sizes such as 6 millimeter, 8 millimeter, and 10 millimeter. Allen wrenches are made to fit on socket drivers as shown in Figure 3-24.

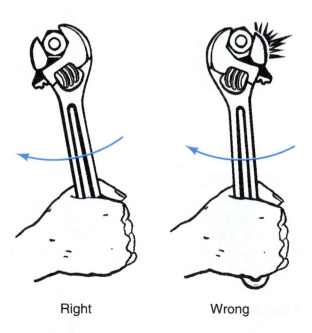

Right   Wrong

Figure 3-22 The right and wrong way to use an adjustable wrench. (Courtesy of General Motors Corporation, Service Operations)

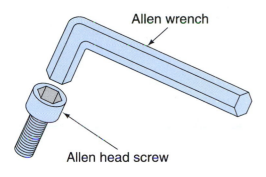

Allen wrench

Allen head screw

Figure 3-23 An Allen head screw and Allen head wrench set.

Figure 3-24 An Allen head wrench for a socket driver. (Courtesy of Snap-on Tools Company)

Torque Wrenches

When some automotive parts are reassembled after repair, the bolts and nuts must be tightened a certain amount. A special socket handle called a **torque wrench,** which measures torque, is used for this purpose. Torque is the turning or twisting force. A common measurement of torque is foot-pounds (ft.-lb.). This means a force, measured in pounds, is acting through a distance of 1 foot. For example, imagine a wrench that is 1 foot long. Then push on the end of the wrench with 50 pounds of force. We have just applied 50 foot-pounds of torque.

There are many types of torque wrenches. One popular type (Figure 3-25) uses a beam and pointer assembly. During tightening, the beam on the wrench bends as the resistance to turning increases. The torque is shown on a scale near the handle. Another type of torque wrench (Figure 3-26) has a ratchet drive head and a breakover hinge. Its adjustable handle and scale allows the mechanic to adjust the wrench to a certain torque setting. When the torque setting is reached, the breakover hinge swivels to signal the technician to stop turning. Another type has a dial on the handle (Figure 3-27). This dial has a needle that points to the amount of torque the wrench is delivering.

Several different torque measurement systems are in use. Specifications are given in metric or U.S. (English) units. Typical U.S. (English) units are inch-pounds and foot-pounds. There are 12 inch-pounds in 1 foot-pound. Metric system torque specifications are most often given in Newton-meters.

A *torque wrench* is designed to tighten bolts or nuts to a certain tightness or torque.

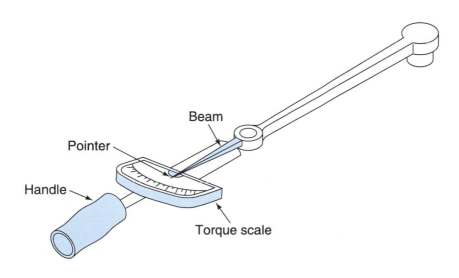

Beam

Pointer

Handle

Torque scale

Figure 3-25 A beam-type torque wrench.

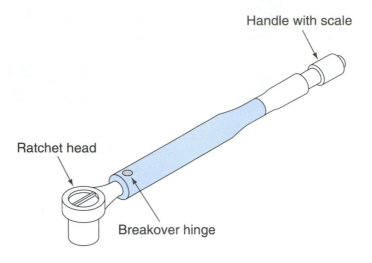

Handle with scale

Ratchet head

Breakover hinge

Figure 3-26 A torque wrench with a breakover hinge.

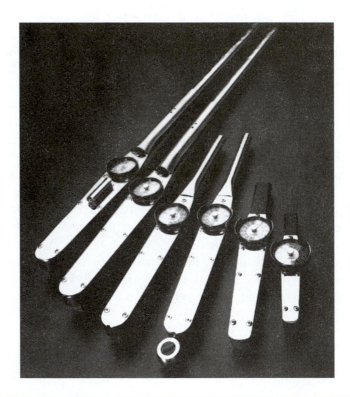

Figure 3-27 A dial readout-type torque wrench.

Special Purpose Wrenches

Most automotive service jobs are done with regular open-end, box-end, combination, and socket wrenches. Some jobs, however, require special wrenches. These are usually variations of the common wrenches that have special shapes to get around obstructions. For example, a *brake bleeder wrench* (Figure 3-28) is a box-end wrench that is bent to fit on the small bleeder valves on brake wheel cylinders. Another example (Figure 3-29) is a *distributor wrench*. This wrench has a long, bent handle that allows it to fit on distributor holddown clamps. Then there is the *flare nut* or *fuel line fitting wrench* (Figure 3-30). This one is a combination of an open-end and a box-end wrench. It is used on fuel line fittings. An open-end wrench might round off the soft metal fittings. A box-end wrench would be impossible to slip over a fuel line. There are, of course, many other special purpose wrenches. Most technicians collect these wrenches for special jobs.

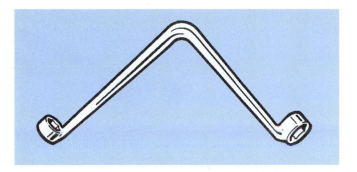

Figure 3-28 A special brake bleeder valve wrench. (Courtesy of U.S. Navy)

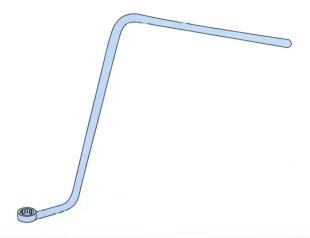

Figure 3-29 A special box-end wrench for distributor holddown fasteners.

Figure 3-30 A special open-end wrench for line fittings. (Courtesy of Snap-on Tools Company)

Pliers

Pliers act as an extension of the technician's fingers. They allow the technician to grip parts with great force. In addition, some pliers are made for cutting things like wire or cotter keys. Good technicians know which pliers to use for every job. They never use pliers when some other tool will do the job better; for example, pliers are never used to loosen or tighten a nut.

Combination Pliers

Combination pliers (Figure 3-31), one of the most commonly used types, have a slip joint where the two jaws are attached. This slip joint can be set for either of two jaw openings: one for holding small objects and one for holding larger objects. These pliers are used for pulling out pins, bending wire, and removing cotter pins. They come in many different sizes. The pliers

Combination pliers are general purpose pliers with a slip joint for two jaw openings.

39

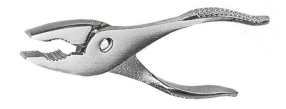

Figure 3-31 Combination pliers. (Courtesy of Snap-on Tools Company)

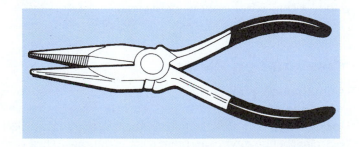

Figure 3-32 Channel-lock pliers.

most commonly used in automotive work is the 6-inch combination. Some combination pliers have a side cutter to allow the cutting of wire and cotter pins. The better grades of combination pliers are made of drop-forged steel and can take a great deal of hard usage.

Channel-Lock Pliers

Channel-lock pliers have channels cut so that the jaws may be adjusted to open wide or narrow openings.

Channel-lock pliers (Figure 3-32) are used to grip large objects. They get their name from the channels that allow the jaws to be set at many different openings. The design of the channels permit quick, nonslip adjustments with practically parallel jaws. These pliers are slim enough, with long enough handles, so that the 45-degree jaws will reach and firmly grip objects that would be out of reach of ordinary pliers. The jaw teeth are sharp and deep to take a firm grip on pipes and hoses. The interlocking design of the slipjoint works well to prevent slipping under load.

A BIT OF HISTORY

Channel-lock pliers are often called "water pump pliers." This is the only tool with which a technician can remove the large nut from the water pump in cars made from the 1920s to the 1940s.

Diagonal Cutting Pliers

Diagonal cutting pliers have cutting edges on the jaw for cutting pins or wire. These pliers are often called "dikes," which is short for "diagonals."

Diagonal cutting pliers (Figure 3-33) are used to cut electrical wire and cotter pins. Their jaws have hardened cutting edges. Because they are made for cutting wire, they have hard cutting edges. Like other pliers, they are made in many different sizes and are grouped by their overall length.

Needle Nose Pliers

Needle nose pliers (Figure 3-34) have long, slender, tapering jaws that are useful in gripping small objects. They are sometimes called "long-nose pliers." Some needle nose pliers are equipped with side cutters; others are bent at right angles for hard-to-get-at places (Figure 3-35).

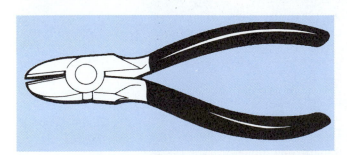

Figure 3-33 Diagonal cutting pliers. (Courtesy of U.S. Navy)

Figure 3-34 Needle nose pliers. (Courtesy of U.S. Navy)

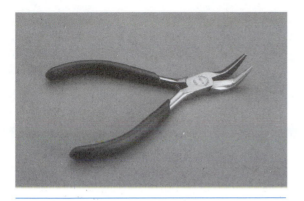

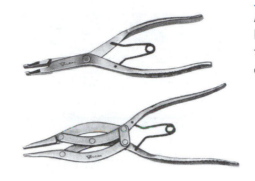

Needle nose pliers have long, slim jaws for gripping small objects.

Figure 3-35 Bent needle nose pliers.

Figure 3-36 Snap ring pliers.

Snap Ring Pliers

Many automotive components are held together with rings that snap into place under tension. Special pliers are required to remove and replace these rings. There are two major types of **snap ring pliers** as shown in Figure 3-36. *Inside snap ring pliers* have jaws that close and grip when the handles are closed like an ordinary pair of pliers. *Outside pliers* have jaws that open when the handles are drawn together. The jaws of each are designed to safely engage the powerful snap rings to be removed or installed.

Snap ring pliers fit into internal- or external-type snap rings and allow them to be removed safely.

> **CAUTION:** Snap rings are powerful springs. They must be handled with the proper tools and with great care. If not, they can fly off their retaining grooves and cause injury. Always use eye protection and the right tool to remove or install snap rings.

Vise Grips

Vise grips (Figure 3-37) are compound-lever pliers that can be locked to the part so the technician can remove her or his hand. The adjustable lower jaw is retracted by an overcenter spring when the locking handle is released. Adjustment of the jaw opening size is made by turning a screw at the end of the primary handle. Vise grips do not take the place of wrenches. They are, however, often used to hold nuts that have had their corners rounded off by the misuse of wrenches.

Vise grips can be locked on a part to hold it tightly like a hand-held vise.

Over center spring   Adjusting screw

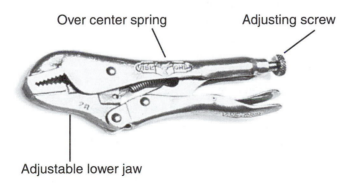

Adjustable lower jaw

Figure 3-37 Vise grip pliers. (Courtesy of Snap-on Tools Company)

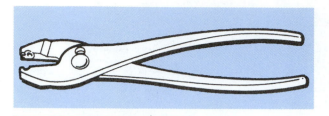

Figure 3-38 Hose clamp pliers. (Courtesy of U.S. Navy)

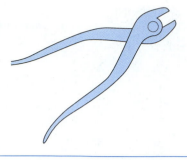

Figure 3-39 Battery terminal pliers.

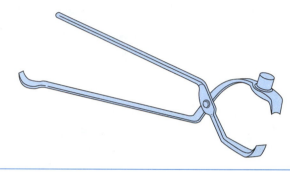

Figure 3-40 Grease cap pliers.

Special Purpose Pliers

There are many special purpose pliers designed for certain repair jobs. The pliers shown in Figure 3-38 are used to remove and replace the spring tension clamps on some coolant hoses. The special groove in the jaws grip the ends of the hose clamps.

The pliers shown in Figure 3-39 are *battery pliers*. The jaws on these pliers are made to grip terminal nuts on battery connections. The tool shown in Figure 3-40 is used to remove hubcaps and grease caps. The round pad on the one jaw is used as a hammer to replace grease caps.

The *crimping tool* shown in Figure 3-41 is used for electrical work. It has cutting edges to cut wire. There are special cutting edges for stripping the insulation off of wire. The jaws of the tool are designed to squeeze or crimp solderless terminals and connectors on the ends of wire.

Screwdrivers

Many automotive components are held together with screws. A screwdriver is used to turn or drive a screw. There are many different types of screws and screw heads. Consequently, there are many different types of screwdrivers to drive them.

Standard Screwdrivers

A *standard screw-driver* is made to drive standard slotted screws.

A **standard screwdriver** (Figure 3-42) is the most common type. It is used to drive screws with a straight slot in the top. The main parts of the standard screwdriver are shown in Figure 3-43. The technician grips and turns the *handle*. The steel part extending beyond the handle is the *shank*. The part that fits into the screw head is the *blade*, which is often called a *bit*. The blade, or bit, is flat and ground at a right angle to the shank. The handle may be made from wood or plastic. The shank is designed to withstand a great deal of twisting force. The length of these screwdrivers is determined by their overall length. The larger the screwdriver, the larger its blade.

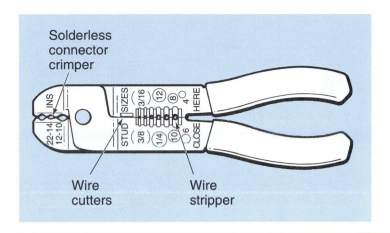

Figure 3-41 Crimping tool for electrical work. (Courtesy of U.S. Navy)

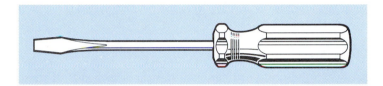

Figure 3-42 A standard screwdriver. (Courtesy of U.S. Navy)

Standard screwdriver bits are available as socket drivers (Figure 3-44). These can be driven with the same drivers and attachments as socket wrenches.

> **CAUTION:** Care must be taken to use a screwdriver whose blade fits snugly into the in the screw. If the fit is loose, the head of the screw may be damaged. More important, the screwdriver could slip out of the screw and stab the technician's hand. The technician should not use a screwdriver to pry on parts. If a screwdriver is used as a prybar, the shank will bend. Then the tip of the blade, which is hardened to keep it from wearing, is brittle and may break. The broken tip could fly into the face and cause injury. Never use pliers to turn the blade of a screwdriver. There are heavy-duty screwdrivers with square shanks that can safely be used with a wrench.

Recessed Head Screwdrivers

There are a number of screws with **recessed-type heads.** Special screwdrivers are required to fit these screws. The most common example is the *Phillips screwdriver* (Figure 3-45). This screwdriver is made to fit Phillips-head screws. Phillips-head screws are used primarily to hold moldings

Recessed head screwdrivers are made to fit recessed head screws like Phillips, clutch, Reed and Prince, and torx.

Figure 3-43 Parts of a screwdriver. (Courtesy of U.S. Navy)

Figure 3-44 A standard screwdriver bit driven by socket drivers. (Courtesy of Snap-on Tools Company)

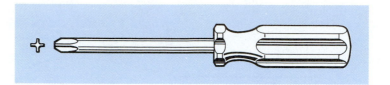

Figure 3-45 A Phillips or cross-tip screwdriver. (Courtesy of U.S. Navy)

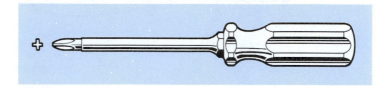

Figure 3-46 A Reed and Prince screwdriver. (Courtesy of U.S. Navy)

and other trim usually on the body or interior of the car. The heads of these screws have two slots that cross at the center and do not extend to the edges of the heads. This design has the advantage that the screwdriver head will not slip out of the slots and scratch the finish. The blades are sized on a numbering system from 0 to 6, with 0 being the smallest and 6 the largest.

The *Reed and Prince* (Figure 3-46) is another common recessed head screwdriver. Reed and Prince cross-slot screws are similar to, and often confused with, Phillips-head screws. But the Reed and Prince slots are deeper and the walls separating the slots are tapered (Figure 3-47). The technician must be careful not to mix up the two types of screws. Using the incorrect screwdriver can result in damaged screw heads.

Clutch screwdrivers are made for turning clutch-bit screws (Figure 3-48). These are also known as figure-eight screws. The screw head they fit looks like the number eight. Like the Phillips and Reed and Prince screws, clutch-type screw heads are designed to keep the blade from slipping off the screw head.

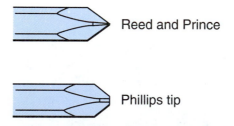

Reed and Prince

Phillips tip

Figure 3-47 The Reed and Prince bit is longer than the Phillips.

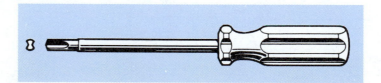

Figure 3-48 A clutch screwdriver. (Courtesy of U.S. Navy)

Figure 3-49 A torx-type screwdriver. (Courtesy of © 1998 Stanley Tools, a Product Group of The Stanley Works, New Britain, CT)

Another common recessed tip screwdriver, called a *torx-type screwdriver*, is shown in Figure 3-49. It has a six-prong tip and is often used to fasten automatic transmission parts.

There are many other types of recessed head screws. Each requires a different type of screwdriver. Screwdrivers are also available that have interchangeable bits on the end of the shanks. Bits are available for each of these recessed head screws.

Offset Screwdrivers

Offset screwdrivers (Figure 3-50) are designed for working in a tight space. The blades at opposite ends are at right angles to each other. The technician can change ends of the offset screwdriver after each swing and continue to turn the screw (Figure 3-51). These screwdrivers come in all the common blade or bit sizes.

Offset screwdrivers have blades arranged at an offset to allow driving screws in small spaces.

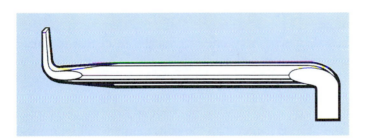

Figure 3-50 An offset screwdriver. (Courtesy of U.S. Navy)

Figure 3-51 The ends of the offset screwdriver are reversed to keep the screw turning. (Courtesy of General Motors Corporation, Service Operations)

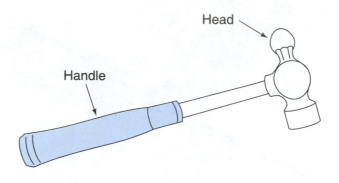

Figure 3-52 Parts of a hammer. (Courtesy of Snap-on Tools Company)

Shop Manual
page 56

Hammers

Hammers are used in many different trades. Every hammer has two basic parts (Figure 3-52): a head and a handle. The handles are made from wood or plastic. The heads are made from different materials for different jobs. The automotive technician will need to know how to use several different types of hammers. A good technician knows when and how to use the correct type of hammer.

> **CAUTION:** Make sure the hammer head is securely attached to the handle before using any hammer. A loose head can fly off and cause serious injury. Never strike a hardened-steel surface with a steel-headed hammer because particles of steel may break loose from the hammer or steel surface and fly off and cause injury.

Ball-Peen Hammers

A *ball-peen hammer* has a head with one round face and one square face.

The **ball-peen hammer** (Figure 3-53) is one of the most common hammers used by automotive technicians. These hammers have one face on the head that is square to the end of the handle. The opposite end is rounded and is called the ball peen. The square end is used for general hammering, such as driving a pin punch or chisel. The rounded ball peen is used to form or peen over the end of a rivet as shown in Figure 3-54.

The hardened head of a ball-peen hammer should not be used to hammer on automotive parts. Because the head is harder than the part, the part will be damaged. Ball-peen hammers of different sizes are listed according to the weight of their head. Small ones weigh as little as 4 ounces; big ones weigh over 32 ounces.

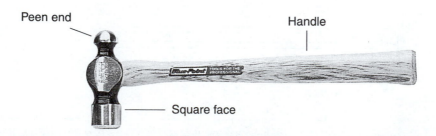

Figure 3-53 Parts of a ball-peen hammer. (Courtesy of Snap-on Tools Company)

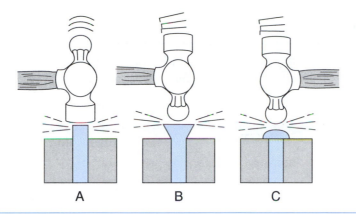

Figure 3-54 Using a ball-peen hammer to set a rivet.

Plastic Tip Hammers

Plastic tip hammers (Figure 3-55) have a head made from plastic. These heads are much softer than the ball-peen, steel-head hammers. They are used to align or adjust parts that might be damaged by a steel-head hammer. Plastic hammers will not damage metal parts because the plastic is softer than metal. Plastic can also absorb the force of the impacts instead of transmitting them to the metal parts. The technician should always remember to use a hammer that is softer than the material being hammered on.

A plastic tip hammer has plastic tips for protecting the surface of parts being hammered.

Brass Hammers

Hammers with brass heads (Figure 3-56) are softer than ball-peen hammers. **Brass hammers** are much heavier than plastic ones. They can be used when more force is required. Brass hammers are often used for driving pilot pins and other materials that will be harmed or distorted by steel hammers.

Brass hammers have heads made from soft brass and are used to hammer on parts without damaging them.

Rubber Mallets

Rubber mallets (Figure 3-57) are softer than brass or plastic. These are used to seat hub caps, wheel covers, and other trim parts. They are soft enough so that they will not damage the finish of parts like wheel covers.

Rubber mallets have rubber heads made to hammer on trim parts like wheel covers.

Figure 3-55 A plastic hammer. (Courtesy of Snap-on Tools Company)

Figure 3-56 A brass hammer. (Courtesy of Snap-on Tools Company)

Figure 3-57 A rubber mallet. (Courtesy of Snap-on Tools Company)

Punches

Punches are used with ball-peen hammers to drive pins or to make the center of a part to be drilled. The most common types are the *pin punch, starter punch, center punch,* and *cotter pin remover.*

Pin and Starter Punches

A *starter punch* has a taper on the end that allows the starting of pin removal.

Some automotive components are held together with pins or rivets. Two different kinds of punches may be needed to remove a pin or rivet. A **starter punch** is used to break the pin loose. Then a **pin punch** smaller than the hole is used to drive a pin out of a hole. A starter punch (Figure 3-58), also called a "drift punch," is made with a shank that is tapered all the way to its end. The tapered shank of a starter punch is stronger and will withstand greater shock from hammer blows than a pin punch. The starter punch is used to start the removal of pins and rivets.

A *pin punch* is a tool used with a hammer to drive out pins from automotive parts.

Drifts are usually made of brass and are used to drive out hard or highly-finished parts without damaging them. They are not always tapered. Long, tapered, steel drifts are often used to align two parts so that a bolt or screw can be installed.

Pin punches (Figure 3-59) have straight shanks and are used to complete the removal of pins and rivets that have been started with a starter punch. The thin, straight shank of the pin punch is more likely to break under heavy blows from a hammer. The starter punch is used only to start a rivet or pin moving. The tapered design of its shank will prevent the starter punch from going all the way through. The pin punch is used to complete the job (Figure 3-60). Always use the largest starter and pin punches that will fit the hole.

A *center punch* is made of steel and used to mark the centers of holes to be drilled.

The **center punch** is a short, steel punch with a hardened conical point ground to a 90-degree angle (Figure 3-61). A center punch is used to mark the centers of holes to be drilled.

Tapered shank

Straight shank

Figure 3-58 A starter punch.

Figure 3-59 A pin punch.

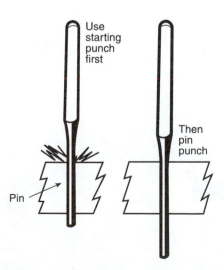

Use starting punch first

Then pin punch

Pin

Figure 3-60 Using a starter and pin punch to remove a pin. (Courtesy of General Motors Corporation, Service Operations)

Figure 3-61 A center punch.

Figure 3-62 A cotter pin puller.

If the center punch is used correctly, the depression it makes in the metal will guide the point of the drill bit and make sure the hole is drilled in the correct spot.

Cotter Pin Pullers

Diagonal cutting pliers are often used to pull out cotter pins after the ends have been straightened out or cut off. The **cotter pin puller** is a special tool that does the job more easily (Figure 3-62). This tool has a hook that engages the loop of the pin and a handle with which to pull it.

A *cotter pin puller* is used to hook and remove a cotter pin.

Power Wrenches

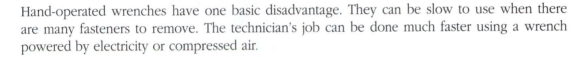

Hand-operated wrenches have one basic disadvantage. They can be slow to use when there are many fasteners to remove. The technician's job can be done much faster using a wrench powered by electricity or compressed air.

Shop Manual
page 39

Electric Wrenches

An **electric wrench** (Figure 3-63) has an electric motor operated by a trigger on the handle. Special heavy-duty sockets are attached to a socket drive at the front of the wrench. Holding down the trigger spins the drive and socket. A reversing switch allows the technician to loosen as well as tighten bolts and nuts. The main advantage of the electric wrench is speed. Its motor drives a socket much faster than it can be driven by hand. Electric wrenches are especially useful for disassembling parts that are held together with many bolts and nuts, such as engines and transmissions.

An *electric wrench* is powered by electricity for fast removal or installation of bolts and nuts.

Air-Operated Ratchet Drivers

An **air-operated ratchet driver** (Figure 3-64) is connected to an air line. Pulling the trigger causes the air to rotate a socket attached to the drive on the wrench. A reversing lever allows the technician to loosen as well as tighten. Many air wrenches are designed with an impact feature.

An *air-operated ratchet driver* is a ratchet handle powered by air for fast removal or installation of bolts or nuts.

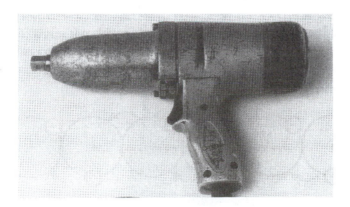

Figure 3-63 An electric wrench. (Courtesy of Snap-on Tools Company)

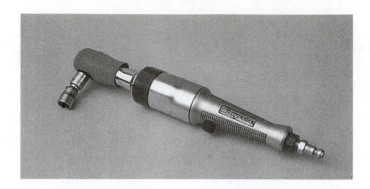

Figure 3-64 An air-operated ratchet. (Courtesy of Snap-on Tools Company)

Figure 3-65 An impact wrench. (Courtesy of Snap-on Tools Company)

Impact Wrenches

An *impact wrench*
is an air-operated
wrench that
impacts as well as
drives a bolt or nut.

An **impact wrench** (Figure 3-65) is an air-operated wrench with a powerful driver. An impact wrench not only drives the socket but also vibrates or impacts it in and out. The force of the impact helps to loosen a bolt or nut that is difficult to remove. They are often used to remove and replace wheel lug nuts.

Most impact wrenches are not torque controlled. This means the technician must use a torque wrench to finish tightening after using the impact wrench.

CAUTION: Special heavy-duty impact sockets must be used with impact wrenches. Impact sockets are thicker and usually not chrome plated as shown in Figure 3-66. Standard sockets cannot withstand the forces; they can break and cause serious injury.

Figure 3-66 Special impact sockets are used with an impact wrench. (Courtesy of Proto Tools)

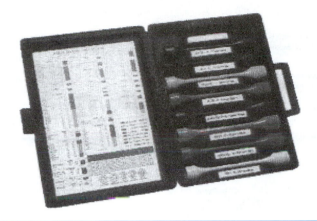

Figure 3-67 Torque sticks are used primarily to tighten wheel lug nuts with an impact wrench. (Courtesy of Snap-on Tools Company)

Wheel Torque Sockets

Wheel torque sockets are commonly known as *torque sticks*. They are used with impact wrenches to deliver a specified torque to a fastener. They are similar to an extension in shape, but are used to prevent overtorquing a fastener while using an impact wrench for its speed. Tire stores are the primary users because of the high number of lug nuts that must be torqued each day.

Each stick is made for a certain torque and most toolmakers color them for that torque (Figure 3-67). The technician need not read the torque printed on the stick: just pick the right color for the desired torque and mount it to the wrench.

The torque stick will deliver the wrench's output up to the stick's limit. Once the limit is achieved, the wrench's output is absorbed by the stick and no torque is delivered to the fastener. However, a torque wrench is still the best method to torque fasteners.

Some tool manufacturers offer adjustable torque impact wrenches. This wrench can be adjusted to provide accurate torque to many types of fasteners. Others offer an impact wrench that is preset to a maximum torque. Impact wrenches of this type work best on larger fasteners such as wheel lug nuts.

Metalworking Tools

Shop Manual
page 51

Many automotive repair jobs require the technician to do some metalworking. Jobs like drilling holes to install accessories or cutting sheetmetal to repair body rust damage are typical examples. The common metalworking tools are described here.

Chisels

Chisels are often used for cutting sheet metal, cutting off rivet heads, and splitting nuts that cannot be removed with a wrench. The most common chisel is called a *flat chisel* (Figure 3-68); these are often called "cold chisels." They are forged from round, square, rectangular, or hexagonal bars of tough carbon steel. A cold chisel is driven with a ball-peen hammer; the heavier the chisel, the heavier the hammer.

The *cape chisel* (Figure 3-69) is another common chisel. The cutting edge of a cape chisel is relatively narrower than the cutting edge of a flat chisel. The cape chisel is forged so that the cutting edge is slightly wider than the shank so that it will not bind when used to cut a narrow groove. Cape chisels are used for cutting narrow grooves such as keyways.

The cutting edge of a *round nose chisel* (Figure 3-70) has only one bevel. It is used mostly for cutting semicircular grooves and inside rounded-off corners.

A *chisel* is a bar of hardened steel with a cutting edge ground on one end, driven with a hammer to cut metal.

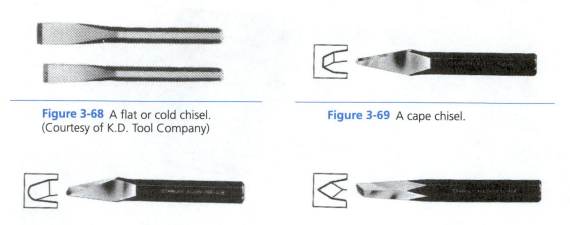

Figure 3-68 A flat or cold chisel. (Courtesy of K.D. Tool Company)

Figure 3-69 A cape chisel.

Figure 3-70 A round nose chisel.

Figure 3-71 A diamond-point chisel.

Diamond-point chisels (Figure 3-71) are forged with a tapered shank of square section that is ground on an angle across diagonal corners. This results in a diamond-shaped cutting edge. Diamond-point chisels are used for cutting V-shaped grooves and for squaring up the corners of slots.

CAUTION: Always wear eye protection when using a chisel. Make sure the part to be cut is held securely in a vise. Hammering sometimes curls over or mushrooms the upper end of a chisel (Figure 3-72). The technician must grind the head smooth because the chips may fly off a mushroomed head and cause injury.

Twist Drills

A *twist drill* is a cutting tool used in a drill motor to cut holes.

The technician will often use **twist drills** mounted or chucked in an electric drill motor to drill a hole. The four parts to a twist drill are shown in Figure 3-73. The end of the drill is called the *point*. The spiral portion is made up of the *body* and the *flute*. The part that fits in the electric drill motor is the *shank*. A straight shank drill is commonly used in portable drill motors.

Twist drills are made in four different size groups: (1) fractional sizes from $1/64$ inch to $1/2$ inch and larger in steps of $1/64$ inch, (2) letter sizes from A to Z, (3) number sizes 1 to 80, and (4) millimeter sizes. Drill sets, or indexes, are sold in each size group. Drill size charts are available that list each drill and give decimal and metric equivalent sizes.

The size of a twist drill is stamped on the shank. But, after a great deal of use, the stamp may be difficult to read. A drill may be measured with a drill gauge. The drill gauge shown in Figure 3-74 is a metal plate with holes identified by size. The drill to be measured is placed in the holes until it is found which size hole best matches the drill.

A twist drill must be sharp to do a good job of cutting. The point of the drill is sharpened with a grinding wheel, usually in a special fixture.

Twist drills are most often used in portable electric drills (Figure 3-75). There are three common sizes of portable drills based on the size of the drill chucks. These are $1/4$, $3/8$, and $1/2$ inch. The size of the chuck determines the largest drill size that will fit inside.

CAUTION: Always wear eye protection when operating a drill. Small pieces of metal to be drilled must always be clamped tightly in a vise.

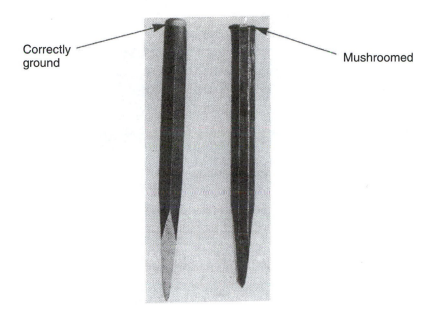

Correctly ground

Mushroomed

Figure 3-72 The mushroomed end of a chisel must be ground off.

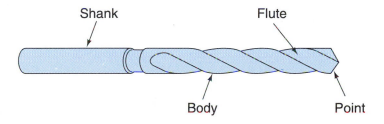

Shank

Flute

Body

Point

Figure 3-73 Parts of a twist drill.

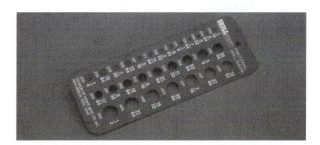

Figure 3-74 A drill gauge for measuring the size of a drill.

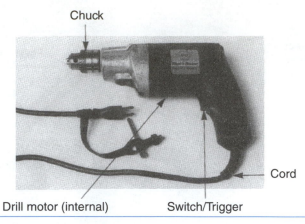

Figure 3-75 Parts of a portable electric drill motor.

Reamers

A *reamer* has cutting edges used to remove a small amount of metal from a drilled hole.

A **reamer** is used when a very precise hole is necessary for a precision fit. A reamer has cutting edges designed to remove a small amount of metal from a drilled hole. Many automotive parts have bushings that must be finished to size with a reamer.

Machine-driven reamers are available, but most automotive jobs require a hand-driven reamer. Like drills, reamers are made in many different sizes, which are stamped on the shank. Reamers are also made that adjust to many different sizes. Some common types of reamers are shown in Figure 3-76.

A reamer should be turned with a wrench or with a tap wrench in a clockwise direction. Turning a reamer backward will quickly dull its cutting edges.

Hacksaws

A *hacksaw* is made to cut metal.

Hacksaws are made to cut metal. A technician may use a hacksaw to cut exhaust pipes and other metal parts that are made during a repair job. They consist of two parts: the *frame* and the *blade* (Figure 3-77). Almost all hacksaw frames are made so that they can use 8-, 10-, or 12-inch blades. The hacksaw blade installed in the frame is the part that does the cutting. Blades are made in different lengths and also with different numbers of teeth per inch (TPI). The

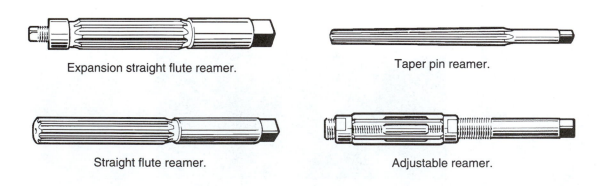

Expansion straight flute reamer.

Taper pin reamer.

Straight flute reamer.

Adjustable reamer.

Figure 3-76 Types of reamers. (Courtesy of U.S. Navy)

Figure 3-77 Two main parts of a hacksaw. (Courtesy of Snap-on Tools Company)

common TPI are 14, 18, 24, and 32. The number of teeth are marked on the hacksaw blade. For almost all automotive work, the 18- and 32-tooth blades are best. The 18-tooth blade is used for all sawing except thin metal, such as sheets or tubing, which should be sawed with 32-tooth blades.

When installing a hacksaw blade in a frame, always have the teeth pointing toward the front of the frame and away from the handle (Figure 3-78). Hold the saw with both hands and push forward and down to cut. Release the pressure to back the saw up for the next stroke. Take about one stroke per second.

Files

A file is used to remove metal for polishing, smoothing, or shaping. The parts of a file are shown in Figure 3-79. The *face* is the larger part of the file with the cutting teeth. The *tang* is the part that goes into the handle. The tapering part between the tang and the face is the *heel*. Files are made in different lengths, measured from the *tip* to the heel. The tang is shaped to fit into the handle. A handle must always be attached when filing to protect the technician from the sharp tang. The handle is set tightly on the file by striking the handle on a workbench, as shown in Figure 3-80.

A *file* is a hardened, steel tool with rows of cutting edges used to remove metal for polishing, smoothing, or shaping.

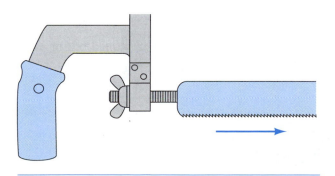

Figure 3-78 The teeth point forward or away from the handle.

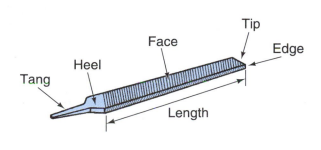

Figure 3-79 The parts of a file.

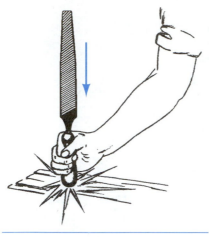

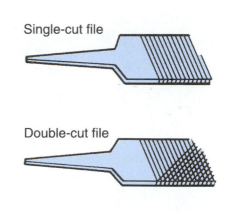

Single-cut file

Double-cut file

Figure 3-80 Strike the file handle on a bench to attach the handle. (Courtesy of General Motors Corporation, Service Operations)

Figure 3-81 Single- and double-cut files.

Files with cutting edges that run in only one direction are called *single-cut files*. Files made with cutting edges that cross at an angle are referred to as *double-cut files* (Figure 3-81). The cutting edges may be spaced close together or wide apart. The wider they are spaced, the faster the file will remove metal. A file whose cutting edges are closer together will remove less metal and can be used to smooth or polish a metal surface. Files are also available in different shapes: flat, half-round, round, triangular, and square (Figure 3-82).

When using a file, grip the handle with one hand. Push down on the file face with the other hand. Because a file is designed to cut in only one direction, raise the file on the return stroke. Dragging it backward dulls the cutting edges. Mount small parts in a vise for filing. The correct way to hold a file is shown in Figure 3-83. When the teeth on the file become clogged with metal filings, remove them by tapping the file handle or brushing the teeth with a file card.

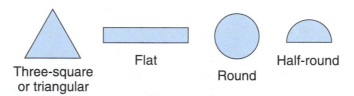

Three-square or triangular

Flat

Round

Half-round

Figure 3-82 Different types of file shapes.

Figure 3-83 Correct way to hold a file.

CAUTION: Always wear eye protection when using a file. Never use a file without a handle. Be sure to mount small parts in a vise when filing. Never use a file as a prybar because it could break and cause injury.

Threading Tools

Many repair jobs involve repairing or replacing the threads on automotive parts. A **tap** is a metal-working tool (Figure 3-84) designed to make or repair inside threads. A tap of the correct size is installed in a holding tool called a tap wrench (Figure 3-85). The tap is then turned in the hole to make a new thread or to repair damaged ones. Taps are available for all the common thread sizes.

A **die** (Figure 3-86) is a metal-working tool used to repair or make outside threads. A die is installed in a tool called a *die stock* (Figure 3-87). The die is then turned down over the part to make new threads or repair damaged ones. Like taps, dies come in a variety of thread sizes.

When a bolt or screw has broken off in a threaded hole, the technician must remove it. One tool for this job is a **screw extractor** (Figure 3-88). A hole of the proper diameter is drilled into the broken screw. A screw extractor is inserted into that hole. A wrench is used to turn the screw extractor counterclockwise. The reverse threads dig into the broken screw and allow the technician to unscrew it as shown in Figure 3-89.

Shop Manual
pages 51–56

A *tap* is used to cut external threads.

A *die* is used to cut internal threads.

A *screw extractor* is used to remove broken screws, bolts, or studs from automotive parts.

Figure 3-84 A tap is used to make internal threads. (Courtesy of U.S. Navy)

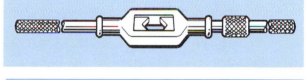

Figure 3-85 A tap wrench. (Courtesy of U.S. Navy)

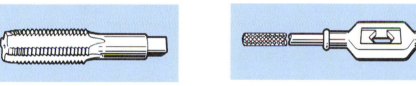

Figure 3-86 A die is used to make external threads. (Courtesy of U.S. Navy)

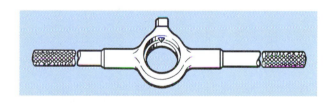

Figure 3-87 A die stock. (Courtesy of U.S. Navy)

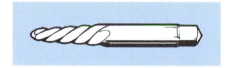

Figure 3-88 A screw extractor is used to remove broken screws or studs. (Courtesy of U.S. Navy)

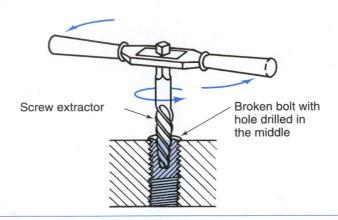

Figure 3-89 Using a screw extractor to remove a broken screw or stud.

Equipment

Shop Manual
page 57

There are many different types of equipment in an automotive shop. They range from large ³/₄-inch and 1-inch drive socket sets to automotive lifts capable of lifting up to 9,000 pounds safely.

Oxyacetylene Welding and Cutting Outfits

The **oxyacetylene welding and cutting outfit** is a common tool used in many automotive repair shops. This equipment can be used to weld parts such as exhaust systems. It can be used to cut metal to make repairs to body panels or to exhaust systems. Technicians often use the oxyacetylene outfit to heat parts in order to make them easier to remove.

The parts of the oxyacetylene welding and cutting outfit are shown in Figure 3-90. The oxyacetylene welding and cutting outfit uses two different kinds of compressed gases—oxygen and acetylene—to create heat. Each gas is stored in a steel cylinder. A regulator on the top of each cylinder is used to regulate the pressure of the gas coming out of the cylinder to the correct amount. Two gauges are attached to each regulator. One shows the pressure in the cylinder. The other shows the working pressure or the pressure going to the torch. Hoses connect the cylinders to the torch. The torch mixes the oxygen and acetylene in the proper proportions to create a heating flame. The flame is created at the tip of the torch and is hot to melt steel for welding or cutting.

An *oxyacetylene welding and cutting outfit* is a system that uses oxygen and acetylene gas as a heating source for welding, cutting, and heating.

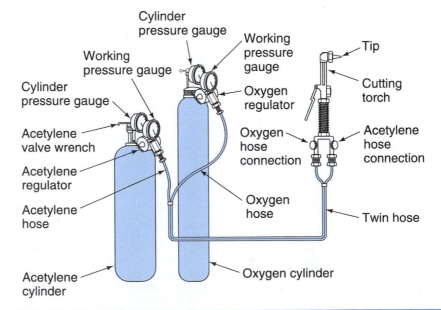

Figure 3-90 Parts of an oxyacetylene welding and cutting outfit.

Jack Stands, Wheel Blocks, and Jacks

Jack stands come in different heights with different load capabilities. The smallest can extend to about 24 inches in height and will support two to four tons of weight (Figure 3-91). Jack stands are used to support the vehicle or one end so the technician can move under it on a creeper. Most under-vehicle work can be performed using this type of stand.

Other stands are much taller and designed to support a portion of a vehicle raised on a lift (Figure 3-92). For instance, a stand may be used to hold a rear-axle housing as the suspension system is being repaired. This type of stand is usually rated at a one-ton capacity.

Jack stands are used to support a vehicle so a technician can move under it with a creeper.

Wheel Blocks

CAUTION: Do not lift one end of the vehicle without placing the transmission in park or in a forward gear (manual), setting the parking brake, and blocking the lowered wheels. Failure to do so could result in injury or damage.

Wheel blocks are placed forward and rearward of the wheels to prevent the vehicle from rolling. Chocks should always be used if only one end of the vehicle is being raised. Place the blocks forward and rearward of at least one wheel left on the floor (Figure 3-93).

Wheel blocks are also known as "wheel chocks."

Jacks

The most common shop jack is a **floor jack** (Figure 3-94). This type of jack may be rated from one ton to ten or more tons. However, the type found in most automotive shops is rated between two and four tons. Floor jacks are designed to roll as the vehicle is lifted. This keeps the

Floor jacks are designed to roll as a vehicle is lifted, thereby keeping the lift plate in position.

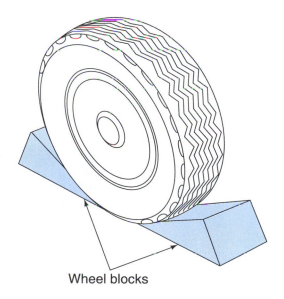

Wheel blocks

Figure 3-91 Jack stands are placed to hold the vehicle securely while under-vehicle work is being done. (Courtesy of MATCO Tool Company)

Figure 3-92 Tall jack stands are used to support portions of a raised vehicle. (Courtesy of Snap-on Tools Company)

Figure 3-93 Blocks should be rearward and forward of the wheels before jacking.

Figure 3-94 Floor jacks are designed to travel under the lift point as the vehicle is being raised. (Courtesy of MATCO Tool Company)

lift plate in position under the vehicle, but it *can* present a problem. On an incline, the jack and vehicle could roll down the slope. This is the reason for using wheel blocks.

There are also **transmission jacks.** They can be tall enough to fit under a lifted vehicle or low enough to be used with the vehicle on jack stands (Figure 3-95). This type of jack has attachments to secure the transmission to the jack.

Transmission jacks can be used under a lifted vehicle or with a vehicle on jack stands.

■ **CAUTION:** Never attempt to move a loaded transmission jack without securing the transmission to the jack. Injury to the technicians or vehicle damage could result.

Lifts

Automotive lifts make the technician's work much easier. The vehicle can be raised so almost any under-vehicle repair can be easily accessed and accomplished. The entry-level technician will use a lift to change oil and inspect or replace tires.

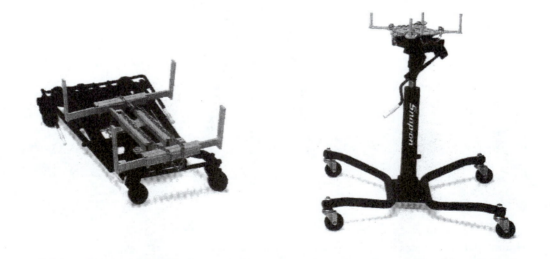

Figure 3-95 Transmission jacks are used to support the transmission during removal and installation. (Courtesy of Snap-on Tools Company)

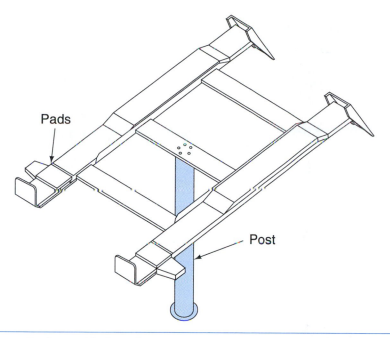

Pads

Post

Figure 3-96 Single-post lifts have the post extending from and retracting into the floor.

Single Posts

Single-post lifts are older types of lifts. The hydraulic cylinder is buried in a floor cavity. The lift arms and the top of the cylinder present a blockage to an otherwise smooth floor (Figure 3-96). In addition, many of the under-vehicle components like the transmission and exhaust system are blocked by the arms when the vehicle is raised. The floorboard and exhaust can also be damaged if the lift arms and pads are not positioned correctly. There is usually only one safety lock positioned at the lift's full extension. Any position between *completely down* or *completely up* relies on the hydraulic and air control valves to hold the vehicle in place.

Double Posts

The **double-post lift** is also known as the "above-ground lift" (Figure 3-97). It is designed to lift a vehicle and allow the technician to perform almost any repairs needed under the car. Many lifts of this type have a crossover between the two posts at the top. This provides a clear work area under the vehicle.

Most lifts of this type use two equalizing hydraulic cylinders to raise the vehicle. Cables or flat chains run over pulleys and attach to each cable, ensuring that the vehicle is lifted evenly. There are locks at different heights so the vehicle can be safely locked in place at several working heights.

Scissors and Drive-Ons

Scissor lifts are best used for brake or tire service. They fit under the vehicle and have limited height (Figure 3-98). As such, there is almost no room to work under the vehicle.

Drive-on lifts allow the technician full access to under-vehicle components (Figure 3-99). Unless equipped with secondary jacks, the vehicle rests on its tires for the duration of the repairs. Drive-on lifts with secondary jacks are generally used for alignment. The vehicle can rest on its tires for alignment or the axles can be raised for access to the wheel and brake assembly.

Single-post lifts are an older type of lift. They can be difficult to use because many under-vehicle components are blocked by the lift arms when the vehicle is raised.

Dual-post lifts may also be mounted under the floor like a single-post lift.

Scissor lifts are used for brakes or tires, fit under the vehicle, and have limited height.

Drive-on lifts provide full access to components under a vehicle and are generally used for alignment.

Figure 3-97 Double-post lifts with an overhead crossover leave a clean floor and work area. (Courtesy of Snap-on Tools Company)

Figure 3-98 A scissor lift works very well for brake and tire repairs. (Courtesy of Snap-on Tools Company)

Figure 3-99 Drive-on lifts are excellent work platforms for lower engine, transmission, and exhaust repair. (Courtesy of Gray)

Gear and Bearing Pullers

All technicians should be able to set up and use a **gear puller** or bearing puller. Most shops have a large selection of these pullers (Figure 3-100). These tools are designed for hundreds of common service jobs on the automobile. Although the number and variety of puller tools is huge, they are all designed to do three simple jobs (Figure 3-101).

Shop Manual
pages 50–51

A *gear puller* is a tool used to remove gears, bearings, shafts, and other parts off shafts or out of holes

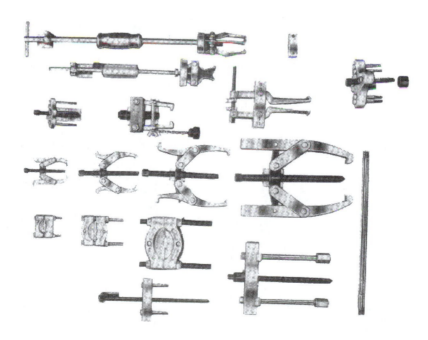

Figure 3-100 Typical puller selection found in most shops. (Courtesy of Owatonna Tool Company)

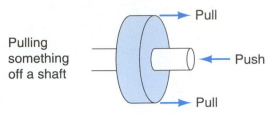

Pulling
something
off a shaft

Pull

Push

Pull

Removing a gear, bearing, wheel, pulley, etc.,
to replace it or get at another part.

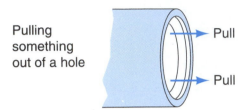

Pulling
something
out of a hole

Pull

Pull

Internal bearing cups, retainers or oil seals are
usually press-fitted and are difficult to remove.

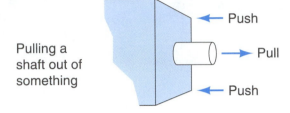

Pulling a
shaft out of
something

Push

Pull

Push

A transmission shaft or pinion shaft is often
hard to remove from a bore or housing.

Figure 3-101 Three common pulling operations. (Courtesy of Owatonna Tool Company)

1. Pulling something off a shaft: A common service job is to pull a gear, bearing, or pulley off a shaft. These parts are made to fit tightly together. Rust and corrosion on the shaft can cause them to be very difficult to remove.

2. Pulling something out of a hole: Parts like bearing cups and seals may be installed inside holes with a press fit. These parts would be very difficult to remove without a puller.

3. Pulling a shaft out of something: Shafts are installed in many transmission components. These parts are very difficult to remove except with the correct puller.

Computerized Data Storage and Retrieval

Shop Manual
page 66

Computerized data banks are ideal for shops that have terminals in each bay. One data bank is available for every technician with a click of a button.

Paper service manuals are rapidly being replaced by **computerized data** storage (Figure 3-102). A series of CD-ROM disks contain all of the information found in many different con-

Figure 3-102 Paper manuals like the ones shown require several different manuals to cover the entire vehicle.

ventional service manuals. The data can be retrieved and printed quickly and easily. Most of the programs are user friendly and require no training other than some hands-on experience. Fax, e-mail, Internet, and direct satellite uplinks can also transmit computerized data. In addition, storage is not required except for the space required for the computer terminal. There are after-market data systems such as Mitchell's On-Demand and Snap-on's Alldata.

The computer data system is less expensive after initial setup than purchasing the required paper manuals. The system is designed for a shop. With the correct software and computer hookup, the shop and even the technician can ask, receive, and give technical assistance over the Internet. Groups like the Coordinating Committee for Automotive Repairs, International Automotive Technician's Network, and the National Institute for Automotive Service Excellence provide on-line tips and conduct forums for automotive representatives to discuss repairs and the business in general. Many automotive businesses have Internet sites along with libraries and schools that provide technical assistance and general information. An Internet search for automotive sites can be made using the keyword "automotive."

Computerized Test Equipment

Shop Manual
page 66

There are many different types of test equipment on the market today. They are designed to communicate with the vehicle's on-board computer or with a specific system. Most are known as *scan tools* or *hand-held lab scopes* (Figure 3-103). Computerized test equipment retrieves data and **diagnostic trouble codes (DTC)** for the technician to use in diagnostic procedures. They will only collect data on electrical troubles. Mechanical problems will not show up as a problem, but may cause a sensor to transmit erroneous information to the computer. This data may show as a sensor problem on the scan tool. Correct interpretation of the information is essential for correct diagnosis.

DTCs are numerical data that can be retrieved from the vehicle's computer with a scan tool. Each code designates a specific fault.

DIAGNOSTIC SCANNER AND DIAGNOSTIC CARTRIDGES

Figure 3-103 Scan tools are used by most shops to retrieve data from the vehicle's computer. (Courtesy of Snap-on Tools Company)

DIAGNOSTIC DIGITAL OSCILLOSCOPE

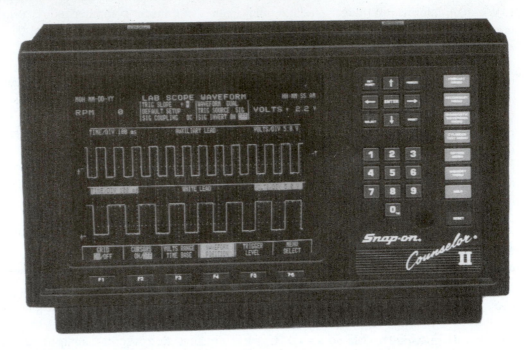

Figure 3-104 Digital oscilloscopes like this one are required for diagnosing almost any new engine or transmission. (Courtesy of Snap-on Tools Company)

The *DSO* is a large, expensive piece of equipment. It is capable of storing and displaying enormous amounts of data in graph, wave, and numerical forms.

A larger, more expensive test instrument is the **digital storage oscilloscope (DSO)** (Figure 3-104). Shops with the customer base and funds to purchase a DSO can diagnose a fault down the last loose screw. The small, hand-held type will perform many of the functions of the larger scopes, but the large unit usually performs a more detailed readout of several data streams from the vehicle's computer at once. They are also quicker, which provides a better readout of what is happening within the vehicle's electronic systems. Most are equipped with pickup tubes to sample the exhaust and provide a total readout of four or five gases. By analyzing the amount of oxygen (O), carbon monoxide (CO), carbon dioxide (CO_2), hydrocarbons (HC), and nitrogen oxides (No_x), gases within the exhaust provide the technician with information on the operation of the engine and its systems.

Shop Manual
page 75

Electrical Test Instruments

Electrical theory is covered in Chapter 4, "Automobile Theories of Operstion." Electrical measurements have become a way of life with today's technician and are essential test devices. Some of the electrical terms used here are also explained in Chapter 4 where they are in context within electrical theory.

Test Lights

 WARNING: Do not use a test light on electronic systems and only use it when directed by the service manual. Damage to electronic circuits may occur.

Even though the *test light* is old technology, it is still valid in many situations (Figure 3-105). One end, the *ground clip,* is attached to a metal portion of the vehicle. The other end, the *probe,* is used to probe the conductor. The test light will indicate voltage, but not the amount.

Figure 3-105 The test light should be used to test an electrical circuit only when directed by the vehicle's service manual. (Courtesy of MATCO Tool Company)

If voltage is present, the lamp will light. It is used on heavy current circuits like the headlight and cooling fan motors.

Multimeters

The multimeter is used to check almost any circuit on the car, but caution must be used when selecting one for purchase. It should be an automotive meter of **high-impedance** and preferably digital (Figure 3-106). A high-impedance meter is used to protect electronic circuits during testing. It will have two regular leads: a black and a red. Automotive multimeters have additional leads for temperature, engine speed, and other measurements exclusive to the automobile. The controls on most multimeters include selections for the three basic electrical measurements: voltage, current, and resistance. Some meters require the operator to select the range of measurements. The better and more expensive ones have automatic ranging. *Ranging* is the scale of measurement. For instance, if voltage is being read, the operator may have to select a 0-to-20-volt scale or a 20-to-100-volt scale. Automatic ranging meters do this automatically.

High-impedance is basically the same as high-resistance. It is required to obtain an accurate measurement of electricity.

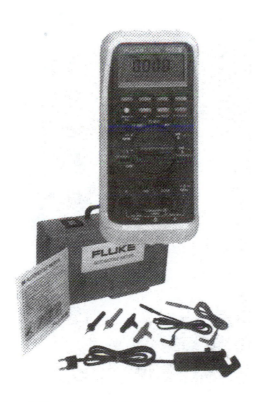

Figure 3-106 Multimeters can measure most vehicles' electrical circuits. (Courtesy of MATCO Tool Company)

Frequency is the number of times an action or repeated action happen within a given time period. It is also known as "duty cycle."

Automotive multimeters have selections for revolutions per minute, temperature, **frequency**, and others depending on the make and model of the meter (Figure 3-107). Additional leads are required to make the most of these measurements.

There are meters that measure only resistance (ohmmeter), voltage (voltmeter), or current (ampere). Meters of this type are not common in the automotive field. There is one meter, however, that is common to almost every tire, parts, and repair facility. It is a *volt/ampere tester* commonly known as the "VAT 40" (Figure 3-108). The VAT 40 is dated technology because it is an *analog* type of meter, but many still exist and work as good as new. An analog meter is similar to the speedometer on a car where a moving needle points to a number for the measurement. Its replacement, the VAT 60, performs the same tests, but the display data is in a digital format. Both are used to test batteries and charging systems. This is one automotive meter that is capable of reading high current. On a starting system, the current may be as high as 400 to 500 amperes. Most multimeters can read no more than 10 amperes.

Figure 3-107 Automotive multimeters can measure many different functions of the vehicle's electronic circuits. (Courtesy of Snap-on Tools Company)

Figure 3-108 This is a typical volt-amp tester commonly known as a VAT 40 or VAT 60. (Courtesy of MATCO Tool Company)

Summary

- Wrench sizes are determined by the size of the bolt head or nut they fit. Sizes are given in either the U.S. (English) or metric system. Common wrench types are open-end, box-end, combination socket, and adjustable.

- Pliers are used as an extension of the technician's fingers. Common pliers are combination, channel lock, diagonal cutting, and vise grips.

- Screwdrivers are used to drive screws. There are many different types of screwdrivers to fit the many types of screw heads. Slotted screws are driven by the standard screwdriver. Recessed head screws—such as Phillips, Reed and Prince, and clutch type—require a different screwdriver.

- Hammers are used to drive tools or position parts. The ball-peen hammer is used for driving punches and chisels. Softface hammers—such as brass, plastic tip, and rubber—are made to protect the parts being hammered on.

- Punches are used to remove parts held together by pins. Removing a pin requires both a starter punch and a pin punch. A center punch is used to make a mark for locating a hole to be drilled.

- Power wrenches are used for rapid removal or installation of bolts or nuts. Common power wrenches are the electric wrench, air-operated ratchet driver, and the impact wrench.

- Many repair jobs require the use of metalworking tools. Chisels are used for cutting metal. Twist drills are used to cut holes. Reamers are used to finish holes to an exact size. Metal is cut with a hacksaw. Files are used for removing, shaping, or polishing metal.

- Threads are made or repaired with several threading tools. Dies are used to make or repair outside threads. Taps are used to make or repair inside threads. If a screw is broken while still in a part, it can be removed with a screw extractor.

- An oxyacetylene welding and cutting outfit mixes oxygen and acetylene gases to create heat at the tip of a torch for welding, cutting, and heating.

- Jacks that come with the vehicle for tire changing should not be used for any other repair work.

- Jacks are designed to lift only one end or one side of a vehicle.

- Lifts may be powered by air rather than hydraulics when air is used to operate under-floor controls for the hydraulic system.

- Service manuals may be on paper or computer disc.

- Computerized data banks can contain data for many years, makes, and models of vehicles.

- Web sites on the Internet can provide assistance on vehicle repair or information on buying a vehicle.

- Computerized test equipment can interface with the PCM to display sensor and actuator data.

- Large DSOs display more data than hand-held scanners and scopes. They can also be equipped to measure certain exhaust gases.

- The most common electrical measuring instrument is the multimeter.

- Multimeters must be high-impedance to prevent damage to electronic circuits.

- VAT 40 and VAT 60 meters are used to test starting and charging systems and the battery.

Terms to Know

Adjustable wrench
Allen wrench
Analog
Air-operated ratchet driver
Ball-peen hammer
Box-end wrench
Brass hammer
Channel lock pliers
Chisel
Combination pliers
Combination wrench
Computerized data
Cotter pin puller
Current
Diagonal cutting pliers
Diagnostic trouble codes
Die
Digital storage oscilloscope
Double-post lift
File
Floor jack
Frequency
Hacksaw
High-impedance
Impact wrench
Jack stands
Multimeter
Needle nose pliers
Offset screwdriver
Open-end wrench
Pin punch
Ratchet handle reamer
Scissor lift
Scan tool
Screwdriver
Screw extractor
Service manuals
Single-post lift
Socket handles, drivers, and extension
Socket wrench
Snap ring pliers
Starter punch
Tap torque stick
Torque wrench
Twist drill
Vise grip pliers

Review Questions

Short Answer Essays

1. Describe how to determine the size of a wrench to use on a nut or bolt.

2. Explain how to safely pull and push on a wrench in a tight space.

3. What determines the size of a ratchet handle?

4. What is the purpose of a socket extension?

5. List the three main types of torque wrenches.

6. Describe how to safely use an adjustable wrench to loosen a nut.

7. Explain how to safely use an adjustable wrench to tighten a nut.

8. Explain why special heavy-duty sockets are used with an impact wrench.

9. What is the purpose of a die?

10. What three basic electrical measurements can be made with a high-impedance multimeter?

Fill-in-the-Blanks

1. The two different measuring systems for wrench sizes are _____ and _____.

2. A wrench with a box-end and an open-end is called a(n) _____ wrench.

3. Torque can be measured in U. S. (English) units called _____-_____. Torque can be measured in metric units called _____-_____.

4. When using an adjustable wrench, care must be taken to apply the load to the _____ jaw.

5. When using an impact wrench, be sure to use _____ sockets.

6. A(n) _____ hammer is used to drive a chisel.

7. The _____ will measure resistance only.

8. A file should never be used without a(n) _____.

9. A drilled hole can be finished to a specific size with a(n) _____.

10. A broken screw can be removed with a(n) _____ _____.

ASE Style Review Questions

1. Two technicians are discussing a wrench with the size 10 stamped on the handle. *Technician A* says this is a metric size wrench.
   Technician B says this is a U.S. (English) size wrench. Who is correct?
   A. A only
   B. B only
   C. Both A and B
   D. Neither A nor B

2. The use of a wrench to remove a tight nut is being discussed. *Technician A* says an open-end wrench is the best choice.
   Technician B says the box-end wrench is a better choice. Who is correct?
   A. A only
   B. B only
   C. Both A and B
   D. Neither A nor B

3. Two technicians are discussing which of two socket wrenches to use on a tight nut. *Technician A* says a 12-point socket is the strongest.
   Technician B says a six-point socket wrench is the strongest. Who is correct?
   A. A only
   B. B only
   C. Both A and B
   D. Neither A nor B

4. Two technicians are discussing a torque reading that is specified in U.S. (English) units. *Technician A* says there are 12 foot-pounds in an inch-pound.
   Technician B says there are 12 inch-pounds in a foot-pound. Who is correct?
   A. A only
   B. B only
   C. Both A and B
   D. Neither A nor B

5. A torque specification calls for a reading of 144 inch-pounds. *Technician A* says this is the same as 12 foot-pounds.
   Technician B says this is the same as 6 foot-pounds. Who is correct?
   A. A only
   B. B only
   C. Both A and B
   D. Neither A nor B

6. A torque specification calls for a reading of 72 inch-pounds. *Technician A* says this reading is equal to 6 foot-pounds.
   Technician B says this reading is equal to 12 foot-pounds. Who is correct?
   A. A only
   B. B only
   C. Both A and B
   D. Neither A nor B

7. Two technicians are discussing an Allen wrench with an 8 stamped on the handle. *Technician A* says the wrench fits metric Allen head screws.
   Technician B says the wrench fits U.S. (English) Allen head screws. Who is correct?
   A. A only
   B. B only
   C. Both A and B
   D. Neither A nor B

8. Two technicians are discussing the use of an impact wrench. *Technician A* says heavy-duty impact sockets must always be used with an impact wrench.
   Technician B says standard sockets can be used with an impact wrench. Who is correct?
   A. A only
   B. B only
   C. Both A and B
   D. Neither A nor B

9. The removal of a pin from a part is being discussed. *Technician A* says a starter punch is used to begin the operation.
   Technician B says the final removal of the pin is done with a pin punch. Who is correct?
   A. A only
   B. B only
   C. Both A and B
   D. Neither A nor B

10. Lifts are being discussed. *Technician A* says dual-post lifts may extend from under the floor.
    Technician B says a drive-on lift is normally used for brake repair. Who is correct?
    A. A only
    B. B only
    C. Both A and B
    D. Neither A nor B

Automobile Theories of Operation

Upon completion and review of this chapter, you should be able to:

❏ Discuss the requirement to apply theories of physics in designing and operating vehicles.

❏ Discuss the first two laws of thermodynamics and their application in an engine.

❏ Define *energy* and the two types of energy.

❏ Discuss how pressure and vacuum help operate the engine.

❏ Discuss Newton's three Laws of Motions.

❏ Define *electricity, voltage, current,* and *resistance.*

❏ Discuss magnetism.

❏ Name the types of electrical circuits.

❏ Discuss the use of hydraulics in vehicles.

❏ Discuss other factors in designing and operating a vehicle.

Introduction

The automobile is made up of eight major interactive systems based on the application of certain physics theories. The systems power the vehicle, deliver power to the wheels, and comfort and protect the passengers. Functional systems working together make it possible for the automobile to be what it is today: reliable, durable, and comfortable. In this chapter, we will discuss some of the theories of operations and examples of how they apply to the major systems. Understanding theories will assist the technician in diagnosing problems that do not respond to the conventional practice of electronic diagnostics and direct testing of suspected failures.

Thermodynamics

Shop Manual
pages 75–77

Thermodynamics is the study of using radiant heat to perform mechanical work.

Engines use the theory of **thermodynamics** to generate power. The first law of thermodynamics is a law of energy conservation: energy cannot be created or destroyed. The amount of internal energy in a system is a direct result of the heat transferred into that system and the work done. Heat and work are means by which natural systems exchange energy with each other. Basically, it is the relationship between heat and mechanical energy (Figure 4-1).

The second law concerns the thermodynamic cycle. In a perfect heat engine, all of the heat produced would be completely converted to mechanical work. Nicolas Leonard Sadi Carnot, a nineteenth-century, French scientist, proved that this idea cycle could not exist. The best engine would lose heat and work because the engine must be exhausted. The exhaust would remove and waste a great deal of heat.

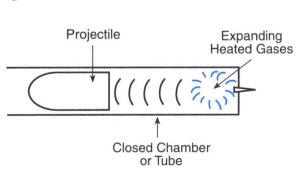

Figure 4-1 The force of hot, expanding gases are transformed into mechanical energy by the movement of the projectile.

Liquids that are heated enough will change to a vapor and expand greatly in volume. A prime example of this is boiling water. As the water is heated, it expands in volume, which is the amount of space it occupies. Somewhere around 212 degrees Fahrenheit, the water changes from a liquid to a gas, or vaporizes. The gas or steam greatly increases its volume. For example, place a lid on top of a container of boiling water. As the water boils more rapidly, the lid will begin to rattle and move about. We can capture the energy of the water by bolting the lid down and not allowing the steam to escape. However, if we do not in some way allow the steam to escape or use its energy, the volume of steam will continue to build until it ruptures the container and an **explosion** occurs.

Most lightweight vehicles use gasoline that is rated at about 110,000 **British Thermal Units (BTU)** per gallon. BTUs are used to measure the heat-producing capabilities of a chemical or mechanism. Burning vaporized gasoline in a closed chamber called a **combustion chamber** will result in high heat and a rapid expansion of gases (Figure 4-2). This rapid heating and expansion of gases creates a very high force that acts against the engine's pistons and produces useable power (Figure 4-3). The chamber is closed but it does have one side that can move: the piston. The energy of the expanding gas is transmitted against the piston, forcing it downward. This, in turn, rotates the crankshaft, producing a twisting force or *torque*. As the torque is transmitted through the vehicle's drive system, it moves the wheels and work is performed. Work in this instance is measured as **horsepower**.

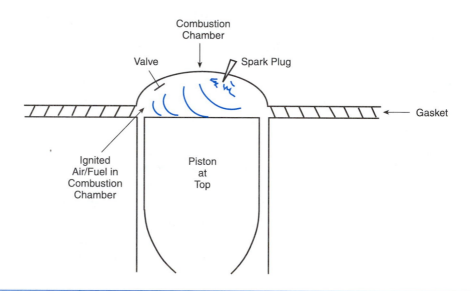

Figure 4-2 The air/fuel mixture burns in a controlled manner from the ignition point across the chamber.

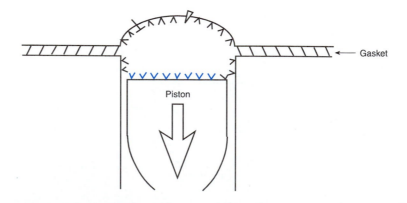

Figure 4-3 The force of the gases acts against the piston the same as it did against the projectile in Figure 4-1.

However, the engine loses *heat* and therefore *work* through the exhaust and cooling system. The exhaust is required to remove spent gases while the cooling system must remove heat from the mechanical components. Roughly two-thirds of the heat in a gallon of gas is lost through exhaust and cooling. Only about 35,000 BTUs are left to operate the engine and vehicle. After friction and overcoming the weight, the mass of components, and the vehicle, only about 19 percent to 20 percent of the energy in a gallon of fuel is actually applied to the drive wheels. Vehicles made in the mid- and late-1990s do somewhat better, but the nature of the beast prevents any large increase in the efficiency of current internal combustion engines.

The application of thermodynamics in an engine requires fuel, oxygen, spark, and a combustion chamber. The engine will not operate or operate poorly if a fault occurs in any of these four elements. Almost nothing will burn without oxygen present. Considering that only about 20 percent of the earth's atmosphere is oxygen and the number of fires that occur worldwide, it is apparent that oxygen does not have to be present in large amounts. By the same token, almost everything will burn if enough oxygen is present. Steel will burn like a match if enough oxygen is available.

The control of the fuel and air (oxygen) to an engine is essential for a safe, controlled burn (Figure 4-4). While an automobile does not and cannot use pure oxygen, too much or too little upsets the **ratio** of air and fuel, resulting in a poorly operating engine. An example of upsetting the air/fuel ratio would be a dirty air filter. The filter will not allow sufficient air to enter the combustion chamber. With insufficient oxygen, the gasoline may or may not fire. If it does fire, it will not produce the necessary energy to perform the work needed.

Ratio is the proportion or relation one element has to another. In this case, it is the relation of air to fuel.

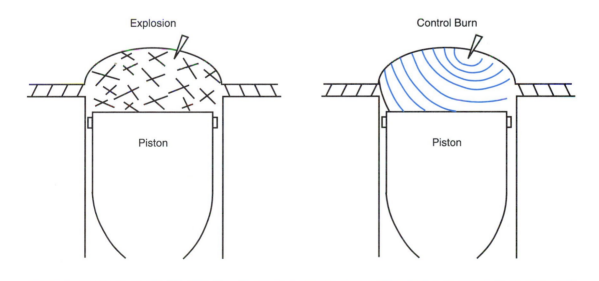

Figure 4-4 An explosion (left) creates uncontrollable energy radiating in unpredictable directions. A controlled burn radiates energy equally in all directions at once.

Shop Manual
pages 75–77

Potential energy available is based on the size, weight, and position of a stationary object. The speed of a moving object plus its weight and size creates the *kinetic energy* of the object.

Foot-pounds is a measurement of force being applied in a rotary motion or a measurement of torque.

Energy and Work

Energy is defined as the capacity of matter to perform work through its motion or position. The motion of an object produces **kinetic energy**. **Potential energy** is available energy based on the position of an object. Consider the action of a baseball. While the pitcher is holding the ball in his or her hand, the ball represents potential energy. When the ball is released toward the hitter, it has kinetic energy based on its weight and speed. The hitter's bat has potential energy until she or he swings it. When the bat and ball connect, the kinetic energy of both result in an expenditure of energy (Figure 4-5). The ball is driven in a different direction with great force. An equal amount of opposing force is absorbed by the bat and the hitter (See Newton's Laws of Motion later in this chapter).

When an object moves and energy is expended, work has been accomplished. This includes raising a hand or moving a house. *Work* is a unit of energy and may be measured in **foot-pounds (ft.-lbs.)**. Foot-pounds measure torque or twisting force. Work is measured by multiplying the weight times the distance, or it is stated as one foot-pound will lift one pound the distance of one foot (Figure 4-6).

Gasoline is potential energy in the fuel tank. Its rapid expansion in the combustion chamber results in great amounts of kinetic energy. This energy is used to transform a resting vehicle with potential energy into a moving vehicle with large amounts of kinetic energy. This is proven by the amounts of kinetic energy released through actions occurring in a traffic accident.

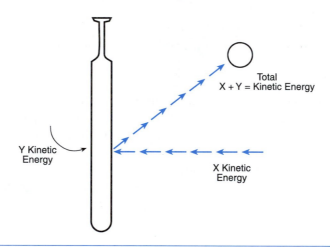

Figure 4-5 The kinetic energy created by merging or connecting a moving ball and bat creates a larger amount of kinetic energy in the ball.

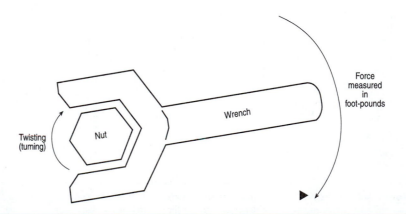

Figure 4-6 Torque is measured by determining the amount of twisting force being applied.

Pressure and Vacuum

Pressure is the amount of force used to hold a gas or liquid in place. The atmosphere applies about 15 pounds of pressure per square inch to every external inch of all objects on the surface of the earth including the human body. **Atmospheric pressure** can be used to achieve work in a vehicle or components and can be used to create pressure within a system to perform work.

Vacuum is defined as the absence of pressure. The earth does not have sufficient natural vacuum to perform automotive work, but it can be created to some extent by certain components. One of those components is the engine.

As discussed in the section on thermodynamics, fuel and air must be present in proper amounts in the combustion chamber to create power. Since the chamber can be sealed, the downward movement of the piston within the chamber can create a vacuum above it. This combination of vacuum and atmospheric pressure forces fresh air through a valve that is timed to piston movement (Figure 4-7). Fuel is pressure-injected into the airflow near the same valve or directly into the chamber during the intake of air. With the proper equipment, the amount of fuel can be regulated in correct ratio to the amount of air entering the chamber. Older systems used the movement of air passing through a ventura to create a low-pressure area and draw fuel from the carburetor fuel bowl (Figure 4-8). Now, almost all engines use electronic devices to control the fuel.

Shop Manual
pages 148–151

Pressure is the amount of force used to hold a gas or liquid in place.

Base *atmospheric pressures* are taken at sea level. The pressure decreases about one pound for each 1,000-foot rise in elevation.

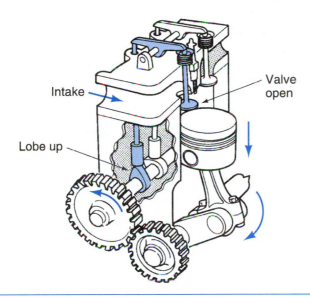

Figure 4-7 This valve opens to let the air/fuel into the cylinder.

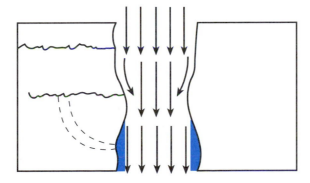

Figure 4-8 An object obstructing the airflow will create a low-pressure area downwind and behind the object.

The force of the driver's foot creates pressure within a brake system (Figure 4-9). The force of the driver's foot is also transferred by pressurized fluid to the brakes at each wheel. Engine vacuum and atmospheric pressure can be used to assist the driver in braking. A steering system uses a pump to supply high-pressure fluid to a power-assist unit controlled by the actions of the steering wheel. In both cases, pressure is used to assist the driver in the operation of the vehicle. Both systems are covered in later chapters.

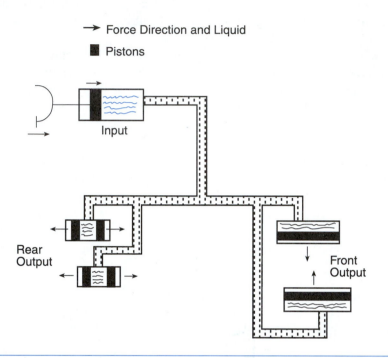

Figure 4-9 Applying force to a liquid in a sealed system will created pressure.

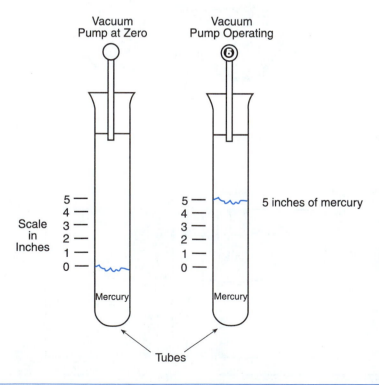

Figure 4-10 A given amount of vacuum will raise mercury or water to a specific height.

Pressure is measured in **pounds per square inch (psi).** The use of pressurized fluid is covered later in this chapter in the paragraph on hydraulic theory. Vacuum is usually measured in **inches of mercury** or **inches of water**.

The most common vacuum measurement used in automotive repair is inches of mercury. Vacuum supplied against mercury in a closed tube will cause the mercury to rise based on the amount of vacuum (Figure 4-10). The amount of rise is measured in inches. However, since mercury is a deadly poison and a glass tube is not practical in the shop, calibrated mechanical gauges are used to register the measurement.

Newton's Laws of Motion

Newton wrote three basic Laws of Motion while proving that the Earth revolved around the sun instead of the sun moving around the Earth. The laws pertain to almost every system on the vehicle.

Newton's First Law deals with movement: An object at rest tends to stay at rest and an object in motion tends to stay in motion. This unchanging action is called **inertia** and it assumes that there is no outside force acting on or against the object. A stationary vehicle will not move unless something forces it to move. Parking a vehicle in neutral with the brakes off on level ground will result in a parked vehicle. However, parking the same vehicle on a hill, in neutral, and without brakes will allow gravity to act on the vehicle and cause it to roll down the hill, thereby overcoming rest inertia (potential energy) and generating motion inertia (kinetic energy).

Newton's Second Law: A moving object will travel in a straight line and in one direction. Try driving a car on a flat, level road. Assuming that the car is set up correctly, it will travel in a straight line. A force (steering) must act upon the car to change its direction. A second example is an inflated balloon. When an inflated balloon that has not been knotted is released, it immediately moves through the air in erratic directions. The balloon is not shaped to travel through air and it changes direction as the air acts against it. When all of the trapped air in the balloon is exhausted, the balloon falls to Earth. The balloon's shape and the loss of pressure are only two actions that work not only on the vehicle as a whole but on each moving component of the vehicle as well.

For every action there is an equal and opposite reaction. That is Newton's Third Law. Again, it is assumed that there are no outside forces in play. If a ball were thrown against a wall in outer space, it would return to the thrower with the same force and same speed. Doing the same on Earth would result in the ball slowing on its return journey. This is due to air resistance and gravity working against the ball so it is slowed, and the direction of travel has changed.

Earlier, we discussed kinetic energy in a vehicle accident. Consider, for example, that a three-ton vehicle moving at 60 miles per hour with 100 tons of kinetic energy was suddenly stopped by an immovable object. The action of stopping 100 tons would immediately result in an opposite and equal reaction of 100 tons against the immovable object. That release of energy would not only affect the object but the vehicle as well. Since energy cannot be destroyed, the 100 tons of suddenly-released kinetic energy must be dissipated or changed in some manner, in this case the destruction of the vehicle and object.

The movement, weight, and operation of all vehicle components are subject to the laws of motion. To move a piston, energy must be used. In turn, the actual movement of the piston in turn creates work. Each movement of every part requires energy and produces work. The engineering of a vehicle requires a balancing of the energy used versus the amount of work performed.

Pounds per square inch (psi) are units by which pressure is gauged. Vacuum is measured in *inches of mercury* or *inches of water*.

Shop Manual page 75

Inertia is the tendency of an object to do what it is already doing: moving or resting.

Electrical Theory, Measuring, and Circuits

Georg Simon Ohm (1787–1854) wrote the law regarding electrical theory and measurement. It states that it takes one *volt* to push one *ampere* of current through one *ohm* of resistance. A change in one of the three affects the other two values proportionally. In order to understand Ohm's Law, we must first look at **electricity**.

The basic element of any material is the **atom**. It is made up of three parts: *neutrons*, *protons*, and *electrons* (Figure 4-11). The center or *nucleus* of the atom is made of the neutrally-charged neutrons and positively-charged protons in equal numbers. An equal number of negatively-charged electrons travel in orbits around the nucleus much as the planets travel around the sun. An atom that has equal numbers of electrons, neutrons, and protons is in balance. A balanced atom really does not produce any outside actions. If the atom can be unbalanced in some way, some type of energy can be created and used. Splitting the atom creates enormous amounts of energy, either uncontrollably, as in the case of nuclear weapons, or in controllable situations like nuclear power plants. However, if only a portion of the atom is moved and it basically remains intact, useable and controllable energy can be acquired.

Conductors

Causing an electron to leave one atom and move to another creates a chain reaction in which an extremely large number of electrons are moving (Figure 4-12). At this point, electricity or electrical current is present and can be used if controlled. However, there are only a few natural elements that can be used. In order for an element to be used for electricity, it must not have more than four electrons in the atom's outer orbit or the *valence ring*. The best element or **conductor** is copper (Figure 4-13). It has one electron in the valence ring. That electron can be forced out of orbit in a relatively easy manner. Elements with two, three, or four electrons in the valence ring can be used but require more voltage. For ease of understanding in this book, we will assume the conductor is copper.

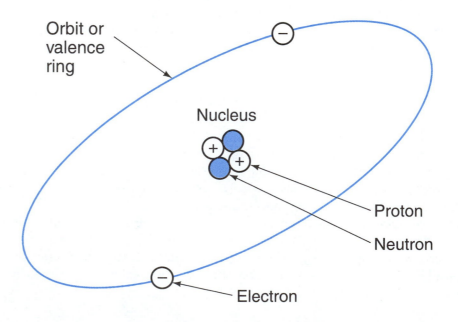

Figure 4-11 A balanced atom has equals numbers of protons, neutrons, and electrons.

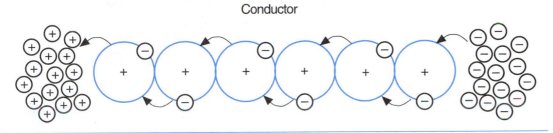

Conductor

Figure 4-12 Electricity is the mass movement of electrons from negative to positive.

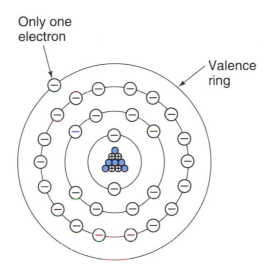

Only one electron

Valence ring

Figure 4-13 An atom with one electron in the valence ring makes an excellent conductor. (Courtesy of Cooper Automotive/NAPA Belden)

Voltage

An electron is moved by voltage. Voltage acts like the pressure in a water hose and is techni-cally known as **electromotive force (EMF)**. Voltage is produced or is available when there is a difference in the numbers of free electrons at one end of a circuit compared the other end (Figure 4-14). This is known as *potential differential* or *voltage*. If there is no potential differ-ence between negative and positive, then there can be no voltage. An example is a driver trapped in a car under a live, 13,500-volt electrical line. Many people might assume that the driver is protected by the rubber tires which act as insulators. The actual situation, however, is that the driver and the vehicle are the same voltage as the electrical line, hence there is no potential differential. When the driver attempts to leave the vehicle and places a foot on the ground, he or she has created a conductor between the line and the ground. Since the ground

Electromotive force is another term for "voltage."

Conductor

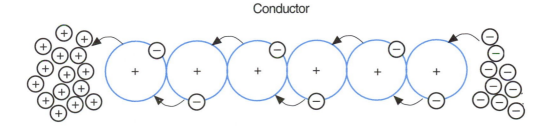

Figure 4-14 The difference between the number of electrons at the positive and negative sides creates voltage.

will have very little voltage, the potential differential is enormous and 13,500 volts of electricity will travel through the new conductor and usually result in death.

As voltage pushes an electron out of orbit, two things happen at once. The atom becomes unbalanced and tries to regain balance by attracting another electron. The moving electron attempts to attach itself to another atom. A chain reaction has started. All of the electrons will move from the **negative** (ground) toward the **positive** because unlike polarities, negative and positive attract each other. The movement of the electron is known as the **Electron Theory of Electricity**. When about 628 billion electrons pass a single point in one second, one ampere of current is available for use.

Current

Current is the amount of moving electrons expressed in an understandable term as *amperes*. A typical automobile has systems that use current ranging from milliamperes up to 30 amperes. Most electronic systems, like the engine computer and radio, use very low amperage while engine cooling-fan motors may use up to 30 amperes. The electrical current will continue to flow toward the positive until it meets resistance.

Resistance

Resistance is anything that slows or stops electricity and is measured in *ohms* (Figure 4-15). Elements that are non-conductive are resistant and classified as **insulators**. The only way to use electricity is to create a resistance or a load within a system.

Resistance may be caused by **corrosion** on wires and connectors, small wire size, or it may be engineered into the system. Unintentional resistances like loose connections or corrosion disrupt the current flow in the system and devices may or may not work. Each electrical system is engineered to have specific resistance using specific amounts of current and voltage. Lamps, motors, and other electrical devices are engineered to use resistance. They are designed

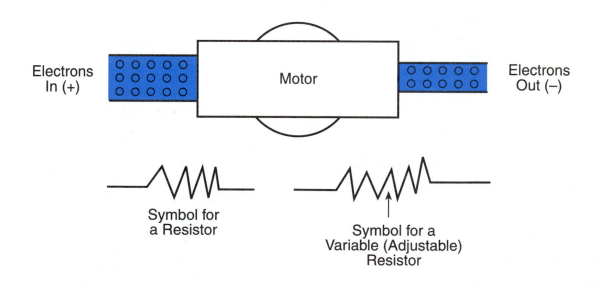

Figure 4-15 Resistance uses or stops current. Note the symbols for a resister.

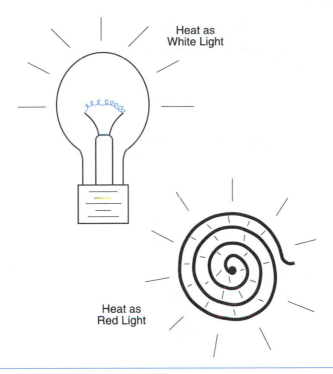

Heat as
White Light

Heat as
Red Light

Figure 4-16 Depending on the type of resistance and circuit design, current can be used for light or heat.

to use the electrical current for tasks. The lamp has a filament with high resistance (Figure 4-16). As current flows through the filament or resistance wire, it produces heat. The heat is seen as white light. An electric stove uses high-resistance coils for heating. The coils are heated to high temperature by current flow, which is usually seen as a red light. Electric motors overcome the resistance of a shaft and attached devices.

Magnetism

Any time current is flowing, a **magnetic field** is created around the conductor (Figure 4-17). If the conductor is wound tightly in a spool fashion around a shaft and each winding is insulated from the others, the magnetic fields overlap and increase the total magnetic field around the shaft. This is known as a **coil** (Figure 4-18). Placing the shaft and its coil inside stationary magnetic fields makes the shaft turn. Stationary magnets produce the second magnetic fields. The positive and negative polarities of the shaft's coil fields attempt to move toward their opposing polarity of the stationary fields (Figure 4-19). The result is a spinning shaft, which can be attached to a fan or other device. The polarity of the shaft's magnetic field can be continuously changed to continue the rotation. Magnetic fields can also be engineered to produce electrical current. Devices that use or produce electricity will be covered in other chapters in this book.

Magnetism may be natural, like a magnetic rock or something electrical. Materials that are magnetized by electricity are known as "electromagnets."

A *coil* is a wire or conductor that is insulated and tightly wound around a center. Each wrap around the center is called a "winding."

Current Flow

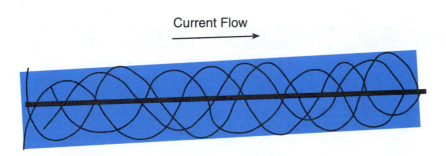

Figure 4-17 Current creates a magnetic field as it moves through a conductor.

Straight conductor, magnetic flux of 1

Conductor folded unto itself, magnetic flux of 2 plus

Generated fields

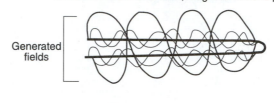

Total field generated by overlapping fields

1 winding

8 windings

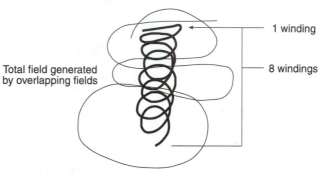

Figure 4-18 Overlapping magnetic fields will create a larger field.

Wire loop (armature)

Field winding

Pole shoe

S

N

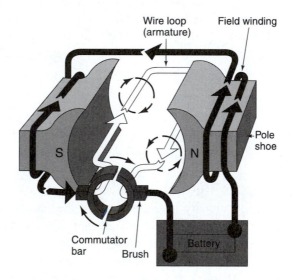

Commutator bar

Brush

Battery

Figure 4-19 Changing the polarity of the center magnetic field will make the center field rotate. (Courtesy of General Motors Corporation, Service Operations)

A BIT OF HISTORY

A *farad* is a unit to measure capacitance.

Michael Faradday (1791–1867) was an English chemist and physicist best known for his experiments with electromagnetism. Using the principle that electricity could be made by moving a magnet inside a coil of wire, he developed the first electric motor, generator, and transformer. He was responsible for several electrical terms in use today including electrode, cathode, anode, and ion. The unit of capacitance, the **farad**, is named in his honor.

Electrical Measurements

Since electricity can be used for definite purposes, there must be a way to measure it for control. Using **electrical measuring instruments** and service data as outlined in Chapter 3 and using Ohm's Law, we can calculate if a system is working properly. Since Ohm stated that one volt will push one ampere through one ohm of resistance, the mathematical formula for measuring and calculating electrical values is E = IR where E is voltage, I is current, and R is resistance (Figure 4-20). If two of the values are known, the third can be calculated (Figure 4-21). Consider the following problem.

A system is using 12 volts to feed a current through 4 ohms of resistance (Figure 4-22). The missing measurement, *current*, can be determined using Ohm's mathematical formula. Since E = IR, then I = E/R or current is equal to voltage divided by resistance. In this case, I = 12/4 or I equals 3 amperes. If the system or device being checked uses 3 amperes to operate, then the system is functional. A device that uses more or less current would not work properly in this system. By the same token, if we measured this system with an instrument and found that only one ampere was being used, then the system is inoperative.

The most common *electrical measuring instrument* in the automotive shop is the multimeter, which can measure volts, resistance, and amperes with the flick of a switch or button.

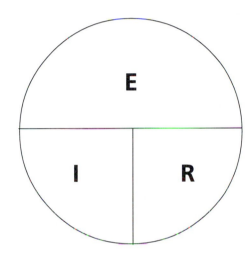

Figure 4-20 A method of remembering the mathematics formula for Ohm's Law.

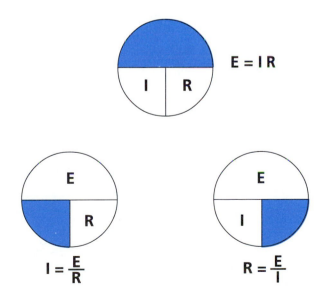

Figure 4-21 The equation can be used to find one unknown factor.

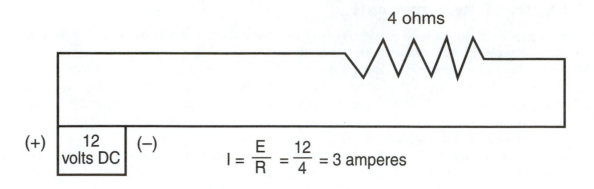

4 ohms

(+) | 12 volts DC | (−)

$$I = \frac{E}{R} = \frac{12}{4} = 3 \text{ amperes}$$

Figure 4-22 If a circuit can be measured and two known factors are found, Ohm's Law can be used to determine the third factor of the circuit.

Consider a second situation.

A technician used a service manual to find that a system is supposed to have 12 volts supplied and 6 amperes of current (Figure 4-23). According to the formula, the working device should have 2 ohms of resistance (R = E/I). After taking measurements, it was found that only 1 ampere of current was present. Applying the formula again (R = E/I), the system appeared to have 12 ohms of resistance. Somewhere in the system there was too much resistance and the device did not work.

Circuits

A *circuit* is the path by which electricity follows from positive to negative.

Up to this point, we have discussed electrical systems. The systems are comprised of one or more electrical circuits. A **circuit** is the path that electricity follows from the negative side to the positive side. A circuit can work with a power source, conductor, and load. However, if those were the only things used, the operator would have to open the hood and connect the

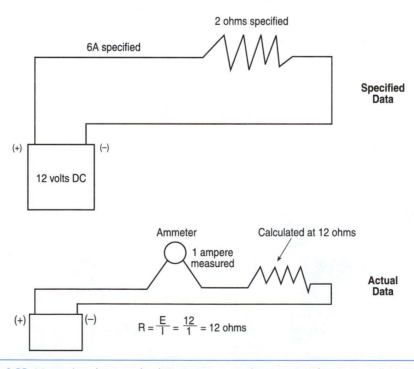

Figure 4-23 Measuring the actual voltage, current, and resistance of a circuit will assist the technician in diagnosing.

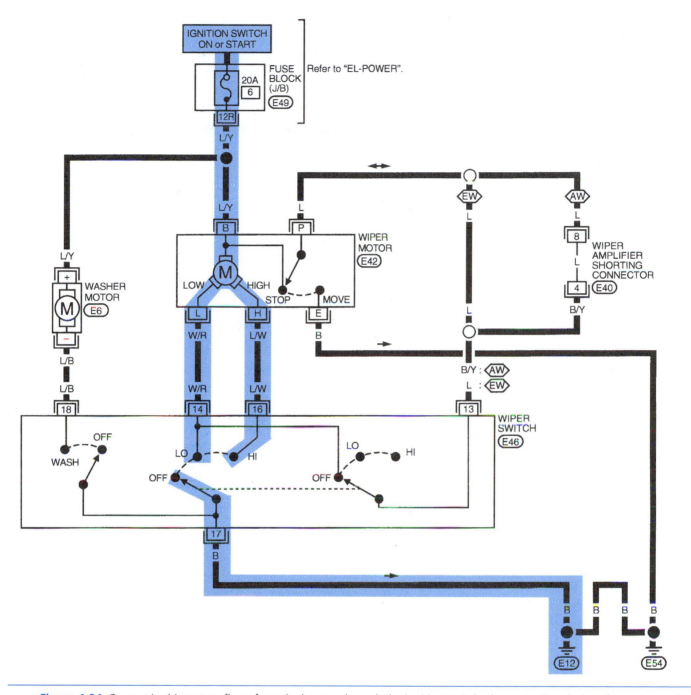

Figure 4-24 Current in this system flows from the battery through the ignition switch, the protection device, the motor, and finally to the wiper switch. When the switch is moved to ON, current flows through the switch to ground.

wires to the battery just to start the car. Such a procedure would have to be done for every device or system to be turned on or off. With this in mind, a viable working circuit must have load, a conductor, a power source, and controls like the ignition or headlight switches (Figure 4-24). There are three different types of circuits.

Series Circuits

A **series circuit** provides one path for the current to travel. In order to work, the circuit must be intact from one end to the other (Figure 4-25). A defect anywhere in the circuit would affect the entire circuit. Older Christmas tree lights used series circuits. If one lamp went bad, then all of the other lamps would not work.

A series circuit provides only one path for the current to travel.

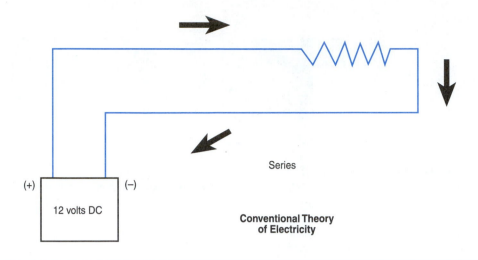

Series

Conventional Theory
of Electricity

(+) (−)

12 volts DC

Figure 4-25 A series circuit has one path for current to follow.

There is only one major series circuit on the typical vehicle and that is the starter circuit. The current must flow from the battery, through the ignition switch, and finally through the park/neutral (clutch switch on manual transmission) to the starter. If the transmission is in a drive gear (clutch pedal up), then the park/neutral (clutch) switch is opened and the starter will not engage.

Parallel Circuits

A parallel circuit
provides more than
one path to ground
and is common on
vehicles.

A **parallel circuit** provides more than one path to ground and is common on vehicles (Figure 4-26). One portion of the circuit can work while the others may not. Typical examples of this type circuit on a vehicle are the exterior lights. In the case of the headlights, from the last control device in the system, the conductors branch into two legs or parallel circuits, one to each headlight. If one light fails, the other will still work. The brake lights, parking lights, and turn signals work in the same manner. Most of a vehicle's safety equipment, like the lights, are parallel wired to some extent. The most common type of electrical circuit on a vehicle is the **series-parallel**.

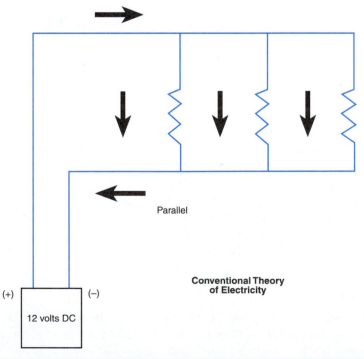

Parallel

Conventional Theory
of Electricity

(+) (−)

12 volts DC

Figure 4-26 A parallel circuit has multiple paths to ground.

Series-Parallel Circuits

Part of the **series-parallel circuit** is in series and other parts are in parallel (Figure 4-27). In effect, the series portion controls the parallel portion. Again, a typical example is the headlight system. As discussed in the last section, separate circuits feed the headlights from the last control device. That control device is usually an electric-mechanical switch known as a **relay** (Figure 4-28). Before the relay is the headlight switch. The current flows from the power supply, through the headlight switch, and finally to the relay. The switch controls the relay, which in turn routes power to the lights. When the switch is moved to the ON position, the relay closes and current moves from the power source through the relay to the headlights.

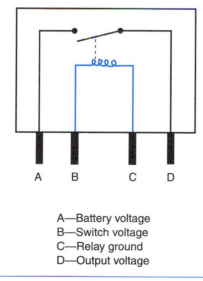

Series-Parallel

Conventional Theory
of Electricity

(+)   (−)

12 volts DC

In a *series-parallel* circuit, the series portion controls the parallel portion. An example is the headlight system.

A *relay* is an electromechanical switch that uses a low current to control a larger current.

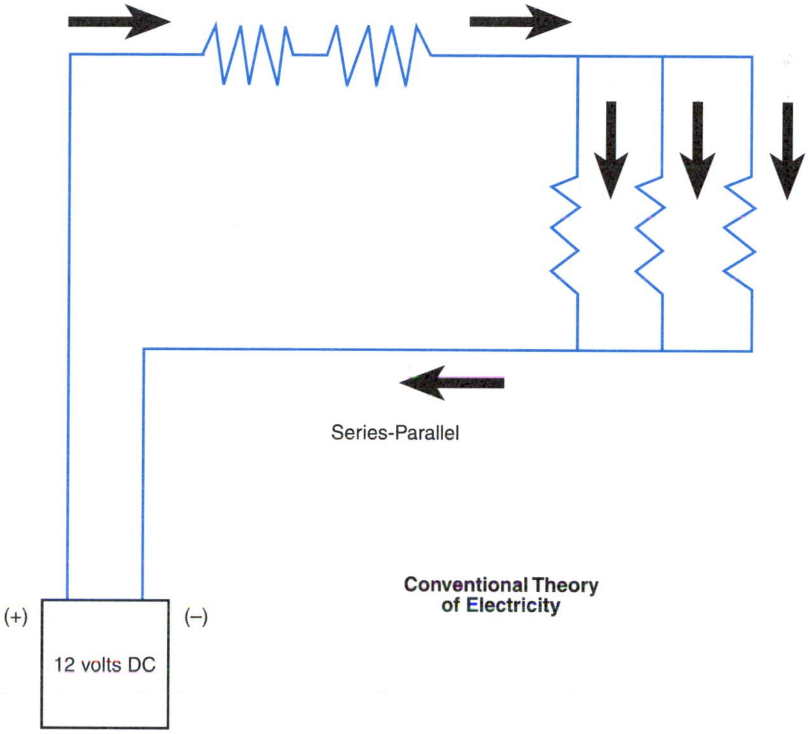

Figure 4-27 The first two resisters in this circuit can be set up to control the parallel portion of the circuit.

A   B   C   D

A—Battery voltage
B—Switch voltage
C—Relay ground
D—Output voltage

Figure 4-28 Current through the coil (B to C) will switch on the high current circuit (A to D).

Circuit Protection

Circuits should be protected by some device that will fail if the circuit is overloaded. This protects other circuits and possibly the whole vehicle. There are three basic protection devices: **fuses**, **circuit breakers**, and **fusible links**. If a circuit protector fails, it is for a reason and the circuit must be checked to find the problem. Simply replacing the protector may damage the circuit or at the very least burn through it again.

 WARNING: Do not replace a circuit protector with one rated at a higher ampere. A larger one will carry more current and may damage electrical components.

A fuse has a small internal wire through which all of the circuit's current flows (Figure 4-29). When a circuit current exceeds the capability of the fuse, the wire burns through, thereby stopping the current. The fuse can be removed from its holder and visually checked for failure. Fuses are rated in different amperes ranging from 5 to 60. The larger ones, 30 to 60 amperes, are used in circuits like charging units, fuel pumps, and cooling fans. The fuse may be a cartridge-type or blade-type. The cartridge has been replaced with the blade for convenience and reliability. A fuse is not serviceable and must be replaced when it fails.

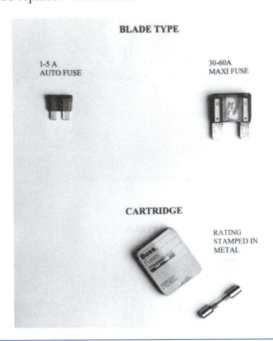

Figure 4-29 Fuses are made to be effective, but they are also inexpensive and easy to replace.

Figure 4-30 A circuit breaker can be used to protect a high-current circuit or used as the flasher for the turn signal.

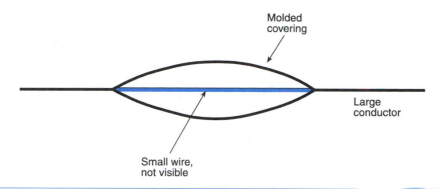

Molded
covering

Large
conductor

Small wire,
not visible

Figure 4-31 The fusible link is being phased out but can be found on trucks and older vehicles.

Circuit breakers open because of high current and are rated like fuses (Figure 4-30). Some circuit breakers will reset after they cool off. Others have to be replaced or manually reset. Breakers are better in some circuits because their design allows them to be slow acting compared to a fuse. This prevents an electrical surge from tripping the breaker. The problem with a circuit breaker is that most have to be checked with an ohmmeter or multimeter. There are usually no visible signs of failure.

A fusible link is used on some heavy-current circuits. The link is a wire within a molded form (Figure 4-31). The wire is sized below the wire of the circuit and usually rated for high amperes. Again, excessive current will burn through the link. A major problem with this type of protection device is the diagnosis and replacement of it. Many times the link fails without any visible signs. They are usually checked with a test light, voltmeter, or multimeter. They also are placed in locations difficult to access. Fusible links must be replaced when they fail. Fusible links have been replaced on newer vehicles with high-rated blade fuses.

Circuit Defects

There are only three possible defects in an electrical circuit. They are **open**, **short**, and **ground**. Any type of electrical device, including ones that use magnetism, can have one or more of the defects.

The three types of circuit defects are open, short, and ground.

Open

An *open* circuit is an incomplete circuit where current will not flow. When a light switch is moved to the OFF position, the circuit is open, but a defect is an unintentional open. A burned-through fuse creates an open circuit the same as a cut conductor.

Short

A *short* is an unintentional connection between two or more conductors (Figure 4-32). This could happen where two conductors rub against each other, thereby weakening the conductors' insulation. In this case, operating a switch could cause two or more circuits to switch on.

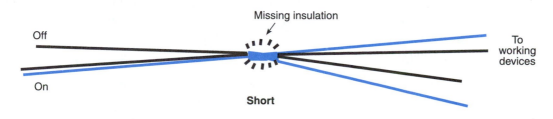

Off

Missing insulation

To working devices

On

Short

Figure 4-32 In this short, current is being directed into a circuit that is actually turned off. The result will be a blown fuse or poorly performing electrical devices.

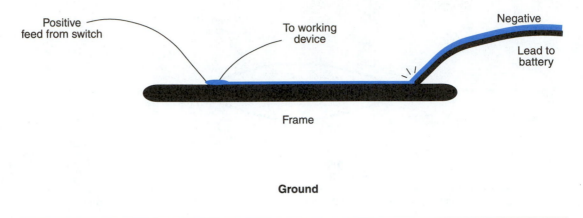

Frame

Ground

Figure 4-33 If the protection device does not blow in this grounded circuit, the conductor's insulator will probably melt and possibly start a fire.

In most cases, none of the affected systems will work correctly and will probably trip a protection device. A short can be found using a test light, voltmeter, ohmmeter, multimeter, or a short finder. A *short finder* is the common name for a tool that can be moved over a conductor for testing. A moving needle indicates current is present and a stationary needle indicates no current. The same tool can be used to locate a grounded circuit.

Ground

A *ground* is an unintentional connection between the feed conductor (positive) and the ground of the vehicle (Figure 4-33). Since the ground is the negative side of the power source, and there will probably be no resistance at this time in the circuit, extremely high current will flow. The protection should trip to protect the circuit. If it does not, heat can damage the circuit very quickly or a fire could erupt, possibly causing damage and injury.

Electrical Service Information

Shop Manual
page 85

In order to diagnose an electrical circuit or system, information must be obtained on how the circuit is placed in the vehicle. It may also be necessary to trace a circuit without tearing the vehicle apart chasing wires. There are two items to help the technician in this area. They are the vehicle's wiring diagram and electrical component locator.

Wiring Diagrams

The *wiring diagram* is a paper or computerized document that shows every circuit on a vehicle (Figure 4-34). It also shows controls, switches, loads, and other devices in each circuit. A single circuit may extend over several pages or screens, so tracing the circuit can be a time-consuming task (Figure 4-35 and Figure 4-35 continued). Wiring symbols are usually shown on the first page or screen of the wiring diagram section of the manual. Wire colors and sizes, along with circuit designators, are also shown on the diagram. In this manner, the technician can trace a circuit by moving from page to page (screen to screen) with the circuit designator instead of trying to trace a black line among many other black lines. Even using a wiring diagram to locate a specific component or connection can be difficult.

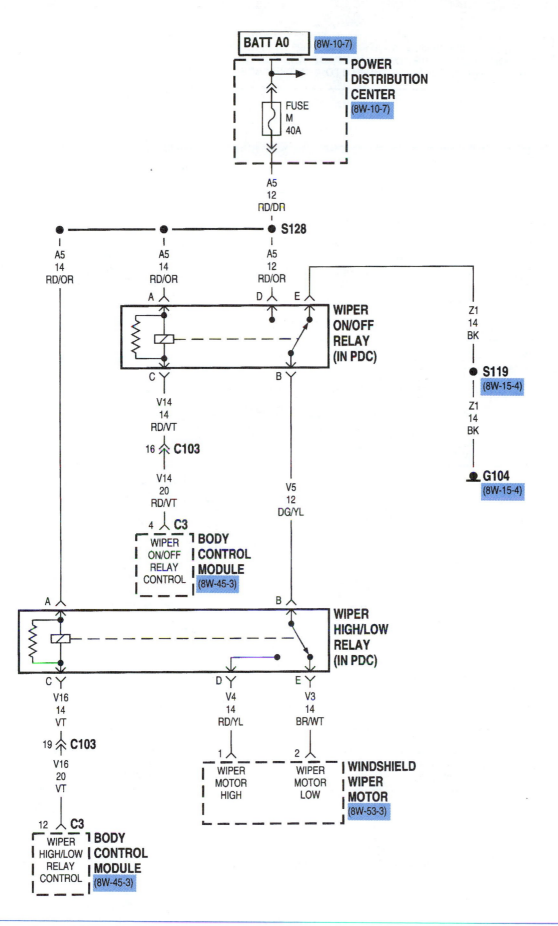

Figure 4-34 The highlighted items indicate other pages or figures that must be used to complete the reading of this wiring diagram. (Courtesy of Chrysler Corporation)

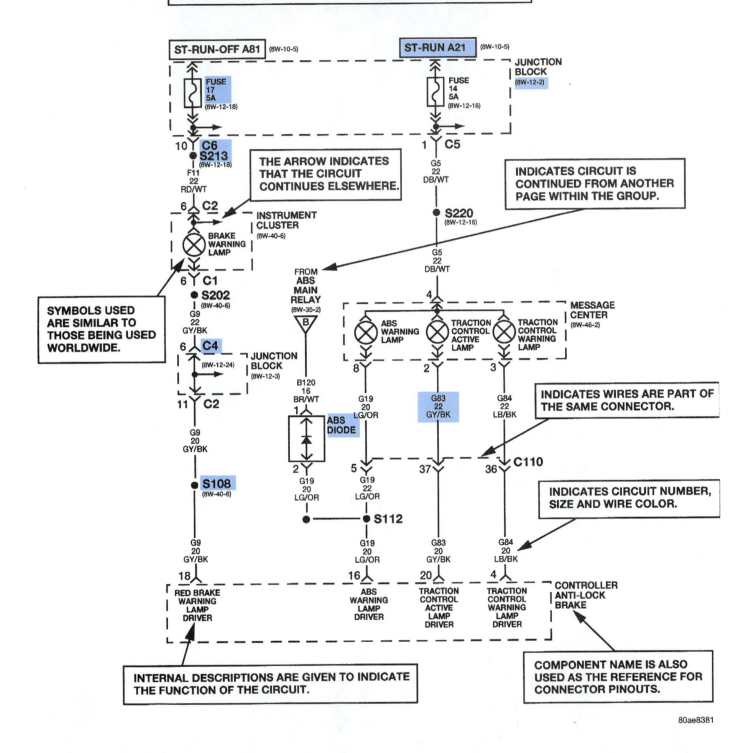

Figure 4-35 Note the amount of information that is included on this manufacturer's information page. (Courtesy of Chrysler Corporation)

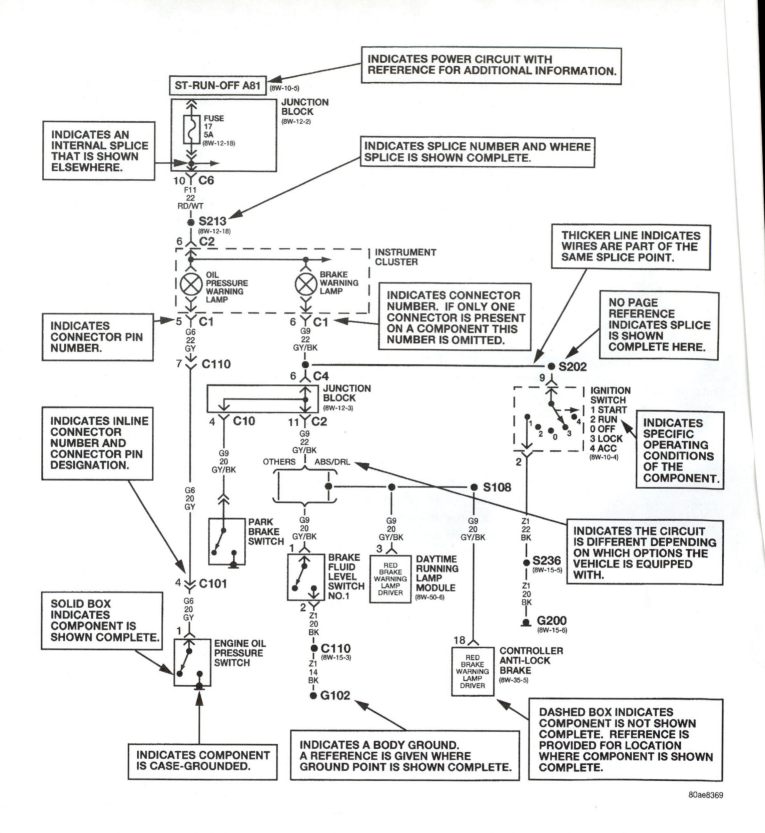

INDICATES POWER CIRCUIT WITH
REFERENCE FOR ADDITIONAL INFORMATION.

ST-RUN-OFF A81 (8W-10-5)

JUNCTION
BLOCK
(8W-12-2)

FUSE
17
5A
(8W-12-18)

INDICATES AN
INTERNAL SPLICE
THAT IS SHOWN
ELSEWHERE.

INDICATES SPLICE NUMBER AND WHERE
SPLICE IS SHOWN COMPLETE.

10 C6

F11
22
RD/WT

S213
(8W-12-18)

6 C2

THICKER LINE INDICATES
WIRES ARE PART OF THE
SAME SPLICE POINT.

INSTRUMENT
CLUSTER

OIL
PRESSURE
WARNING
LAMP

BRAKE
WARNING
LAMP

INDICATES CONNECTOR
NUMBER. IF ONLY ONE
CONNECTOR IS PRESENT
ON A COMPONENT THIS
NUMBER IS OMITTED.

NO PAGE
REFERENCE
INDICATES SPLICE
IS SHOWN
COMPLETE HERE.

INDICATES
CONNECTOR PIN
NUMBER.

5 C1

G6
22
GY

6 C1

G9
22
GY/BK

7 C110

S202

6 C4

9

IGNITION
SWITCH
1 START
2 RUN
0 OFF
3 LOCK
4 ACC
(8W-10-4)

INDICATES INLINE
CONNECTOR
NUMBER AND
CONNECTOR PIN
DESIGNATION.

JUNCTION
BLOCK
(8W-12-3)

4 C10

11 C2

1 2 0 3 4

INDICATES
SPECIFIC
OPERATING
CONDITIONS
OF THE
COMPONENT.

G9
22
GY/BK

2

G9
20
GY/BK

G6
20
GY

OTHERS   ABS/DRL

S108

G9
20
GY/BK

G9
20
GY/BK

G9
20
GY/BK

Z1
22
BK

INDICATES THE CIRCUIT
IS DIFFERENT DEPENDING
ON WHICH OPTIONS THE
VEHICLE IS EQUIPPED
WITH.

PARK
BRAKE
SWITCH

RED
BRAKE
WARNING
LAMP
DRIVER

DAYTIME
RUNNING
LAMP
MODULE
(8W-50-6)

S236
(8W-15-5)

1

BRAKE
FLUID
LEVEL
SWITCH
NO.1

3

Z1
20
BK

4 C101

2

G6
20
GY

Z1
20
BK

G200
(8W-15-6)

SOLID BOX
INDICATES
COMPONENT IS
SHOWN COMPLETE.

1

ENGINE OIL
PRESSURE
SWITCH

C110
(8W-15-3)

Z1
14
BK

18

RED
BRAKE
WARNING
LAMP
DRIVER

CONTROLLER
ANTI-LOCK
BRAKE
(8W-35-5)

G102

INDICATES COMPONENT
IS CASE-GROUNDED.

INDICATES A BODY GROUND.
A REFERENCE IS GIVEN WHERE
GROUND POINT IS SHOWN COMPLETE.

DASHED BOX INDICATES
COMPONENT IS NOT SHOWN
COMPLETE. REFERENCE IS
PROVIDED FOR LOCATION
WHERE COMPONENT IS SHOWN
COMPLETE.

80ae8369

Figure 4-35 *(continued)* Additional information on circuit data is furnished on this second page of manufacturer's information sheets. (Courtesy of Chrysler Corporation)

Component Locators

A *locator* shows the location of each electrical component on the vehicle. It can be a list of the items with their location spelled out as shown in Figure 4-36. Some manufacturers use pictures or drawings of sections of the vehicle with the component highlighted in some manner.

A drawback to locators and wiring diagrams is their publication date. The manuals should be in the dealerships when their corresponding vehicle model is available for public sale. This means that last-minute changes may not be included and either the diagram or locator may be in error. This is not a real issue until the manufacturers update their product during the model year instead of waiting for the next model. Even then, most changes are small and may only be a change in wire color. Experience is the best method of learning how to read wiring diagrams and component locators.

Electricity plays a large part in today's vehicles. It powers the engine controls, accessories, and many options. It can be expected to play a much larger role in the years to come. Electricity made the use of electronics possible on cars built during the last two decades of the twentieth century. The next century will see electronics expand greatly in the automotive industry.

Hydraulic Theory

Shop Manual
page 93

Hydraulics is the use of a liquid to transfer force. Liquids used in hydraulic systems are selected based on the work to be done and the chemical properties of the liquid. Brake fluid will not work well in steering systems.

Blaise Pascal formulated his law on the use of **hydraulics** in 1647. Known as Pascal's Law, it states that pressure in a closed hydraulic circuit will be the same everywhere in the circuit. The law is based on the fact that liquid in a closed or sealed circuit cannot be compressed.

Vehicles use pressure to perform work in almost every system. When combustion takes place in the engine, the expanding gases create extremely high pressure against the interior walls of the chamber and the piston. Since the piston is the only part that should move at this time, it is forced downward. However, the moving piston dissipates the pressure and the force drops with the decrease in pressure. In order to use it as a continual transmitting force, the pressure must be maintained like a closed hydraulic circuit.

A hydraulic circuit is similar to an electrical circuit. There are hoses and lines (conductors), pumps (power source), load, and controls (Figure 4-37). The pump pressurizes the fluid and the control valve directs it into the hoses and lines. At the output end of the line or hose is a working device. It may be a brake component, an electronic fuel injector, or a piston that helps the operator apply force. Remember the inflated balloon from earlier in this chapter? The atmospheric pressure against the outside of the balloon forced the inside gas outward through the balloon's opening. The exhausting gas created a force to move the balloon.

Also like electricity there are three measures normally used in hydraulic circuits: *force, pressure,* and *area.* The force is applied to the contained fluid or delivered by the fluid. Area is the surface on which the fluid acts. Pascal's Law uses a mathematical formula to calculate hydraulic values. It is written as F = PA where F is force (ft.-lb.), P is pressure per square inch (psi), and A is area in square inches (sq. in.) (Figure 4-38). A hydraulic system can deliver force and increase or reduce it. An increase is called a *mechanical advantage.*

In Figure 4-39, the input and output pistons have the same area. Trapped between them is a liquid. Applying force to the input piston will create an equal force being delivered by the output piston because of the equal areas. However, any reaction against the output piston will deliver the same force backward through the circuit to the input piston and anything that is pushing on it.

In Figure 4-40, the output piston has twice the area as the input. In theory, this means the amount of force applied at the input is doubled at the output. We can prove this by using the formula for Pascal's Law. Since the pressurized fluid between the two pistons is the *transfer median,* the amount of pressure generated must first be determined. The figure shows an area of 1 square inch for the input piston and a force of 10 pounds. Using the known values (1 and 10) and the adjusted formula P = F/A (F-PA), we can calculate that the pressure is 10 psi (P = 10/1).

COMPONENT	LOCATION	201-PG	FIG.	CONN
Auxiliary Dome Lamp	Center of windshield header	24	69	
Battery	LH front corner of engine compartment	8	22	
Emergency Vehicle Front Lamp Flasher	Behind RH side of I/P, mounted on relay bracket near A/C module	23	67	
Emergency Vehicle Rear Compartment Lid Lamp	Mounted to inside of rear compartment lid	25	72	
Emergency Vehicle Rear Window Panel Lamp	Mounted to rear package shelf	24	70	
Emergency Vehicle Rear Window Panel Lamp Relay Panel	Mounted below rear package shelf	25	71	
Engine Wiring Harness Junction Block 2	LH side of engine compartment, forward of strut tower	7	19	11-2
Fuse Block	RH side of I/P, in RH front door opening	12	32	
Fusible Links	Front of engine, at starter assembly	6	17	
SEO Circuit Breaker Panel	Behind I/P, on RH side of A/C module	23	67	
Stop Lamp Switch	Behind I/P, on brake pedal support	10	27	202-47
Starter	Lower front of engine			
Turn Signal Switch	In steering column	10	26	
C200 (102 cavities)	I/P harness to crossbody harnesses, mounted behind RH side of I/P, near shroud	18, 19	53, 55	202-10
C201 (48 cavities)	I/P harness to steering column harness, mounted behind I/P on RH side of steering column	10	29	202-15
C208 (2 cavities)	Accessory wiring harness to accessory wiring junction block cable, behind RH side of I/P near blower motor	23	67	
C210 (1 cavity)	I/P harness to accessory wiring harness, behind RH side of I/P near SEO circuit breaker panel			
C400 (8 cavities)	Crossbody harness to rear body harness, below RH sill panel of rear compartment	27	75	202-32
C404 (4 cavities)	Emergency vehicle rear window panel lamp extension harness to body wiring extension harness, behind RH side of rear seat			
C410 (4 cavities)	Emergency vehicle rear lamp harness to emergency vehicle rear window panel lamp extension harness, below LH side of rear package shelf	25	71	
C412 (1 cavity)	Emergency vehicle rear compartment lid lamp harness to emergency vehicle rear window panel lamp extension harness, behind RH side of rear seat			
G204	Below RH side of I/P, right of A/C module, on 6J1 mounting bracket			
G306	At base of RH A pillar	23	68	

Figure 4-36 Note the location of the highlighted fusible links. (Courtesy of Chevrolet Motor Division, General Motors Corporation)

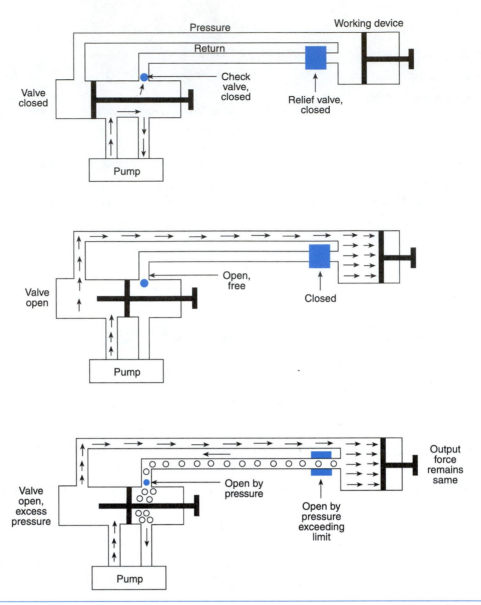

Figure 4-37 Note the actions of the check valve and the relief valve in this hydraulic. Valves of this type are common in hydraulic circuits.

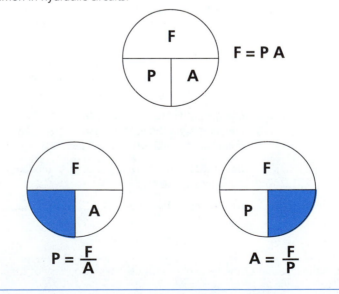

Figure 4-38 The mathematic formula for Pascal's Law is similar to the one for Ohm's Law.

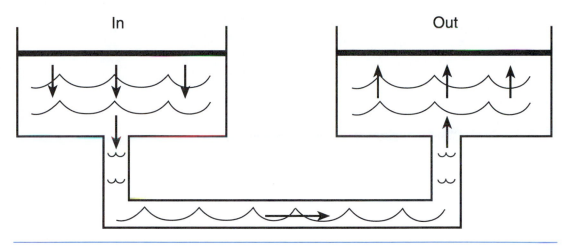

Figure 4-39 If the input and output pistons are equal in size, then the output force will be equal to the input force.

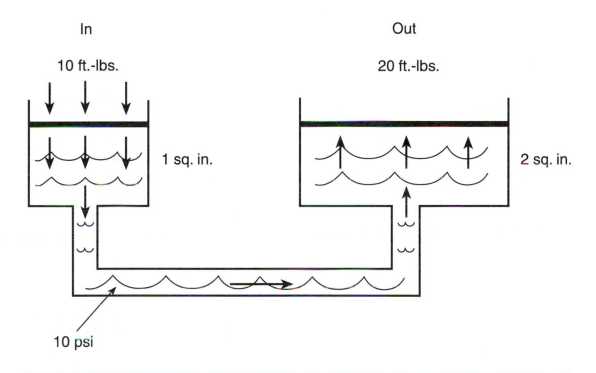

Figure 4-40 Mechanical advantage can be gained or reduced by having input and output pistons of different sizes.

Moving to the output piston and using F = PA, we can calculate the force that can be delivered by the output piston at 20 ft.-lbs. (F = 10 × 2). Remember that pressure is per square inch. The input is 1 square inch and the output is 2 square inches. Since there are 2 square inches on the output and 10 psi on each inch, the delivered force is 20 ft.-lbs. This would give a mechanical advantage of 2:1 (20/10). Calculate the total output force in the following simple braking system.

Refer to Figure 4-41 for the known values for this problem. The figure shows a very simple hydraulic braking system. Notice that the input piston is .75 square inches. The rear brakes have 1-square-inch pistons at each wheel while the front has 2-square-inch pistons at each wheel. First, the system pressure must be calculated. The input piston is being applied with

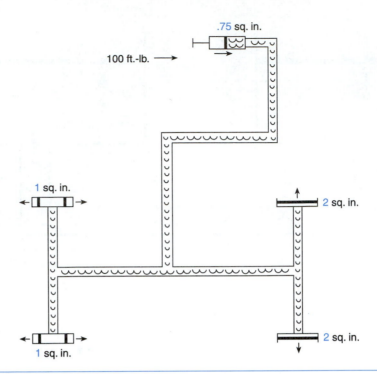

Figure 4-41 This system has 8 square inches of output area versus ³/₄ square inches of input area.

100 ft.-lbs. The known values, force and area, are used in P = F/A or P = 100/.75 resulting in a pressure of 133 psi rounded off. Since the pressure is the same throughout the circuit, the pressure against each square inch of the output pistons is 133 psi. Compute the rear pistons' force using F = PA (F = 1 × 133) or 133 ft.-lbs. at each wheel or a total of 266 ft.-lbs. for the rear. The front is calculated in the same manner. F = 133 × 2 for 266 ft.-lbs. per wheel for a total of 532 ft.-lbs. in the front. The operator is applying 100 ft.-lbs. at the brake pedal and the vehicle is being stopped with a total of 798 ft.-lbs. (front plus rear) of braking effort. This is a 7.98:1 mechanical advantage.

Shop Manual
page 77

Manufacturing treatment of materials includes heating, cooling, and the addition of other materials to increase the properties of the original material.

Aerodynamics is the study of how an object moves through the air.

Vehicle Design

Without going into too much detail, the engineering of an automobile covers not only the theories discussed above, but how different materials can be shaped and formed and their actions under different conditions. The local library has information on how the shape, composition, treatment, and the reactions of metal affect the actions of the component. For instance, aluminum warps at a lower heat then cast iron. In this book, we will not discuss vehicle design and construction at length, but be aware that the material choice, shaping, and the **manufacturing treatment** of each component on the vehicle was researched and designed to satisfy the laws of physics.

Aerodynamics

Until quite recently, the study and use of **aerodynamics** was not considered when building everyday vehicles. Even smaller cars were bulky, heavy, and had the aerodynamic design of a brick! Aerodynamics is the study of how an object moves through the air (Figure 4-42). A vehicle with poor aerodynamics will create high air resistance resulting in higher noise levels and fuel usage. Aircraft and racing vehicles were the first major users of aerodynamic study and application.

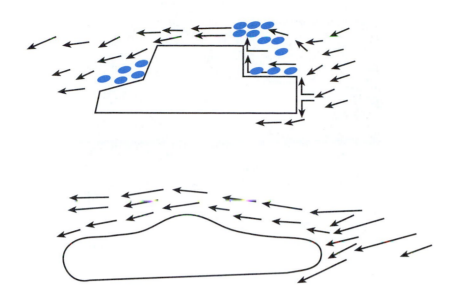

Figure 4-42 On the top vehicle, note how the air is blocked and makes a turbulence area as different air flows meet. The slicker, rounded vehicle at the bottom allows the air to flow freely over and around the vehicle.

Federal regulations and the increased price of fuel required vehicle manufacturers to increase **fuel mileage.** Aerodynamically-designed vehicles incorporated a rounded outside body shape, but all of the different components, including the passenger, had to fit inside this rounded body. Weight also had to be reduced drastically. Newer headlights were redesigned and configured with aerodynamics in mind. The result is the slick body contours of today's vehicle.

One major drawback to smaller-size vehicles with increased accessories is the lack of working area for technicians. Older vehicles provided room under the hood and dash to reach components. On some new vehicles, it is almost impossible to locate or reach some components. The major advantages to smaller, more aerodynamic vehicles are greatly increased fuel mileage and driver protection.

Passenger Protection

Today, almost every car on the road has seat belts (Figure 4-43) and most have air bags (Figure 4-44). While these devices are needed to keep the passenger in his or her seat during an accident, they are minor parts of overall vehicle crash protection.

Fuel mileage is required to meet standards in the federal Corporate Average Fuel Economy (CAFÉ) regulation. A manufacturer's vehicle line must be tested and the average of all models offered must meet CAFÉ.

Figure 4-43 Seat belts hold passengers and drivers in their seats, but do not absorb the kinetic energy of the vehicle. (Courtesy of Chevrolet Motor Division, General Motors Corporation)

Figure 4-44 Air bags absorb the kinetic energy of the moving person to prevent contact with forward parts of the passenger compartment. (Courtesy of Chevrolet Motor Division, General Motors Corporation)

Crumple zones extend from bumper to bumper. The sections of the body and frame interact to absorb as much energy as possible before it reaches the passenger compartment.

While older vehicles had stout bodies and frames, they usually delivered most of the shock of an accident to their passenger compartments. The automotive and insurance industries, along with various government agencies, have destroyed many vehicles to study the effect of accidents on passengers and vehicles. Newton's Laws of Motion and the dissipation of kinetic energy are key points in designing a safe vehicle.

Automotive studies have resulted in vehicles designed and built with material that will crumple on impact and absorb energy. Such areas are called **crumple zones.** More energy absorbed by the vehicle means less shock and damage delivered to the passenger compartment. The first point or crumple zone is energy absorbing bumpers which incorporate devices between the bumper and frame. Under federal law, newer cars must withstand a frontal impact of 30 miles per hour before the passenger compartment begins to absorb kinetic energy from the impact. Entire vehicles have been redesigned to protect passengers and, to some extent, the vehicles themselves.

Materials

Like other things in nature, materials respond to different situations in different ways. The materials used in vehicles are selected to meet specific properties. In the engine, materials must be capable of enduring high pressure, twisting, and heavy loads and still function properly. The most prominent materials in engines are cast iron for the block, aluminum for the cylinder heads, and treated steel for the internal components. Using different materials together presents another problem. The engine produces a great deal of heat and different materials react differently to heat. Aluminum expands much quicker than cast iron. Since they are mated to each other in the engine, engineers had to design some method to compensate for the differences. A gasket was made with materials that would react with the cast iron on one side and aluminum on the other. Gaskets are covered in Chapter 7, "Automotive Bearings and Sealants."

Other systems of the vehicle require different materials to perform their purpose. The material may be high strength and flexible like suspension components, thin sheet metal for the body, or plastics for trim and some no-load components. All of the materials are selected, shaped, treated, and used based on their physical properties and the operational theories involved.

Summary

❑ Thermodynamics is the study of using rapidly expanding gases to produce energy.

❑ The theory of thermodynamics is used to achieve mechanical power from a liquid fuel.

❑ The motion of an object is called work. Work is the use of energy.

❑ Pressure and vacuum are used to perform work on a vehicle.

❑ Vacuum is the absence of pressure.

❑ Newton's Laws of Motion apply to almost every operation on a typical vehicle.

❑ Electricity is made when negatively-charged electrons move from atom to atom in large numbers.

❑ Volts, ohms, and amperes are basic measurements of electricity.

❑ Hydraulic systems use liquids to transfer force.

❑ The measurements of hydraulics are force, area, and pressure.

❑ Vehicle design and manufacturing is based on the capabilities of materials used and their functions within the vehicle.

❑ Aerodynamics is the study of objects moving through air.

Review Questions

Short Answer Essays

1. Describe how the operator's force on the brake pedal is transmitted to the wheels.

2. List the three types of electrical circuits.

3. Discuss the movement of electrons.

4. Explain the importance of the vehicle's electrical system.

5. Discuss the possible defects in a circuit.

6. Explain how force is transmitted with hydraulics.

7. Discuss some of the considerations used in designing and manufacturing a vehicle.

8. Discuss how aerodynamics may affect a vehicle.

9. Discuss the mathematical formula for Ohm's Law.

10. List the three Laws of Motion.

Fill-in-the-Blanks

1. Basic electrical measurements are in _____, _____, and _____.

2. Basic hydraulic measurements are _____, _____, and _____.

3. An electrical system with 12 volts and 5 amperes will have a resistance of _____ _____.

4. The technical name for voltage is _____.

5. Electrons flowing through a conductor will create a(n) _____ _____.

6. A hydraulic circuit consists of_____, _____, _____, and _____.

7. A(n) _____ is the most common electrical measuring instrument in automotive shops.

8. The tendency of an object to stay at rest is called _____ _____.

9. The study of using expanding gases to perform work is called_____.

10. A(n) _____ will not conduct current.

ASE Style Review Questions

1. Electrical systems are being discussed. *Technician A* says the theory is based on 1 volt pushing 1 ampere of current through 1 ohm of resistance.
Technician B says electrons flow from positive to negative. Who is correct?
   A. A only
   B. B only
   C. Both A and B
   D. Neither A nor B

2. Hydraulic systems are being discussed. *Technician A* says the force applied against a piston creates the pressure of the transferring liquid.
Technician B says the pressure of the transferring liquid creates the force on the piston. Who is correct?
   A. A only
   B. B only
   C. Both A and B
   D. Neither A nor B

3. Electrical systems are being discussed. *Technician A* says a circuit is comprised of a load, current, a conductor, and controls.
Technician B says a system may have more than one circuit. Who is correct?
   A. A only
   B. B only
   C. Both A and B
   D. Neither A nor B

4. Work and energy are being discussed. *Technician A* says energy is a unit of work.
Technician B says work is done when an object is moved. Who is correct?
   A. A only
   B. B only
   C. Both A and B
   D. Neither A nor B

5. Thermodynamics are being discussed. *Technician A* says a vehicle uses thermodynamics to produce horsepower.
Technician B says expanding gases is the study of thermodynamics. Who is correct?
   A. A only
   B. B only
   C. Both A and B
   D. Neither A nor B

6. Electricity is being discussed. *Technician A* says electromotive force moves the neutron.
Technician B says electricity produces a magnetic field. Who is correct?
   A. A only
   B. B only
   C. Both A and B
   D. Neither A nor B

7. Newton's Laws of Motion are being discussed. *Technician A* says inertia is the function applied to the change of direction by an object.
Technician B says air resistance and gravity affect objects at rest and in motion. Who is correct?
   A. A only
   B. B only
   C. Both A and B
   D. Neither A nor B

8. Pascal's Law is being discussed. *Technician A* says a liquid is used to transfer force.
Technician B says hydraulics is the study of using liquid to transfer force. Who is correct?
   A. A only
   B. B only
   C. Both A and B
   D. Neither A nor B

9. *Technician A* says current will flow through insulators.
Technician B says current is pushed by voltage. Who is correct?
A. A only
B. B only
C. Both A and B
D. Neither A nor B

10. Electrical circuits are being discussed. *Technician A* says a series circuit provides multiple paths for the current.
Technician B says series circuits and parallel circuits can be combined into one circuit. Who is correct?
A. A only
B. B only
C. Both A and B
D. Neither A nor B

Fasteners

Upon completion and review of this chapter, you should be able to:

❏ Describe the USC threaded fastener measuring system.

❏ Describe the metric threaded measuring system.

❏ Identify the grade markings of threaded fasteners.

❏ List and describe common thread repair tools.

❏ Identify and explain the purpose of common nonthreaded fasteners.

❏ Explain the importance of wire gauge sizes.

❏ Identify common types of electrical fasteners.

❏ Identify common multi-pin, electrical connectors.

Introduction

Automotive components are held together with **fasteners**. As a technician, you will spend much of your day removing and replacing fasteners. These small parts are very important. If fasteners are not used properly, the components they hold together can fail.

Many different types of fasteners are used on an automobile (Figure 5-1). Fasteners can be divided into two basic groups. *Threaded fasteners* use the clamping force from threads to hold parts together. *Nonthreaded fasteners* hold parts together without threads. Both types are discussed in the following sections.

Fasteners are small threaded and nonthreaded parts that hold automotive components together.

Threaded Fasteners

Shop Manual
page 111

Shop Manual
page 111

A BIT OF HISTORY

Threads provide a mechanical advantage to hold parts together. The geometry behind threads was developed a long time ago, around 200 B.C. They were used by the ancient Romans to press grapes for wine.

Threaded fasteners are the most common fastener type. They use spirals called *threads* to wedge parts together. The common types of threaded fasteners are screws, bolts, studs, and nuts and they are shown in Figure 5-1.

Threaded fasteners use spirals called "threads" to hold automotive parts together.

Figure 5-1 Common types of threaded automotive fasteners.

Fastener Sizing and Torquing

For a fastener to work correctly it must have strength and size to handle the task. Fasteners are designed in different grade (strength) ratings for specific loads and applications. Replacement fasteners should meet the same requirement as the original item. A light-duty fastener may be replaced with a heavy-class fastener in most cases. However, some industry and vehicle parts have a certain class fastener that is designed to break or shear to prevent damage to components. A typical example would be the shear pin in a winch. To prevent overloading a cable or chain that may result in very serious injury and great damage, the shear pin is designed to shear apart when its limits are reached. Replacing a shear pin that "broke too quickly" with a stronger pin has resulted in at least one death to this author's knowledge.

 WARNING: Never replace a fastener with a lighter-class fastener. Damage to the associated components can result.

 CAUTION: Never replace safety or protection fasteners with a stronger fastener. Serious injury or death could result to vehicle operators, co-workers, or bystanders.

Screws

A *screw* fits into a threaded hole.

The **screw** is one of the most common threaded fasteners. A screw fits into a threaded hole. Often there is a part without threads between the screw and the part with the threads as shown in Figure 5-2. The threads on the screw engage the threads in the part. As the screw is turned, the two sets of threads work together to clamp the two parts together with a great deal of force.

Hexagon or hexagonal-shaped fasteners are usually just called "hex" for short.

Many different types of screws are used to hold automotive parts together. The most common types of screws used in automotive parts are shown in Figure 5-3. The hex head cap screw is one of the most common types. It has a six-sided or **hexagonal head** and is driven with common automotive wrenches such as the box, open-end, combination, and socket. Hex head cap screws are used to hold large parts together.

Many other screws are driven with a screwdriver. They are used to hold small automotive parts together. These screws may have a slotted head or a recessed head like those discussed under screwdriver types in Chapter 3. They are often called *machine screws* because they are used to hold machined parts together. In addition, they are often classified by the shape of their head. The main head shapes are round head cap screw, round head (with threads that go all the way to the head), flat head, fillister head, and oval head.

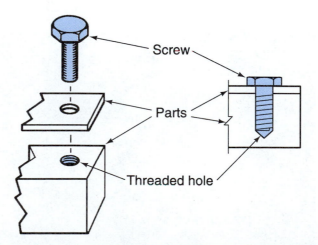

Figure 5-2 A screw fits through one part and into a threaded hole in a second part.

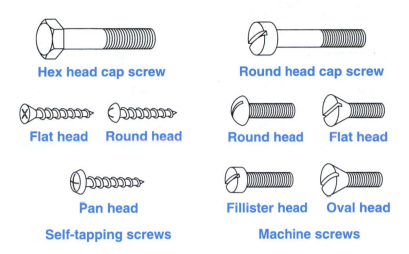

Hex head cap screw   **Round head cap screw**

Flat head **Round head**   **Round head** **Flat head**

Pan head   **Fillister head** **Oval head**

Self-tapping screws   **Machine screws**

Figure 5-3 Common types of automotive screws.

Self-tapping screws do not fit into threaded holes. As they are driven into the part, they use their hard external thread to form, or tap, their own threads. Self-tapping screws are sometimes used in automotive body sheet metal parts.

Bolts

Many automotive parts are held together with **bolts**. Bolts use a nut instead of a threaded hole to hold automotive parts together. Usually the bolt fits through parts that do not have threads. The bolt and nut are driven together to clamp the two parts together as shown in Figure 5-4. Most bolts for automotive use have a hex head (Figure 5-5). The difference between this bolt and a hex head cap screw is that the bolt is used with a nut. The hex head cap screw is used in a threaded hole. There are bolts with heads that are not hex shaped, but they are not often used in automotive parts.

A *bolt* is a threaded fastener used with a nut to hold automotive parts together.

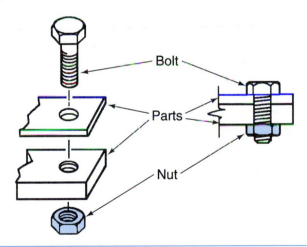

Bolt

Parts

Nut

Figure 5-4 A bolt fits thought both parts and into a nut.

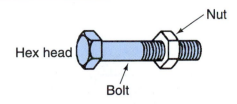

Nut

Hex head

Bolt

Figure 5-5 The hex head bolt is the most common fastener for automotive parts.

Studs

A **stud** is a common type of automotive fastener that has no head. It has threads on both ends. One end of the stud fits into a threaded hole in the part as shown in Figure 5-6. A second part fits over the stud. Then the two parts are forced together with a nut on the other end of the stud.

Studs are used when getting the two parts in perfect alignment is important. A stud may have threads that run all along its length. The threads may be formed only on the ends as shown in Figure 5-7.

Nuts

Nuts are used with bolts and studs. Unlike other fasteners, they have their threads on the inside. We call these *internal threads*. Many different types of nuts are used on automotive parts (Figure 5-8). The most common automotive nuts are hexagonal or *hex nuts*. Hex nuts have six sides and are made to fit box-end, open-end, or socket wrenches.

Several specialized nuts are available when parts may be subject to a great deal of vibration. Specialized nuts are also used where the part held by the nut is critical to safety. These parts might include the *steering-wheel retaining nut* or *wheel spindle nut*. The slotted or *castellated nut* is another important specialized nut. After the castellated nut is tightened onto the bolt or stud, a metal cotter pin is inserted through its slots and also through a hole that has been drilled in the stud or bolt. After the cotter pin is installed, it is bent around the nut.

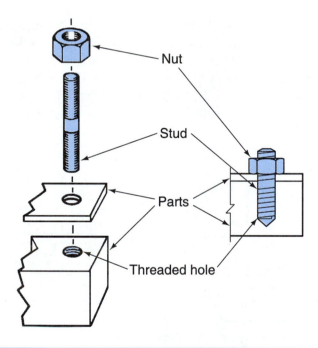

Figure 5-6 One end of a stud fits in a threaded hole and the other end has a nut.

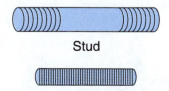

Stud

Figure 5-7 Stud threads may be on each end or along the entire length.

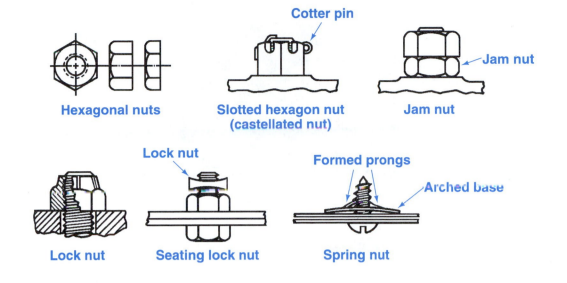

Figure 5-8 Common types of automotive nuts.

Jam hexagon nuts are thinner than regular hex nuts. These are used in pairs and provide double protection against loosening. The lock nut has inserts of fiber or plastic. The inserts jam into the threads as the nut is tightened. This prevents the nut from getting loose during vibration. A *seating lock nut* is a thin nut with a concave surface that flattens when it contacts the top of the regular hex nut. The concave action binds the lock nut to the threads and prevents it from coming loose. *Spring nuts* are made from thin spring metal. They have prongs that fit into a sheet metal screw thread and an arched base that pushes down on the part to be retained. These nuts are used in trim and sheet metal parts.

Washers

Washers (Figure 5-9) are often used with other threaded fasteners for several reasons. First, they can help a nut distribute the clamping load on a part. They can also prevent a nut from getting loose. Finally, they can prevent a nut from digging into a machined surface.

A *flat washer* is often used between a nut and an automotive component or under the head of a screw or bolt. This washer helps spread out the clamping force over a wider area. It also prevents machined surfaces from being scratched as the bolt head or nut is tightened.

Lock washers are used to prevent nuts from vibrating or working loose. The *spring lock washer* has a sharp edge that will dig into a fastener or component surface. This prevents the fastener from working loose. The *tooth washer* has either internal or external teeth that dig into a surface and prevent the nut from getting loose.

Washers are used with bolts, screws, studs, and nuts to protect machined surfaces and prevent nut loosening.

Figure 5-9 Common types of automotive washers.

Figure 5-10 Fine and coarse thread bolts with the same diameter.

U.S. (English) Thread Sizes

Threads on threaded fasteners are made to sizes specified in either the U.S. (English) system or the metric system. Older automobiles manufactured in the United States use only U.S. (English) system threads. Late-model U.S.-manufactured cars and import cars use metric system threads.

> **CAUTION:** Bolt head sizes in the metric and U.S. (English) systems can be very close in size. Remember that using the wrong wrench can cause it to slip off and injure someone, so be sure to use the correct wrench size.

> **WARNING:** Some cars manufactured in the United States in the 1980s had a mix of U.S. (English) and metric fasteners. Be sure not to mix up fasteners when you work on these cars. Using the incorrect fastener in a threaded hole or nut can cause thread damage and part failure.

The *Unified System* is a system for thread classification based on the U.S. (English) system of measurement.

The U.S. (English) threads are manufactured to specifications called the **Unified System.** This system has two common types of threads: coarse and fine. *Course* means there are few threads in a given length of a bolt or screw; *fine* means there are many threads in a given length of a bolt or screw. They are designated as NC for National Coarse and NF for National Fine as shown in Figure 5-10. The difference between fine and course threads is easy to see. Coarse threads are used in aluminum parts because they provide greater holding strength in soft materials. Fine threads are used in many harder materials, such as cast iron and steel.

> **WARNING:** If coarse and fine threaded fasteners are intermixed, you will damage the threads within the parts and may cause early part failure.

The *shank* is the part of the bolt between the head and the threads.

U.S. (English) threaded bolts and hex head cap screws are described by a number of measurements. Three important measurements are head size, shank diameter, and overall length as shown in Figure 5-11. The **shank** is the part of the bolt betwen the head and the threads.

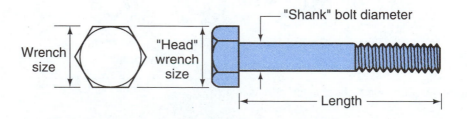

Figure 5-11 Typical measurements for a bolt or hex head cap screw.

Common U.S. (English) Head Sizes	
Wrench Size	Wrench Size
$3/8''$	$1''$
$7/16''$	$1^{1}/16''$
$1/2''$	$1^{1}/8''$
$9/16''$	$1^{3}/16''$
$5/8''$	$1^{1}/4''$
$11/16''$	$1^{5}/16''$
$3/4''$	$1^{3}/8''$
$13/16''$	$1^{7}/16''$
$7/8''$	$1^{1}/2''$
$15/16''$	

Figure 5-12 Common U.S. (English) bolt head sizes.

Its **diameter** is part of the bolt's identification specification. Shank diameter sizes are available in most of the fractional divisions of an inch. Common sizes include $1/4$-, $5/16$-, $3/8$-, $7/16$-, and $1/2$-inch diameters.

The length of the bolt or cap screw is measured from the bottom of the head to the end of the bolt or screw. Bolts and hex head cap screws are commonly manufactured in $1/2$-inch increments. Typical length sizes would be $1/2$, 1, $1^{1}/2$, 2 inch, and so on.

Head size determines what wrench fits the bolt or hex head cap screw. The shank diameter is not the wrench size. The head size is determined by the distance across the hexagonal flats on the head, as shown in Figure 5-11. A bolt with a $1/4$-inch shank diameter has a head size of $7/16$ inch. The common head sizes are shown in Figure 5-12.

The number of **threads per inch** is another important measurement in the Unified System. Each bolt and screw is identified by the number of threads in an inch as shown in Figure 5-13. When a ruler is placed along the bolt shown in Figure 5-13, you can see that there are 20 threads in 1 inch. This bolt would be described as having 20 threads per inch (abbreviated to 20 TPI). When applied to a nut, these measurements refer to the size of the bolt that nut fits.

These measurements are used to give a callout or designation for a threaded fastener. If you were to go to an automotive parts store, you would find the fasteners in drawers with labels like $3/8 \times 16$ UNC $\times 1^{1}/2$ as shown in Figure 5-14. The first measurement is the shank diameter of $3/8$ inch. Each $\times$ means "by." The second measurement is the 16 threads per inch.

The *diameter* is part of the bolt's identification specification.

Threads per inch is a unit of measurement by which bolts and screws are identified.

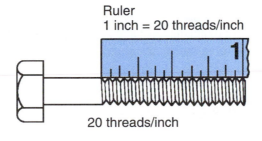

Ruler
1 inch = 20 threads/inch

20 threads/inch

Figure 5-13 A bolt with 20 threads per inch.

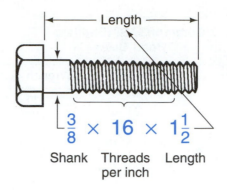

Figure 5-14 A complete bolt size designation.

The UNC means that 16 threads per inch fits in the Unified National Coarse category. Finally, the 1¹/₂ means that the bolt is 1¹/₂ inches long.

Metric Thread Sizes

Fasteners manufactured in metric sizes have the same basic measurements as those described for the U.S. (English) system, but they are given in metric units. A common metric system fastener might be M12 × 1.75 as shown in Figure 5-15. The M indicates that the fastener has metric threads. The first number is the outside diameter of the bolt, hex head cap screw, stud, or the inside diameter of a nut in millimeters. The second number, after the sign ×, is the pitch. This is the distance between each of the threads measured in millimeters.

As you can see, there are no abbreviations such as NC or NF to identify fine and coarse threads. In the metric system, the pitch number distinguishes between fine and coarse. For example, a fine thread metric bolt may be M10 × 1.0. A bolt of the same diameter with a coarser thread might be labeled M10 × 1.25. The larger pitch number indicates wider spacing between threads. The length of metric fasteners is measured the same as in the U.S. (English) system, but the measurements are given in metric system units.

Like the U.S. (English) system, the metric fastener head size is not part of the designation or callout. The head sizes are manufactured in the common metric units. The common head sizes are made to fit common metric wrench sizes. The common head sizes for metric fasteners are shown in Figure 5-16.

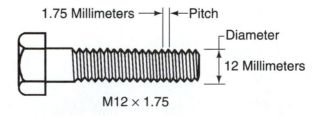

Figure 5-15 A metric fastener designation of M12 × 1.75 mm.

Common Metric Head Sizes	
Wrench Size	Wrench Size
9 mm	19 mm
10 mm	20 mm
11 mm	21 mm
12 mm	22 mm
13 mm	23 mm
14 mm	24 mm
15 mm	26 mm
16 mm	27 mm
17 mm	29 mm
18 mm	30 mm
	32 mm

Figure 5-16 Common metric head and wrench sizes.

Hardness and Strength

Bolts and hex head screws used for different parts of the automobile have different strength requirements. A hex head cap screw used to hold on an engine flywheel must be much stronger than one used to hold on a headlight. On the other hand, the fasteners made to hold on the flywheel are much more expensive than the ones used on the headlight.

The strength or quality of fasteners is identified by **grade markings** on the fasteners. The markings are different for U.S. (English) and metric fasteners. The U.S. (English) grade markings for bolts and hex head cap screws are shown in Figure 5-17. The standards are set by the **Society of Automotive Engineers (SAE).** The system uses marks on the head of the bolt or the hex head cap screw. Unfortunately, the system is confusing because the number of marks and the grade number do not agree. A grade 0, 1, and 2 bolt has no markings on the head and is not very strong. A bolt with three marks is called a grade 5 bolt and is much stronger than a grade 1. The strongest bolt for automotive use has six marks and is called a grade 8. The chart in Figure 5-17 also describes the material each grade is made from and the **tensile strength.** Tensile strength is the amount of pressure the fastener can take before it breaks.

Metric fasteners use property class numbers to indicate their strength. These numbers are stamped on the head of the bolt or hex head cap screw. Typical metric bolt strength numbers are shown in Figure 5-18. The higher the property class number, the stronger the fastener. A 10.9 fastener is much stronger than one marked 4.6. A metric fastener without a number would be the same as a grade 0 U.S. (English) fastener. Metric studs have a marking system on the end of the stud that shows their property class number, as shown in Figure 5-18.

 WARNING: Always replace fasteners with the same strength (grade markings) as the one they replace. Failure to do so may cause the part to fail.

Nuts are also graded according to the same grading systems as shown in Figure 5-19. Nuts must have the same grade as the bolts or studs with which they are used. English nuts have dots that represent grade markings. Three dots represent a grade 5. Six dots represent a grade 8. Metric nuts have numbers that represent strength. The number on the nut is its property class number. In both systems, the higher the number or the more dots, the stronger the nut.

Grade markings on fasteners are used to identify the strength or quality of the fasteners.

Society of Automotive Engineers (SAE) is an organization of engineers, worldwide, that advances the engineering of mobility systems and sets technical standards.

Tensile strength is the amount of pressure a fastener can tolerate before it breaks.

SAE Grade Markings					
Definition	No lines: unmarked indeterminate quality SAE grades 0-1-2	3 lines: common commercial quality Automotive and AN bolts SAE grade 5	4 lines: medium commercial quality Automotive and AN bolts SAE grade 6	5 lines: rarely used SAE grade 7	6 lines: best commercial quality NAS and aircraft screws SAE grade 8
Material	Low carbon steel	Med. carbon steel tempered	Med. carbon steel quenched and tempered	Med. carbon alloy steel	Med. carbon alloy steel quenched and tempered
Tensile Strength	65,000 psi	120,000 psi	140,000 psi	140,000 psi	150,000 psi

Figure 5-17 Grade marking system for U.S. (English) bolts and hex head cap screws.

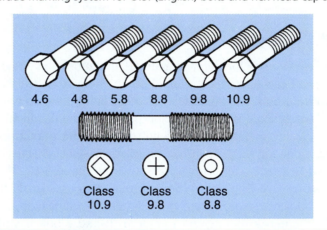

Figure 5-18 Grade marking system for metric bolts, hex head cap screws, and studs.

Inch System		Metric System	
Grade	Identification	Class	Identification
Hex nut grade 5	3 dots	Hex nut property class 9	Arabic 9
Hex nut grade 8	6 dots	Hex nut property class 10	Arabic 10
Increasing dots represent increasing strength.		Can also have blue finish or paint dab on hex flat. Increasing numbers represent increasing strength.	

Figure 5-19 Grade marking system for U.S. (English) and metric nuts.

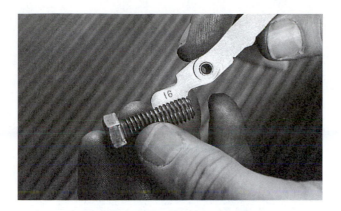

Figure 5-20 A pitch gauge is used to match threads for identification.

Fastener Torque

Each threaded fastener on an automobile must be tightened just the correct amount. Untightened fasteners can loosen and cause parts to fail. Fasteners that are tightened too tightly can damage the part or the fastener and cause the parts to fail. Overtightening causes fasteners to stretch. These no longer have the strength they had before the overtightening.

To avoid these problems, fasteners must be tightened with a torque wrench. There are torque specification charts for each important fastener in the service manual for the car you are working on. Always follow these specifications.

Pitch Gauges

▲ **WARNING:** Always start threaded fasteners several turns into their mating threads by hand. Never start a fastener with a wrench. If the fastener is the wrong thread size, the threads can be damaged by the force of the wrench.

There are two different thread systems in use, each with fine and coarse threads, which can cause a great deal of confusion. Metric threads cannot be used with English threads. Fine threads cannot be used with coarse threads. These different types of threads can make it difficult for the technician to tell one thread from another.

A thread **pitch gauge** (Figure 5-20) is used to identify fasteners. It has a number of blades with teeth. The thread size is written on the blade. By matching the teeth on the blade with threads on a fastener, the thread size can be determined. Pitch gauges are made for both metric and English threads.

Thread Repair Tools

There are several types of thread repair tools or tool sets on the market today. The tools come in metric and USC measurements and can help repair both internal and external threads.

Tap and Die Sets

Tap and die sets are designed to make threads instead of repairing them. However, careful use of **taps and dies** can be used to correct some minor thread damage without reducing the fastener's efficiency. A tap and die set is a combination of many different-size thread repair

Never use a regular impact wrench to tighten a fastener. The impact wrench can quickly exceed the required torque and damage the fastener or part.

When threads are damaged by using the wrong sizes or starting a fastener incorrectly with a wrench, we call this "cross threading" or "stripping."

A *pitch gauge* has toothed blades used to match with threads for identification.

Shop Manual
page 113

Taps and *dies* are used to correct minor thread damage without reducing the fastener's efficiency.

Figure 5-21 A typical tap and die threading set. (Courtesy of Snap-on Tools Company)

tools (figure 5-21). The set generally includes special wrenches, an external die, internal taper taps, internal flat taps, and a pitch gauge. Sometimes special dies and taps for pipes are also included.

A *tap* is used for making or repairing internal threads. Internal threads are located inside a hole. The taper tap is used for holes that extend completely through the material. Flat taps are designed with a flat tip to allow the tap to thread to the bottom of a hole (Figure 5-22).

A *die* is used to repair or make external threads on a shaft or rod (Figure 5-23). Special wrenches are used to turn the tap and die.

Figure 5-22 Flat or bottoming taps. (Courtesy of Snap-on Tools Company)

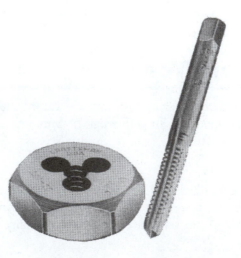

Figure 5-23 A die (left) and a taper tap (right). (Courtesy of Sears Industrial Tools)

118

Helicoils

Basically, a **helicoil** is a threaded device that can be threaded into a hole. Once positioned, it will provide internal threads for a bolt (Figure 5-24). The helicoil is used to make a major thread repair but retains the same diameter size and thread type. It must be used in conjunction with a drill and tap that match the size of the helicoil. A small, single-size helicoil set can be purchased for one-time jobs (Figure 5-25). It consists of a drill bit, the tap, and five or ten helicoils. Larger sets can be purchased with several different-size helicoils and matching drill bits and taps.

A *helicoil* is a device that can be threaded into a hole to provide internal threads for a bolt.

Thread Restorers and Chasers

Thread restoring files are similar to other files in their appearance and use (Figure 5-26). The cutting edges run over and between the damaged threads which are smoothed and straightened by the file's actions. Thread files will only *repair* damaged threads.

    Thread chasers fit over the damaged thread like a threaded fastener (Figure 5-27). The chaser is tightened onto the damaged threads and then turned back and forth over them until they can be used. Like the file, chasers can only *repair* threads, not replace them. This type of thread repair tool does a better job and will correct more damage than a thread file. Chasers are also easier to use.

Thread restoring files are used for repair purposes to smooth and straighten damaged threads.

Thread chasers fit over damaged threads like fasteners and are turned back and forth until the threads can be used. Chasers cannot replace threads but only repair them.

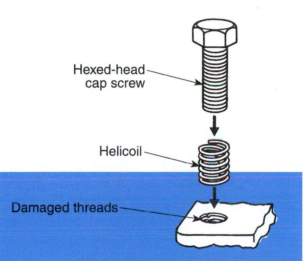

Figure 5-24 A helicoil is used to repair damaged threads.

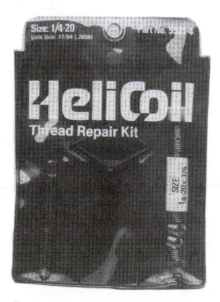

Figure 5-25 A helicoil repair kit. (Courtesy of POP Fasteners)

Figure 5-26 Thread restoring files come in U.S. (English) and metric sizes. (Courtesy of Snap-on Tools Company)

Figure 5-27 A thread restoring tool with insertable blades (lower). (Courtesy of Snap-on Tools Company)

Pins

A *pin* is a small, round length of metal that is driven into two parts to hold them together.

A **pin** is often used to hold automotive parts together when they are not disassembled. A pin is a small, round length of metal. It is driven through holes in two parts to hold them together. There are several common types of pins as shown in Figure 5-28. A *dowel pin* is straight for most of its length but tapered on the very end. This tapered part helps to position two parts that fit together. *Tapered pins* are tapered for their entire length. The taper locks into the part for a tight fit. A *roll pin* is not solid, but rolled from a thin piece of metal. It is used where strong holding power is not required.

Snap Rings

A *snap ring* is an internal or external expanding ring that fits in a groove and works to hold parts together.

A **snap ring** is often used to hold parts in place on a shaft. Because snap rings are used to retain parts in position, they are sometimes called "retaining rings." They are made from a high quality steel that allows them to be expanded or contracted by a tool for installation. Once in place the snap ring "snaps" back to its original size.

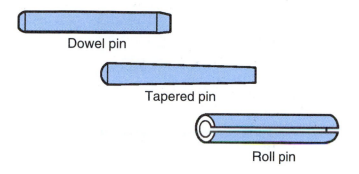

Dowel pin

Tapered pin

Roll pin

Figure 5-28 Three common types of pins.

There are two basic types of snap rings as shown in Figure 5-29. The *internal snap ring* fits in a machined groove inside a hole. The *external snap ring* fits in a machined groove on the outside of a shaft. The external snap ring is expanded with retaining ring pliers for installation. The spring tension from the ring holds it in the groove. The internal snap ring is compressed with retaining ring pliers to fit it in the groove. When the tool is released, its spring tension will hold it in position.

Many snap rings have small holes on the ends as shown in Figure 5-30. The ends of snap ring pliers are made to fit into these holes. Most snap ring pliers come with an assortment of ends to fit different sizes of snap rings (Figure 5-31).

CAUTION: Both types of snap rings are made of spring steel and can cause severe injury if they are allowed to get out of control when removing or replacing them. Always use the right tool and eye protection when working with snap rings.

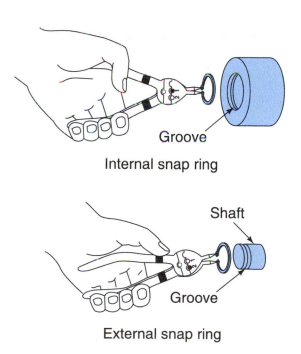

Groove

Internal snap ring

Shaft

Groove

External snap ring

Figure 5-29 Internal and external types of snap rings.

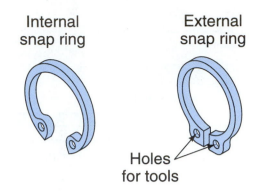

Internal snap ring

External snap ring

Holes for tools

Figure 5-30 Internal and external snap rings with holes for a snap ring tool.

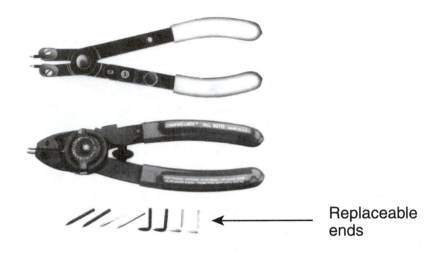

Replaceable ends

Figure 5-31 Snap ring pliers with replaceable ends for snap ring holes. (Courtesy of MAC Tools and Lisle Tools Corporation)

Keys

A *key* is a small, hardened piece of metal used with a gear or pulley to lock it to a shaft.

Another common way to retain a part on a shaft is to use a **key**. A key is a small, hardened piece of metal. They are often rectangular or half-moon shaped. As shown in Figure 5-32, the key fits into a slot called a *keyway*. When it is in position, half the key sticks up past the shaft. There is a matching keyway in the part to be retained on the shaft. The part to be retained in Figure 5-32 is a pulley. The keyway in the pulley fits over the key holding the two parts in position.

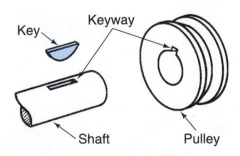

Key

Keyway

Shaft

Pulley

Figure 5-32 A key fits in a keyway on a shaft and pulley to hold the two parts together.

Splines

Splines are another way to hold parts together. These are long teeth that can be cut on the inside or outside of a part. A set of splines is shown in Figure 5-33. The shaft in Figure 5-33 has external splines. The matching part has splines on the inside. The part with the inside splines will slip over the splines on the shaft. The mating splines hold them together.

Rivets

Rivets are used to hold parts together that are rarely disassembled. They are made from soft material such as aluminum. A rivet fits through a hole drilled in the two parts to be joined. The small end of the rivet is formed into a head with a rivet set or a ball-peen hammer. A rivet before and after installation is shown in Figure 5-34. A rivet is taken out by first removing the head with a drill or chisel. It may then be driven out with a punch.

Splines are internal or external teeth cut in a part and used to hold it in place in another part.

A *rivet* is a soft, metal pin with a head at one end that is used to hold two parts together.

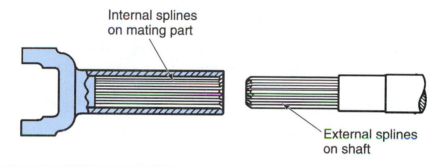

Figure 5-33 Splines can be used to hold a part on a shaft.

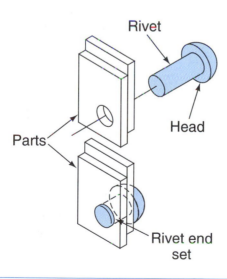

Figure 5-34 A rivet before and after installation.

Electrical Wire and Fasteners

As a technician, you will be required to make repairs to the car's electrical system. This often means replacing damaged wires. Numerous types of wire and electrical fasteners are available to make these repairs. Wires, terminals, and connectors are considered to be the conductors of the electrical current within a vehicle.

When replacing a wire, always use a replacement that is at least as heavy as the one to be replaced. Automotive wire has two basic parts: an inside core of metal that conducts current flow and an outside of insulation material (Figure 5-35).

Wire sizes are based on two systems. The oldest system is the American Wire Gauge (AWG). When wire is made, it is assigned an American Wire Gauge number ranging from 0 to 20. These numbers specify the size of the wire center core diameter and cross-sectional area. The higher the number, the smaller the cross-sectional area of the center core. The largest cross-sectional core wire is 0 and the smallest is 20.

The newer system is based on the metric system. The cross-sectional area of the wire is measured in cubic millimeters. The wire size is then designated by a number such as 0.5 mm. This means the cross-sectional size of this wire core measures 0.5 cubic millimeters. A comparison of common AWG and metric wire sizes is shown in Figure 5-36.

The current-carrying ability of a wire is determined by its cross-sectional area. As shown in Figure 5-37, the greater the cross-sectional area, the more current can flow through the wire. The smaller the cross-sectional area, the more resistance exists and the less current can flow. The wire size of a circuit is carefully determined by automotive electrical engineers to meet

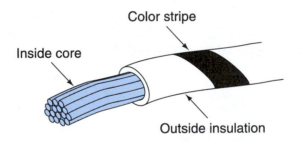

Figure 5-35 Automotive wire has an inside core and an outside insulation material.

Wire Gauge Sizes	
Metric Size (cubic millimeters)	Wire Size
0.5	20
0.8	18
1.0	16
2.0	14
3.0	12
5.0	10
8.0	8
13.0	6
19.0	5

Figure 5-36 Comparison of AWG and metric wire core sizes.

Current increases
(AWG wire size gets smaller)

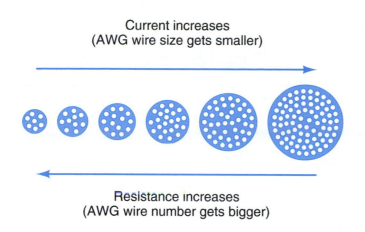

Resistance increases
(AWG wire number gets bigger)

Figure 5-37 Current flow increases and resistance decreases as the wire core gets bigger.

specific requirements. For example, large battery cables are often no. 1 or no. 2 gauge because a larger flow of current is needed. On the other hand, a no. 16 gauge (1.0 mm²) wire is often used with a headlight because less current is needed. As you can see, when a wire is replaced, it must be the same size as the one it is replacing.

New wire is sold in spools that are clearly identified by AWG or metric size. Wire size in various circuits on the automobile are often identified on the wiring diagrams. Two procedures can be used to discover the size of an unidentified wire. First, a wire gauge is available that has a series of graduated holes. Numbers are printed on the wire gauge to correspond with the hole sizes. The center core of the wire to be identified is inserted into the various holes to determine the closest fit. If a wire gauge is not available, the diameter of the center core can be measured with an outside micrometer and then matched to a wire table to determine the size of the wire.

Wires are colored so that they can be traced on circuits on the automobile. Colors on the wire are also printed on the wiring diagrams for the car. You should always try to use the same color as the original when replacing wire.

 WARNING: The diameter of the wire is always specified by the diameter of the center core, not by the outside diameter of the insulation.

Connectors

Soldering

Connecting two wires or a terminal to a wire is best accomplished using electrical-type **solder**. This type of connection adds little or no resistance to the conductor and is resin based. Regular solder is acid based so the soldered area is cleaned as the solder is being applied. Acid based solder will damage the surface of the wire and may add resistance to the circuit. Once soldered, the connection should be covered with heat shrink material instead of tape whenever possible.

Shop Manual
page 121

Soldering is a way to connect two wires or a terminal, adding little or no resistance to the conductor.

Figure 5-38 Butt connectors are used to splice two wires together. (Courtesy of Cooper Automotive/NAPA Belden)

Butt Connectors

A *butt connector* is an electrical connector used to connect two pieces of wire together.

Butt connectors are used to connect two wires together (Figure 5-38). They are quick and easy to install but may add resistance to the total wire. Very low amperage circuits such as computer systems should be soldered.

Terminal Connectors

A *terminal connector* is an electrical connector installed on the end of a wire to connect it to an electrical component.

The **terminal connector** is a device fastened to the end of a wire which allows the wire to connect to or disconnect from a component (Figure 5-39). This allows components to be unplugged from the circuit without damaging the conductor or component. Most positive conductor terminals are covered in some type of insulator or installed in connectors to prevent shorting the circuits.

Molded and Shell Connectors

Crimp-on termnal and butt connectors are called "solderless connectors."

Molded and shell connectors come in many different sizes, shapes, and material depending upon their designed use. In areas of the automobile where there are several connections in one place, the designers have, in many cases, installed connectors that can be matched according to shape, size, and color (Figure 5-40). This reduces the possibility of crossing circuits. Connectors can be made for a single wire or any number of wires installed into one shell or mold.

A molded connector imbeds wires inside a formed plastic or rubber coating (Figure 5-41). The wires cannot be removed from this connector. Replacement connectors with short

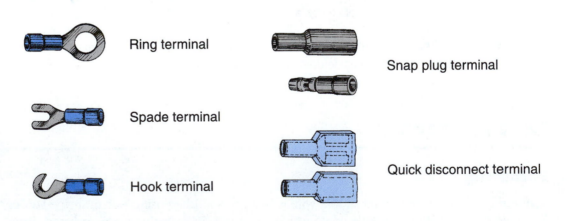

Figure 5-39 The terminals on the right allow a quick hookup and disconnection of the conductor while the terminals on the left are for more permanent connection. (Courtesy of Cooper Automotive/NAPA Belden)

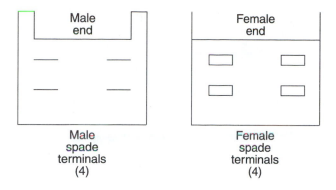

Male end

Female end

Male spade terminals (4)

Female spade terminals (4)

Figure 5-40 The male end will slide and lock into the female end.

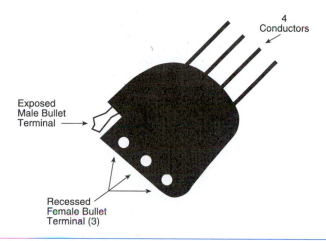

4 Conductors

Exposed Male Bullet Terminal →

Recessed Female Bullet Terminal (3)

Figure 5-41 This is a typical molded connector. Note the bare terminal on the left. This is a four-wire trailer connector.

wires can be purchased to replace the damaged connector. This part is called a *pigtail* and is made to exactly replace the original part, including the color of the wires.

Shell connectors are usually made of hard-formed plastic (Figure 5-42). With the right tool, damaged terminals can be removed, replaced, and reinserted into the shell. Shell connectors may have some type of seals to block moisture from entering the system. This is called a *weather-tight connector*. Most electronic circuits are connected this way.

Weather-Tight Connectors

Most of the electrical connections on late-model vehicles use the weather-tight design. This type of connector may be molded or a shell with rubber seals. As the two the connectors are plugged together, the seal on one side is trapped and clamped between the two (Figure 5-43). The clamped seal prevents moisture and dirt from entering the connector.

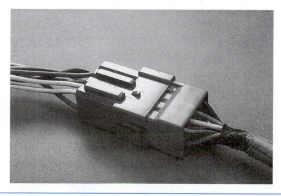

Figure 5-42 Multiple-circuit, hard shell connector.

127

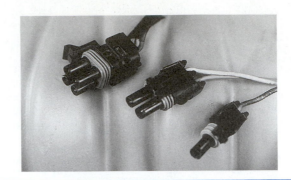

Figure 5-43 The seal is trapped between the mating connectors when they are plugged together.

Summary

❑ Threaded fasteners use threads to hold automotive parts together. Common threaded fasteners include bolts, screws, studs, nuts, and washers.

❑ Threads are measured and classified according to the U.S. (English) and metric systems. U.S. (English) threads are made to standards of the Unified System.

❑ Unified System threads are classified according to the bolt shank diameter, length, and number of threads per inch. Threads in this system can be classified as fine or coarse.

❑ Metric system threads are classified according to shank diameter, length, and pitch.

❑ Both metric and U.S. (English) fasteners have grade markings to show fastener strength. U.S. (English) fasteners use marks on the bolt head to indicate grades. Metric bolts use property numbers to indicate grades.

❑ A pitch gauge can be used to determine the size of a fastener.

❑ Fasteners must be torqued to ensure they are not over- or undertightened. Overtightened fasteners loose their strength.

❑ Damaged threads may be repaired with a tap, die, or by installing a helicoil.

❑ Many automotive parts are held together with nonthreaded fasteners. Common nonthreaded fasteners include keys, snap rings, rivets, splines, and pins.

❑ Electrical system repair involves the use of automotive wire and electrical connectors. Wire is sized according to American Wire Gauge sizes. The smaller the cross section of the wire core, the larger the AWG number. Replacement wires must be the same gauge and should be the same color.

❑ Electrical terminal connectors fit on the end of the wire. Butt connectors join two wires together. Connectors must be the correct shape and wire gauge size.

❑ Weather-tight connectors are used to protect the interior of the connector from moisture.

Review Questions

Short Answer Essays

1. Describe the difference between a bolt and a screw.

2. Explain the difference between a bolt and a stud.

3. Describe how the length of a bolt is measured.

4. Explain how to tell the strength of a U.S. (English) bolt.

5. Describe how to tell the strength of a metric bolt.

6. Explain how to tell the strength of a metric nut.

7. Explain the purpose of a pitch gauge.

8. Explain why a torque wrench should be used when tightening fasteners.

9. Why must the correct wire gauge size wire be used when replacing wire?

10. Explain the purpose of soldering wires and terminals.

Fill-in-the-Blanks

1. A measurement across the head of a bolt gives a(n) _____ size.

2. A measurement from under the head of a bolt to the end of the threads gives a _____ measurement.

3. Grade markings are found on the _____ of a bolt.

4. The number of threads on a metric fastener is called a(n) _____ measurement.

5. A replacement thread installed in a tapped hole is called a(n) _____.

6. A key fits in a(n) _____.

7. The two types of snap rings are _____ and _____.

8. The two types of electrical connectors are _____ and _____.

9. The _____ the inner core of a wire, the greater the resistance.

10. The _____ connector may add resistance to the circuit.

ASE Style Review Questions

1. A ³/₈-inch bolt is being discussed. *Technician A* says a ³/₈ wrench fits the bolt.
Technician B says the diameter of the bolt is ³/₈ inch. Who is correct?
   A. A only
   B. B only
   C. Both A and B
   D. Neither A nor B

2. A ¹/₄ × 28 UNF bolt is being discussed. *Technician A* says this fastener is ¹/₄ inch in diameter.
Technician B says this fastener has 28 threads per inch. Who is correct?
   A. A only
   B. B only
   C. Both A and B
   D. Neither A nor B

3. The grade markings on a U.S. (English) bolt head are being discussed. *Technician A* says the more marks, the stronger the bolt.
Technician B says the more marks, the weaker the bolt. Who is correct?
   A. A only
   B. B only
   C. Both A and B
   D. Neither A nor B

4. The grade markings on a metric bolt are being discussed. *Technician A* says the larger the number, the weaker the bolt.
Technician B says the larger the number, the stronger the bolt. Who is correct?
   A. A only
   B. B only
   C. Both A and B
   D. Neither A nor B

5. Coarse and fine thread bolts are being discussed. *Technician A* says a fine thread bolt has more threads per inch than a coarse one. *Technician B* says a fine thread bolt has fewer threads per inch than a coarse one. Who is correct?
   A. A only
   B. B only
   C. Both A and B
   D. Neither A nor B

6. The installation of a hex head cap screw is being discussed. *Technician A* says always start the screw by hand. *Technician B* says it is acceptable to start a screw with a wrench. Who is correct?
   A. A only
   B. B only
   C. Both A and B
   D. Neither A nor B

7. The installation of a hex head cap screw is being discussed. *Technician A* says it can be installed with an impact wrench. *Technician B* says it should be tightened with a torque wrench. Who is correct?
   A. A only
   B. B only
   C. Both A and B
   D. Neither A nor B

8. The use of washers are being discussed. *Technician A* says washers distribute the load on the part being tightened. *Technician B* says washers protect machined surfaces. Who is correct?
   A. A only
   B. B only
   C. Both A and B
   D. Neither A nor B

9. Automotive wire size is being discussed. *Technician A* says the larger the center core, the larger the AWG number. *Technician B* says the larger the center core, the smaller the AWG number. Who is correct?
   A. A only
   B. B only
   C. Both A and B
   D. Neither A nor B

10. Connectors are being discussed. *Technician A* says terminals are used to connect two wires. *Technician B* says a butt connector can do the same job as soldering two wires. Who is correct?
    A. A only
    B. B only
    C. Both A and B
    D. Neither A nor B

Mechanical Measurements and Measuring Devices

Upon completion and review of this chapter, you should be able to:

❑ Describe United States Customary (USC) system of measurement.

❑ Describe the metric system of measurements.

❑ Explain how to convert measurements between the two measuring systems.

❑ Explain when and how to read rulers.

❑ Discuss the need for feeler gauges and their general use.

❑ Describe how to read measurements made with micrometers.

❑ Explain the use and function of bore gauges.

❑ Describe dial indicators and calipers.

❑ List pressure and vacuum measuring devices.

Introduction

In the latter section of Chapter 3, we discussed collecting repair data for the vehicle. Many times, the technician must collect information about the vehicle for comparison with manufacturer data. The method of collecting the data depends on what is being measured and what measuring devices are used. This chapter will discuss some of the most common automotive measuring devices. Your shop may send components out for machining based on your diagnosis. If you do not check for excessive wear or damage, you may be wasting the customer's money. Also starting in this chapter, you may see vehicle terms, such as *valve guide,* used to explain when a tool may be used. Definitions of some of the terms will be held until a later chapter when the terms can be placed in context with other vehicle components.

English versus Metric

The **measurement** commonly used in the United States comes from the English system and is known as United States Customary (USC). Its scale is based on the inch (2.54 centimeters). USC measurements have been given terms for a certain number of inches. A *foot* equals 12 inches, a *yard* equals 36 inches or 3 feet, and on up the scale.

Shop Manual
page 133

When recording *measurements,* always include the whole number before the decimal or fraction, for example, 0.002 or 1.05.

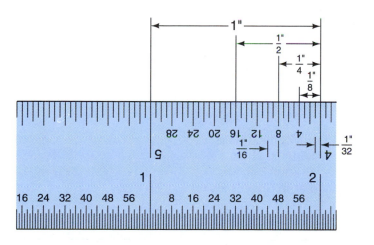

Figure 6-1 USC uses fractions of an inch to measure distances.

Linear measures are measurements of distance such as a mile or a meter.

Decimal numbers are expressions of whole numbers in hundreds, thousands, or ten thousands. *Fractional numbers* are shares of the whole, such as 3/8 or 1/2.

Tolerances set the limit that a component can be different from a perfect size.

Meshed components touch or engage each other all the time, such as two meshed gears.

A *meter* is a unit of the metric measuring system that equals 39.37 inches.

Automotive USC measures **linear** distances in **decimals** (0.000) or **fractions** (1/2, 7/8) of an inch (Figure 6-1). Fractional measurements are usually used for overall vehicle length, tire or brake sizes, and other components where the **tolerances** are not required to be so precise. Engine measurements such as *oil clearance* and *valve guide size* are expressed in thousandths (.000) or ten-thousandths (.0000) of an inch. The tolerances in tightly fitted and **meshed components** must be extremely precise to reduce wear and prolong the life of the vehicle.

The metric system of linear measurements is based on the **meter** (39.37 inches) or a portion of a meter. A kilometer (1,000 meters) is roughly five-eighths (5/8) of a mile (5,280 feet) (Figure 6-2). The meter can also be divided into smaller units. The most common units used on the automobile are *centimeter (cm)* or one-hundredth (1/100) of a meter and *millimeter (mm)* or one one-thousandth (1/1,000) of a meter. Technicians and automotive designers use the millimeter to measure close tolerances the same as using the decimal units of inches.

Within the last ten to twenty years, there has been an effort to change the United States from the USC system to the metric system. Almost the entire world uses the metric system, and with global trade growing, it is becoming necessary to use a single system of measurements. Presently, most automotive measurements show the USC and its matching metric measurement together (Figure 6-3) Another point that makes metric measurements a viable alternative to USC is the math required to change from one unit of measure to the other. With the USC system, the technician must divide or multiply by some hard-to-use numbers like 12, 3, or 5,280. In addition, inches can be expressed as decimals or fractions. The metric system uses 10 for dividing

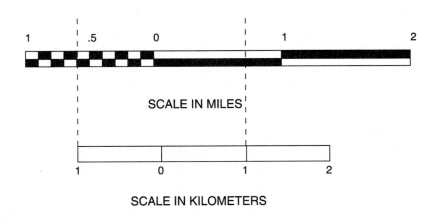

Figure 6-2 The metric system uses tenths as the base divider or multiplier.

DESCRIPTION	TORQUE
Brakes:	
Support Plate Mounting Bolts	109 N·m (80 ft. lbs.)
Brake Hose Mounting Bolt	45 N·m (35 ft. lbs.)
Brake Hose Bracket Bolt	23 N·m (17 ft. lbs.)
Caliper Adapter To Spindle Bolts	115 N·m (85 ft. lbs.)

Figure 6-3 Most service manuals give USC and metric data. (Courtesy of DaimlerChrysler Corporation)

or multiplying to change units of measure and the system does not use fractions. As an example, see which of the following can be calculated the quickest.

One mile (5,280 feet) = _____ inches
One kilometer = _____ centimeters

The answer to the first statement is:
5,280 × 12 (inches in one foot) = 63,360 inches

The second statement?
1,000 × 100 (centimeters in a meter) = 100,000 centimeters

The first equation requires some math. The second one requires the addition or subtraction of zeros.

In the automotive field, almost every vehicle and part is made to metric specifications. However, there are many vehicles on the road that are based on the USC system and those vehicles will be here for years to come. Eventually, the USC will be dropped and the metric system will be used worldwide. Until then, it is necessary to convert some measurements from USC to metric or the reverse to properly perform daily tasks.

Measurement Conversion

At times, the data found may not be in the measurement format needed. This may happen because a technician who has been in the field for awhile does not like the metric system or the tool being used only measures in one system. A prime example is brake disc thickness. Most new brake discs have the minimum thickness marked in millimeters (Figure 6-4). Most older shops use a brake disc caliper that measures in USC. To properly check the disc, the technician must first convert the metric data to a USC measurement. Sometimes a chart is readily available for quick conversion or the technician may have to perform some minor math. The following section will give a brief idea of converting measurements.

Shop Manual
page 133

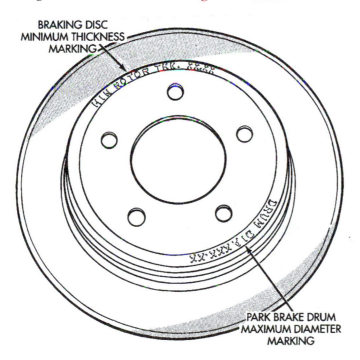

Figure 6-4 Brake discs have their minimum thickness stamped in millimeters. (Courtesy of DaimlerChrysler Corporation)

Conversion factors are numbers or statements that set the method of changing to and from different databases.

End play is the amount of movement a component can move within its mount.

Figure 6-5 shows a chart to help make conversions. Notice that there is a **conversion factor** for going from the metric system to the USC and another for USC to metric. When making conversions, ensure that the correct conversion factor is being used. One of the most common conversions done in automotive repairs is converting inches to millimeters or centimeters or the reverse.

A measurement that may require conversion is shaft **end play** (Figure 6-6). This may be measured with a dial indicator, which may be marked in thousandths of an inch. However, the specification may be listed as metric. Using the data below, compute the measured end play into a metric unit.

Measured end play is 0.0008 in.
Specified end play is .2 mm to .3 mm.

To Find		Multiply	x	Conversion Factors
millimeters	=	inches	x	25.40
centimeters	=	inches	x	2.540
centimeters	=	feet	x	32.81
meters	=	feet	x	0.3281
kilometers	=	feet	x	0.0003281
kilometers	=	miles	x	1.609
inches	=	millimeters	x	0.03937
inches	=	centimeters	x	0.3937
feet	=	centimeters	x	30.48
feet	=	meters	x	0.3048
feet	=	kilometers	x	3048.
yards	=	meters	x	1.094
miles	=	kilometers	x	0.6214

Figure 6-5 Conversion charts make switching between measuring systems easy.

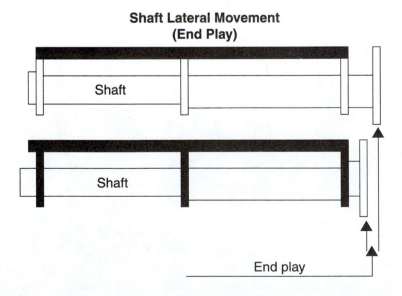

Figure 6-6 End play is the amount of travel a component can move within its mount.

Is the end play within tolerances? If computed correctly, the answer is yes. Multiply the measured end play (0.008 in.) by the conversion factor (25.40 mm). This equals 0.2034 mm, which falls within the .2 mm to .3 mm specified range. Try going in the opposite direction with the following information:

The measured valve lifter bore is 21 mm.
The specified valve lifter bore is 0.7 in.

Is the **bore** worn too much? Multiply the measured bore (21 mm) by 0.03937 inch. The answer is 0.826 inch. In this case, the bore is 0.126 inch too large and outside of tolerances. Notice the decimal-inch conversion factor (0.3937). It is one one-thousandth ($^1/_{1,000}$) of the total inches (39.37) in a meter.

A *bore* is usually considered to be a precise, machined hole. A hole is not machined.

Rulers and Feeler Gauges

Rulers

Rulers are simple devices used to measure straight-line distances where tolerances are not a major factor. They may be marked in USC or metric units. The ruler may have measurements on one side, both sides (Figure 6-7), or may be in the shape of a triangle with each side representing a different scale (Figure 6-8). This triangular ruler is used by drafters and engineers to change the scale of their drawings.

Shop Manual
page 133

A *ruler* can be used to measure the distance across an object or the distance between two or more objects.

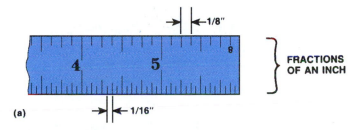

Figure 6-7 One scale has division of $^1/_{16}$ of an inch. The other is $^1/_{32}$ of an inch.

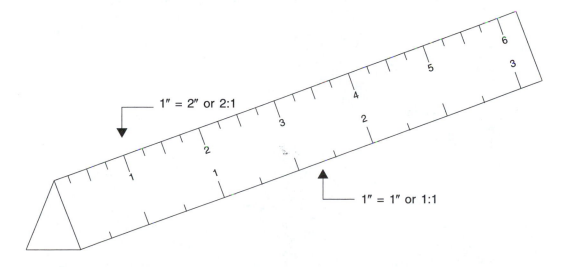

Figure 6-8 This type ruler is used by drafters and engineers for blueprinting.

Rulers can be made of wood, plastic, or metal. The flat, wooden ruler is the type most of us use. The typical USC ruler is 6 inches or 12 inches long. A single-sided ruler may have measurements along one or both edges. The scale will be subdivided or graduated in fractions of an inch. The graduations are normally in $\frac{1}{8}$ or $\frac{1}{16}$ of an inch. Rulers with scales on both edges may have their second scale in $\frac{1}{32}$ of an inch (Figure 6-9). USC rulers may have graduations along both sides, however, this is not a common practice. Most rulers have a thin, metal strip embedded in the graduated edge to act as a straightedge for drawing or marking the work (Figure 6-10).

A metric ruler is made very similar to the USC version. The only real differences are the measuring system and overall length. Metric rulers are usually 20, 25, or 30 centimeters long. Metric rulers are graduated in millimeters, making it easier to interpret the measurement (Figure 6-11). There are no fractions on a metric ruler. Sometimes, a metric ruler will have the 0.5-millimeter marks labeled.

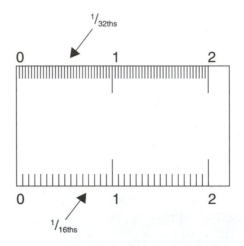

Figure 6-9 Rulers with $\frac{1}{32}$-inch readings can be very hard to decipher.

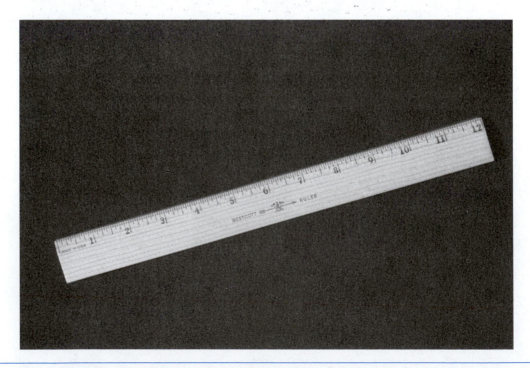

Figure 6-10 The metal edge is provided to help draw a straight line.

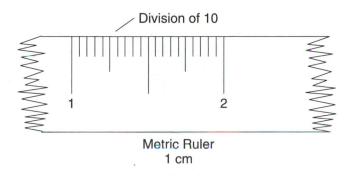

Division of 10

Metric Ruler
1 cm

Figure 6-11 Metric measurements are easier to transfer because they use no fractions.

Feeler Gauges

At one time, **feeler gauges** were one of the most common measuring devices in an auto repair shop. Ignition points and spark plug gap were set using feeler gauges, as were other precision measurements.

A feeler gauge is a flat or round precision-machined piece of metal (Figure 6-12). USC sizes range from 0.001 inch (0.02 mm) upward to about 0.080 inch (1.6 mm). Metric gauges are graduated in tenths of a millimeter. Feeler gauges may be marked in USC or metrics, but the common practice is to stamp each gauge with both measurements (Figure 6-13).

Some *feeler gauges* are so thin, they must be used with extreme care to get an accurate reading and prevent damage.

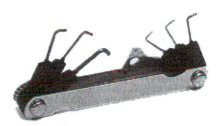

Figure 6-12 Both types of feeler gauges are used to measure distances. (Courtesy of Snap-on Tools Company)

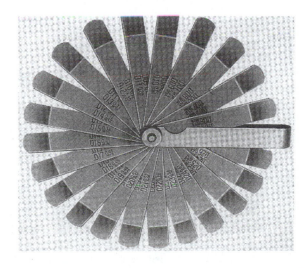

Figure 6-13 A feeler gauge with USC and metric measurements. (Courtesy of A & E Manufacturing Company)

Figure 6-14 A set of feeler gauges has enough individual gauges to meet the measurement requirements of most automobiles. (Courtesy of Snap-on Tools Company)

A *round feeler gauge* is used on spark plugs because of the rounded tip of a plug's rounded center electrode.

Usually the term "feeler gauge" refers to a set of ten to twenty individual gauges (Figure 6-14). Feeler gauges are used to measure the distance between components. Flat gauges are used to check the gap between adjacent parts or measure the air gap between electronic components (Figure 6-15) **Round feeler gauges** are used almost exclusively to measure spark plug gap. Often, the gauges are made from steel, but special gauges may be made from a non-metallic material so they can be used to measure around magnetic components (Figure 6-16). Some gauges are built with an angle to make it easier to measure in tight places (Figure 6-17).

The proper use of a feeler gauge is fairly simple. However, the method of determining the point at which the measurement is correct relies on the technician's touch and feel. A feeler gauge is placed between two adjacent components. The object is to adjust the components until they are exactly x-thousandths of an inch apart.

Instructions for the use of a feeler gauge state that it must slide back and forth between the two components until a *slight drag* is felt (Figure 6-18). Technicians must develop the touch to find the slight drag required for a correct measurement.

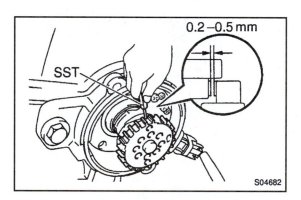

Figure 6-15 Air gaps should be checked with a flat feeler gauge.

Figure 6-16 A non-magnetic feeler gauge must be used to check the air gap between a magnetic sensor and its trigger. (Courtesy of Snap-on Tools Company)

Figure 6-17 Angle feeler gauges can be used to adjust valve lash. (Courtesy of Snap-on Tools Company)

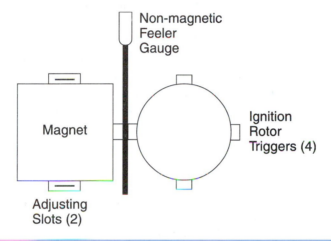

Figure 6-18 The gap must be adjusted until there is a slight drag on the feeler gauge.

In addition to regular feeler gauges, there are go/no-go gauges or **stepped feeler gauges**. This gauge has a tip that is 0.002 inch (0.05 mm) smaller then the rest of the gauge. If the tip slides into the space between the components and the remainder of the gauge will not slide in, the gap or space is correct (Figure 6-19). A larger-size gauge blade can be used to do the same check.

A shaft's end play tolerances may be too large to make use of a *stepped feeler gauge.* Use two different feeler gauges to act as a go/no-go gauge.

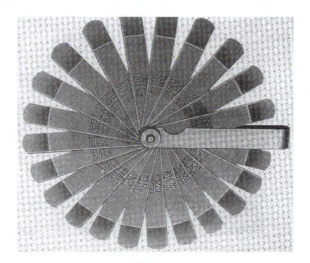

Figure 6-19 Two different-sized feeler gauges may be used in place of a stepped feeler gauge. (Courtesy of A & E Manufacturing Company)

Micrometers

Micrometers are used to precisely measure the different shapes of a component. There are outside, inside, and depth micrometers, each built with a specific purpose. Each type has a measuring scale, an adjustable measuring face, a fixed measuring face, and a frame (Figure 6-20). Because of the limited space on the scales, micrometers are usually set up to measure either USC or metric. However, with the introduction of electronic and digital micrometers, the user can select either USC or metric measurements or the scale (Figure 6-21).

Outside Micrometers

An **outside micrometer** resembles a clamp that can measure linear distances. The *frame* supports the working components and is sized to measure within a limit. Metric scale will be discussed later. A one- to two-inch USC micrometer will measure a component that is between 1 and 2 inches in diameter or thickness (Figure 6-22). The *fixed face* or *anvil* provides a point from which the measurement is made (Figure 6-23). The other measuring face is on a *spindle* that can be extended to or retracted from the anvil by turning the *thimble* (Figure 6-24). The outer portion of the thimble is knurled for grip. At the outer end of most timbles is a smaller, knurled protrusion that acts as a force release (Figure 6-25). The protrusion has a ratchet device that slips when the anvil and spindle are tight enough to make an accurate measurement.

Shop Manual
page 135

A *micrometer* may be referred to as a "mike."

An *outside micrometer* is normally used in the automotive business to measure crankshaft and camshaft bearing journals.

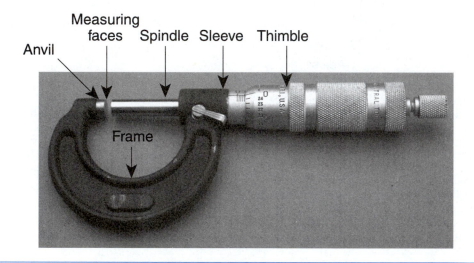

Figure 6-20 A typical micrometer resembles a clamp. (Courtesy of L. S. Starrett Company)

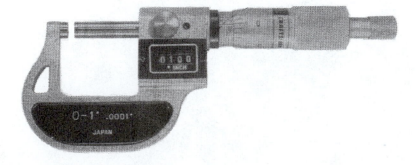

Figure 6-21 Electronic micrometers eliminate the need to add measurements. (Courtesy of Sears Industrial Sales)

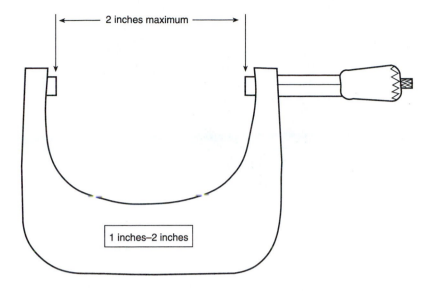

2 inches maximum

1 inches–2 inches

Figure 6-22 The micrometer frame shows the limits of its measurement.

Figure 6-23 The anvil's measuring face is the base point for making measurements. (Courtesy of Snap-on Tools Company)

Figure 6-24 Turning the thimble moves the spindle to and from the anvil. (Courtesy of Snap-on Tools Company)

Figure 6-25 The protrusion is a ratchet that prevents the spindle from being overloaded. (Courtesy of Snap-on Tools Company)

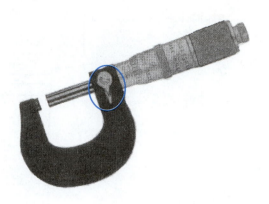

Figure 6-26 The lock holds the spindle in place for reading. (Courtesy of Sears Industrial Sales)

The ratchet also prevents excessive force from being applied to the spindle. The thimble rotates around a sleeve and extends from the frame in the opposite direction of the spindle. Most micrometers use some type of lock to hold the spindle and thimble in place (Figure 6-26). This allows a technician to remove the micrometer from the work to better see the measurement markings.

The sleeve has an **index line** along its length. Above the line are numbers ranging from 0 to 10 at $1/100$-inch graduations (Figure 6-27). Below the lines are markings at 0.025 inch. The thimble markings are laid rotationally around its lower edge (Figure 6-28). Thimble graduations are 0.001 inch each and are numbered each 0.005 inch up 0.025 inch (Figure 6-29). One complete rotation of the thimble will move the spindle 0.025 inch. Some micrometers have an additional scale on the sleeve called a **vernier scale**. The vernier scale is laid rotationally around the base of the sleeve starting at the index line. Each line parallels the index line and is equal to 0.0001 inch (Figure 6-30). Vernier scales are used to read measurements in one ten-thousandths ($1/10,000$) of an inch.

Reading an USC Outside Micrometer

Reading any micrometer is easy after a little practice. Place the anvil against the work. Rotate the thimble to move the spindle to contact the opposite side of the work (Figure 6-31). Use the ratchet knob to ensure sufficient contact is made with the work for an accurate reading. When the ratchet slips, lock the spindle and **remove the micrometer** from the work.

For this example, we will say the component is somewhere between 2 and 3 inches thick. This means a 2- to 3-inch micrometer is selected (Figure 6-32). With the micrometer removed from the work, the index line is checked first. Notice that 0 and 1 are exposed (Figure 6-33). Since 1 is the highest number exposed, the component is between 2.100 and 2.200 inches thick. Below the index line there are three graduations past the 0.100 mark that are visible (Figure 6-34). Since each of these lines is equal to 0.025 inch, 0.075 inch is added to the 2.100 inches computed previously. The component is now at least 2.175 inches thick. Following the index line to where it disappears under the thimble, it is found that the third mark above zero on the thimble aligns with the sleeve's index line (Figure 6-35). This is the fourth number in the measurement. Since each line on the thimble equals 0.001 inch, 0.003 inch is added to the 2.175 inches from the last calculation. The final measurement for the thickness of this component is 2.178 inches.

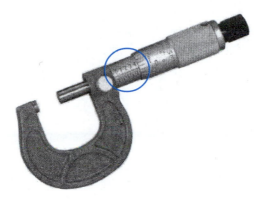

Figure 6-27 Each number represents $1/100$ of an inch. Each of the lower marks represents 0.025 inch. (Courtesy of Sears Industrial Sales)

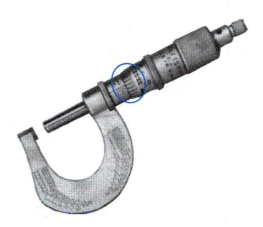

Figure 6-28 The thimble markings are read by finding the closest mark to the index line. (Courtesy of Sears Industrial Sales)

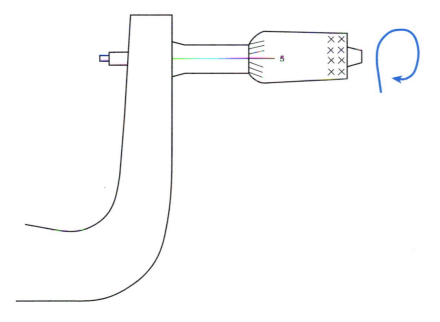

Figure 6-29 Each mark represents 0.001 inch and are numbered every 0.005 inch.

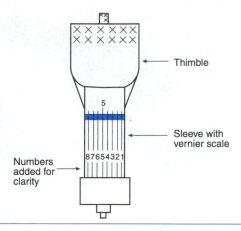

Thimble

Sleeve with vernier scale

Numbers added for clarity

Figure 6-30 The vernier scale measures 0.0001 inch per line up to 0.0010 inch.

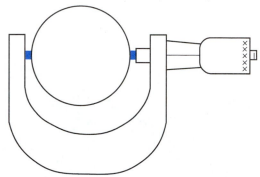

Figure 6-31 Ensure that the two measuring faces are opposite each other.

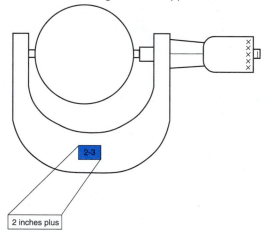

2 inches plus

Figure 6-32 The frame contains the measurement limits of the micrometer.

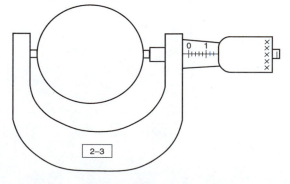

2 inches + .100 = 2.100 inches

Figure 6-33 Use the highest visible number to start your decimal measurement.

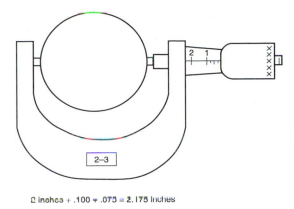

Figure 6-34 The thimble is just past the 0.075-inch mark.

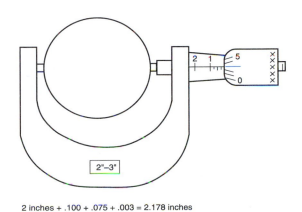

2 inches + .100 + .075 + .003 = 2.178 inches

Figure 6-35 The 0.003-inch line is in direct alignment with the index line.

Reading a Metric Outside Micrometer

A metric micrometer is laid out the same as a USC micrometer except for the size of the scales. Above the index line are 1-millimeter markings ranging from 0 to 25. They are numbered each at 5 mm (Figure 6-36). Below the index line are markings designated 0.5 mm each (Figure 6-37). The thimble is marked from 0 to 50 or half of a millimeter (Figure 6-38). Each graduation is equal to 0.01 mm and is numbered each at 0.05 mm (Figure 6-39).

Reading a metric micrometer is very similar to reading a USC micrometer. After placing the measuring faces properly on the work, lock the spindle and remove the micrometer for better visibility of the scales. On the index line, find the last whole number uncovered. In Figure 6-40, that number is 5. Note that the 0.5-millimeter mark is also exposed below the index line. This shows the work is at least 5.5 mm thick (5 mm + 0.5 mm). Note the thimble reading in Figure 6-40. The twenty-eighth line is aligned with the sleeve's index line. This means that the actual thickness of the work is 0.28 mm larger then 5.5 mm. Add 0.28 mm to the 5.5 mm for a total measurement of 5.78 mm (Figure 6-41).

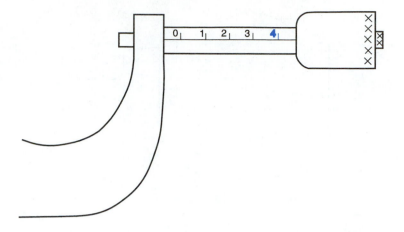

Figure 6-36 Each numbered line above the index line represents 5 millimeters.

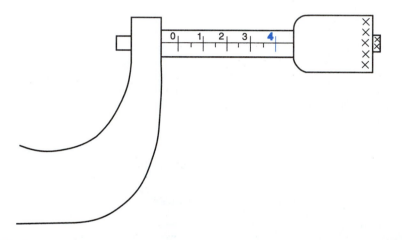

Figure 6-37 The lines below the index equal 0.5 millimeters.

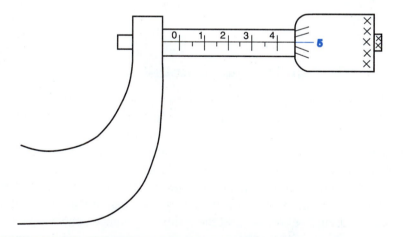

Figure 6-38 One rotation of the thimble equals half (0.5) of a millimeter.

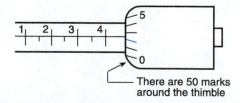

Figure 6-39 Each line on the thimble equals 0.01 millimeter for a total of 0.5 millimeter per thimble revolution.

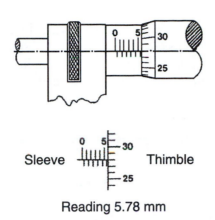

Sleeve ┃┃┃┃┃ Thimble

Reading 5.78 mm

Figure 6-40 The measurements on a metric micrometer are read and totaled the in the same manner as an USC micrometer. (Courtesy of L. S. Starrett Company)

1. Add the last visible number above the index line	5 mm
TO	
2. The last visible 0.5 mm mark below the index line	0.5 mm
TO	
3. The nearest thimble 0.01 mm mark align to the index line	0.28 mm
4. Equals	5.78 mm

Figure 6-41 Add the three scale readings to find the total measurement.

Telescopic Gauges

Telescopic gauges (Figure 6-42) and outside micrometers can be used to measure inside dimensions. Telescopic gauges come in various lengths and are T-shaped with the cross section consisting of two spring-loaded extensions. The leg of the gauge contains a rotatable handle

Telescopic gauges have rounded tips at the ends to better match the curve of the bore.

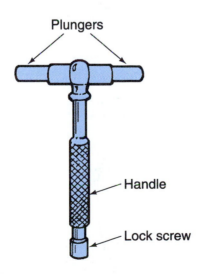

Plungers

Handle

Lock screw

Figure 6-42 The telescopic gauge has two spring-loaded extensions.

that locks the extensions in place. The gauge is placed inside the bore, the handle is twisted to release the extensions, and the extensions spring out to the bore's walls (Figure 6-43). Once the technician is satisfied with the gauge's position, the handle is turned to lock the extensions. The gauge is removed from the bore and its extended length is measured with an outside micrometer (Figure 6-44). The technician must ensure that the telescopic gauge is properly positioned within the bore and the micrometer to guarantee an accurate measurement.

Inside Micrometers

Inside micrometers are USC or metric tools that are used to measure the diameters of large holes up to 1 inch (25 mm) wide. Attachments must be added to measure larger holes.

Inside micrometers are used to measure the inside diameter of large holes (Figure 6-45). They can be USC or metric and the markings are the same as those on an outside micrometer. An inside micrometer looks like the outside micrometer minus the frame. An inside micrometer can only measure up to 1 inch or 25 millimeters without attachments (Figure 6-46). To make larger measurements, the technician must use extension spindles and a half-inch spacer supplied with the inside micrometer set (Figure 6-47). The spindles are made in different lengths and are attached to the micrometer as needed. To measure a hole or bore that is 4 inches in diameter, attach a 3- to 4-inch spindle and spacer to the micrometer. If the bore to be measured is between a whole inch and $1\frac{1}{2}$ inches (3.5 inches for example), the spacer would not be used. Once positioned in the bore, the measurement would be read the same as an outside micrometer.

The critical point in using an inside micrometer is its placement within the bore (Figure 6-48). The micrometer must be at a right angle to the vertical centerline of the bore. Because of the curved surfaces of the bore, the technician must shift the spindle back and forth until the largest measurement is shown on the micrometer. This skill comes with a good deal of practice, and all measurements should be repeated several times to ensure accuracy.

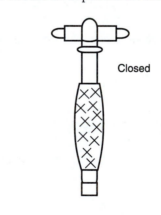

Closed

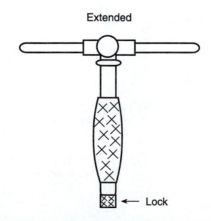

Extended

← Lock

Figure 6-43 Springs push the extensions out to the cylinder wall.

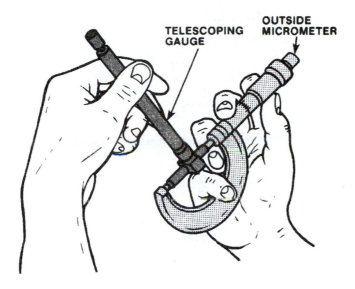

TELESCOPING GAUGE   OUTSIDE MICROMETER

Figure 6-44 The removed telescopic gauge is measured with an outside micrometer.

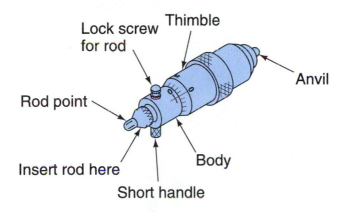

Lock screw for rod   Thimble

Anvil

Rod point

Insert rod here   Body

Short handle

Figure 6-45 Inside micrometers have all the components except for the frame.

Figure 6-46 Anything over 1 inch or 25 millimeters requires attachments. (Courtesy of Snap-on Tools Company)

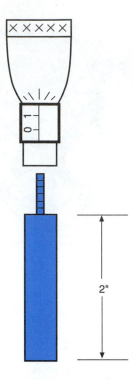

Figure 6-47 Extension spindles are attached to the micrometer to measure a large bore.

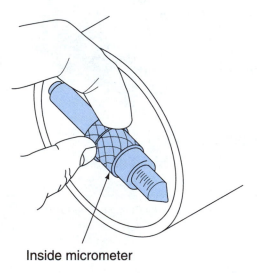

Inside micrometer

Figure 6-48 The micrometer must be carefully placed within the bore for accuracy.

Depth Micrometers

Depth micrometers are used by race engine builders to better position the valves and pistons for additional speed and torque.

Depth micrometers look similar to inside micrometers. This type of micrometer has a base and a depth rod that can be extended like a spindle (Figure 6-49). The scales are the same as those of other micrometers. Place the base on a firm edge of the bore or hole and extend the depth rod to the work. This will provide an accurate measurement of how deep the work is within the bore (Figure 6-50). It should be noted that depth micrometers are used to make small measurements. Anything over an inch will require add-on spindle extensions.

Figure 6-49 The depth rod can be extended like a micrometer's spindle. (Courtesy of Central Tools, Inc.)

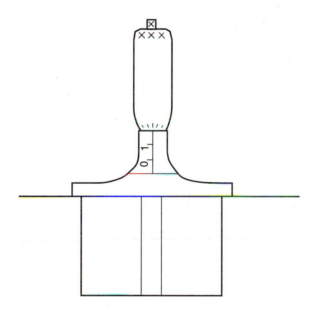

Figure 6-50 Place the base at the edge of the hole and extend the depth rod.

Small Hole Gauges

Small hole gauges do what their name implies: they measure small holes. Technicians do not routinely use large bore gauges because micrometers ("mikes") are more practical. However, micrometers are not practical for small bores, so small hole gauges were built.

> **INSTRUCTOR'S NOTE: NASCAR** and other automotive racing associations use fixed-size bore gauges. During race inspections, inspectors use the gauges to quickly check bores of carburetor air horns, cylinders, and other bored or circular components of the racecar.

Shop Manual
page 135

Small hole gauges are used to measure smaller holes.

NASCAR is the National Association for Stock Car Automobile Racing.

A small hole gauge is straight with a screw handle on one end and a **split ball** on the other end (Figure 6-51). A wedge fits between the two halves of the split ball and extends into the handle. Turning the handle moves the wedge in and out of the split ball, causing it to expand or contract (Figure 5-52). The split ball end is slid into the bore where it is expanded by the wedge. Once the technician is satisfied with the fit, the gauge is withdrawn and the ball diameter is measured with an outside micrometer (Figure 6-53). Small hole gauges are used mostly in the automotive repair business to check valve guide bores.

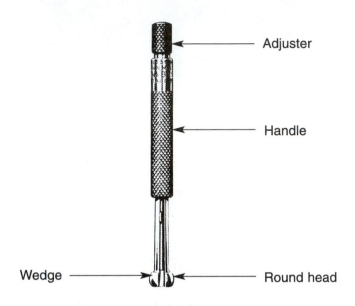

Adjuster

Handle

Wedge ————→ ←———— Round head

Figure 6-51 A small hole gauge. (Courtesy of L. S. Starrett Company)

Wedge ————

Figure 6-52 A wedge expands the split ball. (Courtesy of L. S. Starrett Company)

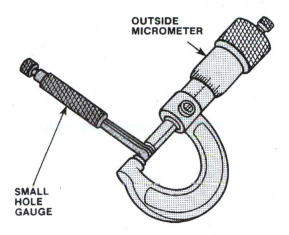

Figure 6-53 The split ball is measured with an outside micrometer.

Dial Indicators and Calipers

The **dial indicator** is used to check how far a component can move. It can also be used to measure the distortion of a component like flywheel **runout**. A dial indicator must be mounted to a solid object for support during its use. The dial has a circular face with a scale around its edge (Figure 6-54). A moveable needle indicates the measurement much like a speedometer needle. Turning a knurled ring on the outside of the dial rotates the scale. There is a lock to hold the scale in place (Figure 6-55). The need is geared to a moveable plunger extending from the bottom of the dial.

The scale can be in USC or metric measurements. The USC scale will use either $1/1{,}000$ or $1/10{,}000$ inch. The scale size will be noted on the face. Metric scales are marked in 0.01 millimeter. The face may be **balanced**, meaning the readings extend from zero counterclockwise and clockwise until they meet at the midpoint of the face (Figure 6-56). A **continuous** type has the markings start at zero and extend clockwise around the scale back to zero (Figure 5-57). Care must be taken with both types when taking large measurements. The plunger may be designed to move the needle more that one revolution around the scale.

Shop Manual
pages 144–148

A *dial indicator* uses a plunger and gear to move a needle in proportion to plunger movement.

Runout is the amount of side-to-side movement in a rotating device. Runout may be referred to as "wobble."

Balanced scale indicators are usually used to measure runout or radial movement. *Continuous scale indicators* are normally used to measure end play or lateral movement.

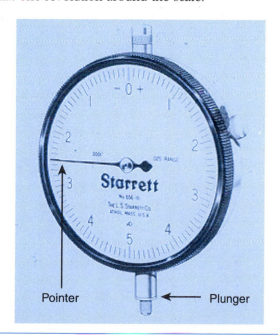

Figure 6-54 The measurement face of a dial indicator. (Courtesy of L. S. Starrett Company)

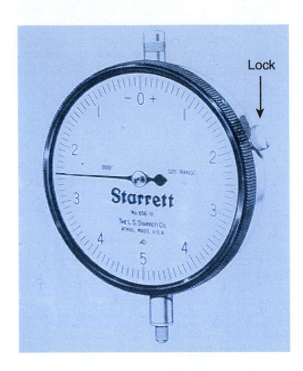

Figure 6-55 The lock holds the scale in place while the measurement is made. (Courtesy of L. S. Starrett Company)

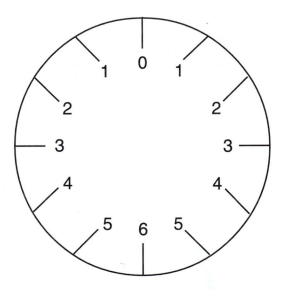

Figure 6-56 The balanced scale numbers meet at the mid-point of the face.

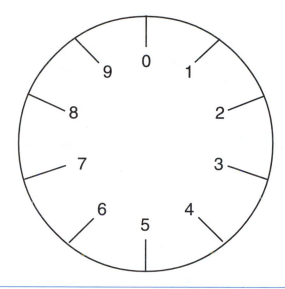

Figure 6-57 Numbers extend upward from zero on a continuous scale.

Dial indicators can also be used to check shaft end play. Before setting up the dial, move the shaft as far as it can go in one direction. Mount the dial to indicate a zero reading at this point. The final step is to move the shaft as far as it will go in the opposite direction. The total movement will be shown on the dial scale.

Dial Calipers

The **dial caliper** is a measuring device that can be used in place of almost all other automotive measuring devices (Figure 6-58). It can measure inside, outside, and depth dimensions. It has two pairs of jaws or calipers that fit inside or outside a component. The top half of each pair is fixed to the bar scale. The bottom half moves along the bar scale when the roll knob is turned. At the opposite end of the bar scale is the depth scale. The bar scale indicates the gross measurement of the caliper's movement. A dial indicator attached to the moveable caliper measures the small movement within its limits.

A typical USC dial caliper can measure between 0 and 6 inches. Its bar scale is divided in 0.100-inch graduations. The USC dial indicator measures in 0.001-inch segments (Figure 6-59). One complete rotation of the dial's needle equals 0.100 inch on the bar scale. A metric bar scale is in 2-millimeter divisions while its dial scale is in 0.02-millimeter graduations. One complete revolution of the needle is equal to 2 millimeters on the bar scale. There are electronic calipers on the market which will allow technicians to change between the two measuring systems.

A *dial caliper* can be used in place of most automotive measuring devices to measure inside, outside, and depth dimensions.

Figure 6-58 A dial caliper. (Courtesy of Sears Industrial Sales)

Figure 6-59 The dial indicator portion of a dial caliper works the same as other dial indicators. (Courtesy of Sears Industrial Sales)

Shop Manual
page 148

Pressure and Vacuum Measurements

Various systems of the vehicle use pressure and vacuum theories to operate. Pressure and vacuum are used to control engine and transmission operations, brakes, and other systems. A brief review of pressure and vacuum theory follows with an explanation of the measuring devices.

Pressure Gauges

Pressure is used in brake, heater, engine cooling, and other vehicle systems.

Pressure is the amount of force being applied as measured in pounds per square inch (psi). If pressure is applied to a liquid in a closed, sealed system, the liquid can be used to transfer force. A hydraulic brake system is an example of pressure and liquid transferring force.

Pressure gauges are fitted into the pressure line using adapters. The scales of a pressure gauge are marked every 1 psi and numbered every 5 psi (Figure 6-60). When the scale is connected and the system energized, the scale will record the pressure being applied to the system.

Vacuum Gauges

Vacuum can be used as brake booster, heater/air conditioning controls, and other vehicle systems.

Vacuum properly defined is the absence of pressure and is measured in inches of mercury (in. Hg). This measure is based on how high (in inches) a vacuum source can raise a column of liquid mercury in a tube. Obviously, technicians cannot carry tubes of mercury around their shops. The vacuum gauge's mechanical system is geared to move a needle a specified distance based on the amount of vacuum in its inlet tube. Vacuum gauges are usually marked in the same manner as pressure gauges except the markings indicate inches of mercury (Figure 6-61).

Pressure scale inside

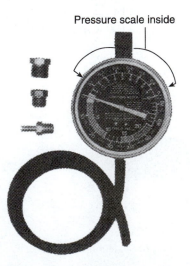

Figure 6-60 Pressure gauges measure pounds per square inch (psi). (Courtesy of Acron Manufacturing Company)

Vacuum scale inside

Figure 6-61 Vacuum gauges measure in inches of mercury (in. Hg). (Courtesy of Acron Manufacturing Company)

Summary

- ❏ The most common measurement system in the United States is the United States Customary or the English system.
- ❏ The base USC measurement is the inch.
- ❏ The USC uses fractions and decimals to measure items less that a whole inch.
- ❏ The metric system is used in almost every nation in the world.
- ❏ The base metric measurement is the meter.
- ❏ The metric system units are based on the number 10 and no fractions are used.
- ❏ USC and metric measurements can be converted to each other using a conversion chart and math.
- ❏ Rulers may be in USC or metric measurements.
- ❏ Feeler gauges are used to measure small gaps or distances between two parts.
- ❏ Micrometers are made for measuring the size of objects.
- ❏ Micrometers may be USC or metric.
- ❏ Small bores can be measured with small hole gauges.
- ❏ Dial indicators are used to measure very small measurements.
- ❏ Shaft end play may be measured with a feeler gauge or a dial indicator.
- ❏ Dial calipers can measure depth, outside diameters, and inside diameters.
- ❏ Pressure is measured in pounds per square inch (psi).
- ❏ Pressure and liquid can be used to transfer force.
- ❏ Vacuum is the absence of pressure.

Terms to Know

Air gauge

Conversion factor

Decimal

Dial caliper

Dial indicator

End play

Feeler gauge

Fractions

Index line

Inside micrometer

Linear

Meshed

Meter

Micrometer

NASCAR

Outside micrometer

Pressure

Ruler

Runout

Small hole gauge

Spindle

Split ball

Stepped feeler gauge

Telescopic gauge

Thimble

Tolerances

Vacuum

Vernier scale

Review Questions

Short Answer Essays

1. Explain the measurement markings on a USC micrometer.

2. List and describe the different components on a micrometer.

3. Explain how stepped or go/no-go feeler gauges are machined.

4. Explain how to convert 3 inches to millimeters.

5. Discuss the components of a dial caliper.

6. Explain how pressure and vacuum are used to move air into the engine's cylinders.

7. List and discuss the differences between a USC and a metric dial caliper.

8. Describe the standard for measuring vacuum.

9. Explain how liquid can be used to transfer force.

10. Explain the metric system of measurement.

Fill-in-the-Blanks

1. The calipers on a dial caliper are moved by a(n) _____ _____.

2. Some micrometers have an additional scale called a(n) _____ _____.

3. One rotation of the spindle on a USC micrometer moves the spindle _____ _____.

4. A 3- to 4-inch micrometer can measure _____ inch.

5. A typical dial caliper will measure up to _____ inch(es) in _____-inch segments.

6. A stepped feeler gauge may be known as a(n) _____ gauge.

7. Large bores may be measured with a micrometer and_____ or a(n) _____.

8. A dial indicator may have a(n) _____ or a(n) _____ scale.

9. A micrometer has a(n) _____ to prevent an over torque of the spindle.

10. The vernier scale is used to read _____ of an inch.

ASE Style Review Questions

1. *Technician A* says pressure and liquid are used to make hydraulic brakes systems work.
   Technician B says vacuum helps the engine operate. Who is correct?
   - **A.** A only
   - **B.** B only
   - **C.** Both A and B
   - **D.** Neither A nor B

2. Dial calipers are being discussed. *Technician A* says the bar scale may be graduated in 0.100 inch.
   Technician B says a bar scale may be graduated in 2 millimeters. Who is correct?
   - **A.** A only
   - **B.** B only
   - **C.** Both A and B
   - **D.** Neither A nor B

3. Feeler gauges are being discussed. *Technician A* says a feeler gauge blade may be machined in two different sizes.
   Technician B says a standard feeler gauge may be used to check the air gap between a magnet and a pickup coil. Who is correct?
   - **A.** A only
   - **B.** B only
   - **C.** Both A and B
   - **D.** Neither A nor B

4. *Technician A* says 2.54 centimeters is equal to 1 inch.
   Technician B says 1 inch is equal to 0.3937 millimeter. Who is correct?
   - **A.** A only
   - **B.** B only
   - **C.** Both A and B
   - **D.** Neither A nor B

5. *Technician A* says a micrometer must be used to measure a gear's runout.
   Technician B says a micrometer can be used to measure a gear's backlash. Who is correct?
   - **A.** A only
   - **B.** B only
   - **C.** Both A and B
   - **D.** Neither A nor B

6. Small hole gauges are being discussed. *Technician A* says an engine's valve guides are best measured with this gauge.
   Technician B says a wedge is used to split two balls apart on this gauge. Who is correct?
   - **A.** A only
   - **B.** B only
   - **C.** Both A and B
   - **D.** Neither A nor B

7. *Technician A* says pressure is measured in inches of mercury.
   Technician B says vacuum is the absence of pressure. Who is correct?
   - **A.** A only
   - **B.** B only
   - **C.** Both A and B
   - **D.** Neither A nor B

8. Pressure and vacuum theories are being discussed. *Technician A* says a vacuum is created when the volume of a sealed bore is increased.
   Technician B says pressure is created when the volume of a sealed bore is decreased. Who is correct?
   - **A.** A only
   - **B.** B only
   - **C.** Both A and B
   - **D.** Neither A nor B

9. Dial indicators are being discussed. *Technician A* says a dial indicator is part of a dial caliper.
   Technician B says a dial indicator is used to measure a small hole gauge. Who is correct?
   - **A.** A only
   - **B.** B only
   - **C.** Both A and B
   - **D.** Neither A nor B

10. *Technician A* says spindles are added to an inside micrometer to make shaft measurements.
    Technician B says telescopic gauges can be measured with a dial caliper. Who is correct?
    - **A.** A only
    - **B.** B only
    - **C.** Both A and B
    - **D.** Neither A nor B

Automotive Bearings and Sealants

Upon completion and review of this chapter, you should be able to:

❑ Explain the purpose of bearings.

❑ Identify the different types, construction, and uses of automotive bearings.

❑ List the types of grease used in a typical light vehicle.

❑ Explain the different types, construction, and uses of gaskets.

❑ Explain the types and uses of chemical automotive sealants.

❑ Explain the different types, construction, and uses of seals.

Introduction

We now have some idea of the theories of operations and how the various components are held together in a vehicle. There are many moving pieces in a typical vehicle, many of which work against or with others. Each moving part requires some type of lubricant and a means to retain that lubricant on or in the assembly. In this chapter, the use of bearings and assembly sealing will be discussed. In addition, common usage of each will be covered. Use and lubrication will also be covered in various system chapters in this book.

Bearings and Bushings

When two components move against or with each other, **bearings** are used. Bearings are used between a moving part and a stationary part or between two adjacent moving parts. A bicycle wheel spins on two bearings that are mounted on and around a shaft or an axle. Bearings are used in the engine, driveline, suspension, steering, and in almost every system in a vehicle. Some components of the electrical system even use bearings.

Bearing Loads

Bearing loads are usually computed by the weight and movement of the load. The weight is used to determine the size, placement, and type of bearing to be used. The movement is either **radial load** or thrust load. Thrust load is also knows as **axial load.** Radial load is the up-and-down movement of the load while axial is the front-to-rear (in and out) movement of the load. It should be noted that the term "load" not only applies to the *weight* being supported but also the *direction* and *amount of force* against that weight. The force can be thermodynamics in the engine or centrifugal force acting against a turning vehicle.

Shop Manual
page 161

Bearings reduce the friction, heat, and wear of moving components.

Bearing load is computed by the weight and movement of the load.

Radial load is produced by the component pushing directly against the bearing from inside out or vice versa.

Axial load is produced by the component pressing on the sides of the bearing.

Bearing Journals and Races

Bearing journals are machined areas on a shaft (Figure 7-1). They have fine, smooth finishes to prevent damage to the bearing. Usually the area is hardened more than other parts of the shaft to withstand the loads. In most bearing failure cases, the journal will have to be machined and a smaller bearing used. There is a limit to the amount of machining before the shaft has to be replaced.

Races

Bearing races serve the same purpose as the journal. However, races are usually purchased as part of the replacement bearing (Figure 7-2). A race can be assembled as part of a bearing assembly or it can be a loose part packed and shipped as part of the assembly. Many wheel bearings have a race that must be press-fitted into the wheel hub before the bearing is installed. If either a bearing or race fails, replace both.

Bearing Inserts

A BIT OF HISTORY

Crankshaft bearing inserts on the Ford Model T could be made from strips of leather.

Figure 7-1 The highlighted areas are the bearing journals on this shaft.

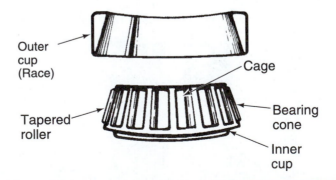

Figure 7-2 A tapered roller bearing is normally used to support a vehicle wheel. Note the separation of the outer cup or race from the bearing assembly. (Courtesy of General Motors Corporation, Service Operations)

Bearing inserts are known as "plain bearings" and are most commonly used in the engine as rod bearings or **main bearings** (Figure 7-3). The piston rod's lower end wraps around a portion of the crankshaft (Figure 7-4). Both the rod and crankshaft move. The bearing allows them to work with reduced friction and increased life. The main bearing is fitted between the rotating crankshaft and the engine block.

A bearing insert is so named because it is fitted into a mounting component. It is made from high-strength steel that is machined to a fine-surface finish, coated, and cut or stamped to the correct length and width. Bearing coating may be babbit, copper-lead, or aluminum alloy.

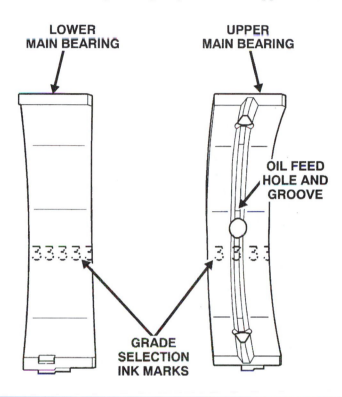

Figure 7-3 This is one main bearing for a crankshaft. (Courtesy of DaimlerChrysler Corporation)

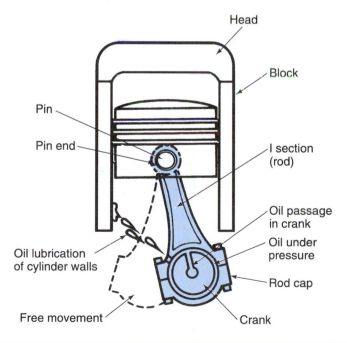

Figure 7-4 The lower end of the rod and rod cap holds a rod bearing in place.

The area called *crush* is needed to allow for tolerances during bearing and rod manufacturing.

Two insert pieces are needed to make one complete bearing which is then curved to fit inside the rod and rod cap (Figure 7-5). One end is notched so the insert will lock into its mount. If the two halves of an insert are placed together like a circle and measured, it will be determined that the circle is not perfectly round (Figure 7-6). This is called *spread*. But when the two halves are fitted into a rod and rod cap, they are forced into a perfect circle. If they do not, the insert will damage the crankshaft and fail quickly. Rod bearings are machined to fit a specified diameter on the crankshaft. A small length at the end of each insert half sticks over the rod and cap. This is called **crush** and is used to properly fit the insert tightly within its mount. The bearings are lubricated by pressurized oil pumped through the crankshaft.

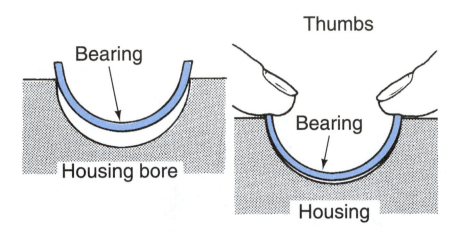

Figure 7-5 The bearing insert is pressed into the mount, usually a rod or main bearing cap, a rod, or the engine block.

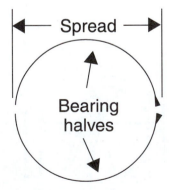

Figure 7-6 The two bearing halves are not in a perfect circle until installed. This is called "bearing spread."

Main bearings are very similar to rod bearings in construction. They are usually a little wider and a little longer. One half fits into the main bearing cap and the other into the engine block. The top half of a main bearing has a lubrication hole that must be aligned with the oil galley hole in the block (Figure 7-7). In this manner, the bearing can be lubricated with pressurized oil. Main bearings also have a lock to keep them in place during operation.

Two other types of bearing inserts are the **camshaft bearing** and **thrust bearing**. Camshaft bearings are shaped like short, highly-machined metal tubes (Figure 7-8). They are press-fitted into cavities of the engine block and the camshaft is slid through from one end. There are usually four or more bearings per camshaft. The camshaft operates the engine's valves.

Camshaft bearings for an engine come with different outside diameters. This assists the technician during bearing installation. The inside diameters are all the same.

Thrust bearings are fitted at the crankshaft and allow only a limited amount of end-to-end travel of the crankshaft.

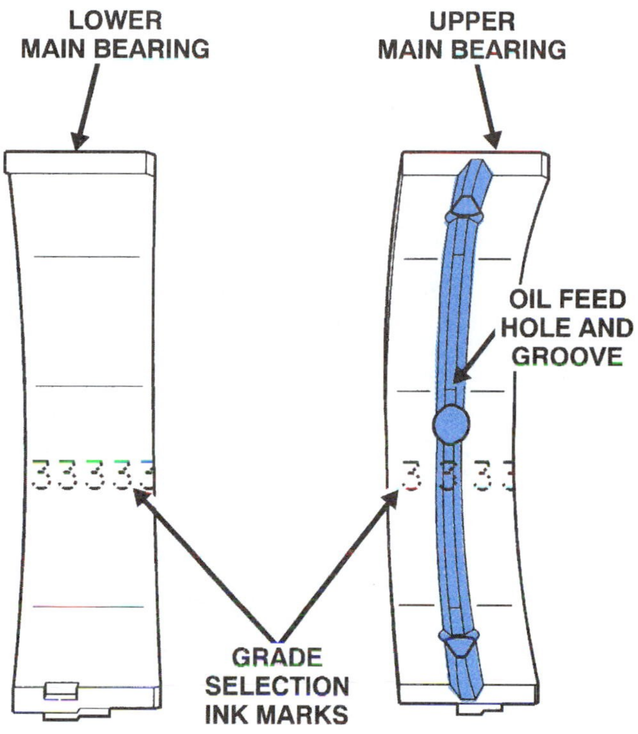

LOWER
MAIN BEARING

UPPER
MAIN BEARING

OIL FEED
HOLE AND
GROOVE

GRADE
SELECTION
INK MARKS

Figure 7-7 The main bearing half with the oil hole and slot is fitted into the block. (Courtesy of DaimlerChrysler Corporation)

Figure 7-8 Note the oil hole in each of the three camshaft bearings shown.

Thrust bearings are fitted at the crankshaft and are used to limit the amount of end-to-end crankshaft travel. The bearing may be part of one of the main bearings or separate pieces (4) that are fitted into the block next to a main bearing (Figure 7-9).

Main, rod, and cam bearings mainly support radial loads. Thrust bearings are made to support axial or thrust loads. However, during normal operation, an engine will experience both types of loads.

The areas on the crankshaft where the rod and main bearings fit is called the *bearing journals*. A typical crankshaft for a four-cylinder engine will have five main and four rod bearing journals (Figure 7-10). Most V-6 engines usually have four mains and three double-rod journals. A double-rod journal means that two piston rods are connected to the crankshaft side by side with a small space between them.

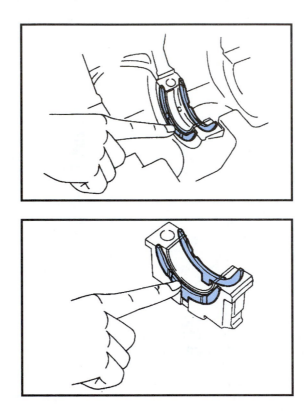

Figure 7-9 This four-piece thrust bearing is fitted to the sides of the bearing journal. (Reprinted with permission)

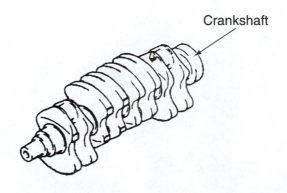

Figure 7-10 A typical four-cylinder crankshaft. (Courtesy of DaimlerChrysler Corporation)

Roller Bearings

A **roller bearing** is a set of machined shafts of steel. The length and diameter of the shafts or rollers are determined by the size of the load that the bearing is expected to carry. The typical roller bearing in a vehicle is fairly small and has several rollers trapped in a cage or separator (Figure 7-11). The cage holds the rollers in a circle. An inner race is fitted through the circle's center and over a shaft. The roller bearing assembly can now rotate around the shaft with each individual roller turning within the cage. An outer race is fitted around the outside of the bearing and into the rotating component like a gear. The contact area between the roller and race extends completely across the length of the roller. Roller bearings can be found in transaxles, and electrical alternators, and other vehicle components. They are used to support radial loads and a small amount of thrust.

Many roller bearing assemblies are made with the same diameter at each end. Tapered roller bearings are made a little differently and usually support a wheel assembly. One end of the circle created by the cage is smaller than the other. This places the rollers at an angle and forms a cone shape (Figure 7-12). The inner race is commonly known as the *cup* and fits into a wheel hub. This race can be separated from the bearing assembly and fitted into the hub. The inner race or cone is part of the assembly and cannot be separated without destroying the bearing. It is made to slide onto the spindle or shaft around which the wheel turns. Tapered roller bearings can support axial and radial loads. When two sets of tapered roller bearings are used in the same component (like a wheel assembly), the bearings compliment each other in handling the loads applied against the wheel assembly.

All *roller and ball bearings* provide a rolling surface for the moving components so friction and wear are reduced.

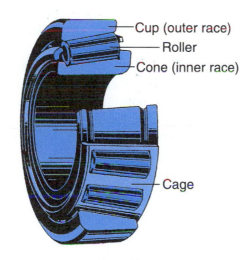

Figure 7-11 A roller bearing is a machined shaft of high-strength steel placed in a cage. (Courtesy of Chicago Rawhide)

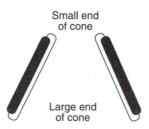

Figure 7-12 A tapered bearing forms a cone with the small end installed toward the inside of the component.

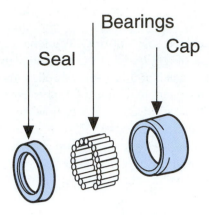

Seal Bearings Cap

Figure 7-13 Normally, needle bearings fit inside a cup but may be placed in a cage similar to the roller bearing.

A third type of roller bearing is known as the **needle bearing**. Needle bearings are very small in diameter and may not be held within a cage. They fit into a bearing cup such as a driveline component or into some older alternators (Figure 7-13). Improper disassembly of a component with needle bearings usually results in many needle bearings on the bench or floor, or they just disappear. If this happens, all of the bearings must be replaced. There can be no empty spaces between the needles. A needle bearing usually support only radial loads.

Ball Bearings

The construction of a **ball bearing** is similar to that of a roller bearing. The primary difference is the use of highly-machined steel balls in place of rollers (Figure 7-14). The balls fit into a machined groove that keeps them aligned and distributes the load evenly to each ball. The two grooved sections form the cage that holds the bearing together and makes the inner and outer races. Contact between the ball and race is very small but the bearing can support axial and radial loads well. Some bearings use two rows of balls to support loads better.

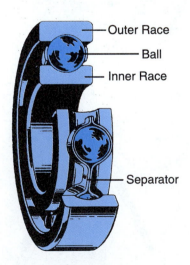

Outer Race
Ball
Inner Race

Separator

Figure 7-14 The balls are separated by a cage within the bearing's grooves to keep the balls apart during component operation. (Courtesy of Chicago Rawhide)

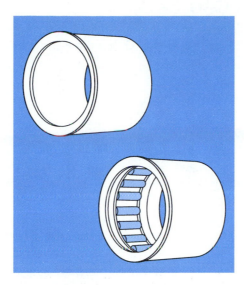

Figure 7-15 The upper item is a bushing. Note that there are no rollers or balls installed. The bottom item performs the same task as the bushing, but has needle bearings installed for better operation. (Courtesy of Federal-Mogul Corporation)

Bushings

Bushings are made of thin steel tubing cut to specific lengths. Usually, they are copper coated, but other materials may be used for special situations. The bushing will not axially support loads very well since their purpose is to limit the radial movement of a shaft. Rotating shafts must be kept in a straight line to prevent damage to the shaft and component. Bushings form a journal on which the shaft can ride (Figure 7-15). However, bushings are not used in high thrust or radial load areas like the camshaft or crankshaft. They can support a transaxle shaft or the shaft in an alternator or pump. The bushing is press-fitted into a housing and the shaft slides through it.

Bushings are more commonly used on transmissions and transaxles where there is not much axial load.

Grease

Grease is a lubricant used on roller and ball bearings and suspension system components. Many bearings, like those used in the engine, transmission, and differential, are lubricated by the oil stored and used within the component. Other components, like the wheels and suspension, require the parts or bearings to be packed with grease.

The National Lubricating Grease Institute (NLGI) (Figure 7-16) classifies automotive grease into two main categories: L for suspension components and U-joints on the drive shaft, and G

Shop Manual
page 165

Grease is a lubricant that reduces friction and smoothes movement between parts of components.

Figure 7-16 G denotes wheel bearing grease. L is for steering and suspension components. (Reprinted with permission from the National Lubricating Grease Institute)

for wheel bearings. The L classification is further broken into an A class for mild duty and B for mild to severe duty. Wheel bearing grease, G, has three sub-classes:

A for mild duty
B for mild-to-moderate duty
C for mild-to-severe duty

A typical passenger vehicle or light truck would use LA for suspension and GB for wheel bearing. The difference between L and G grease is the chemical makeup that is designed to meet certain operating conditions. L grease has a very high adherence to the part that is lubricated. It has a high resistance to being washed off by water. Suspension or chassis grease will not break down from the movement of the suspension parts. Wheel bearing grease is thicker than L grease and is formulated for high resistance to heat. The grease also resists being thrown off by the spinning bearing.

While there are specific greases for the suspension, bearings, and U-joints, the most commonly used grease in an automotive shop is multipurpose grease. As it name implies, it can be used for different purposes. On a vehicle, multipurpose grease is used for suspension, U-joints, and wheel bearings. However multipurpose grease should not be used in all situations, and the service manual must be consulted for specific applications.

Shop Manual
pages 182–183

Gaskets usually seal high-pressure areas, retain lubricants, and seal the space between two stationary components.

Gaskets

Gaskets are used to retain lubricants or to seal a chamber. In most cases, the lubricants are under low pressure but the sealed components may be stationary or moving. A gasket usually seals high-pressure areas and the space between two stationary components or parts.

Head Gaskets

 WARNING: Always replace a head gasket with an exact match in quality and type. Failure to use the proper gasket could lead to engine failure.

The gasket that is under the most force is the *cylinder head gasket* (Figure 7-17). This gasket seals the extremely high pressure of the combustion chamber and seals water and oil passages between the engine block and the head. It must accomplish this task under high and low temperatures, expanding and contracting components, different materials, and exposure to chemicals including acids.

A typical head gasket is composed of layers of different materials bonded together (Figure 7-18). Older head gaskets were stamped or embossed steel sheets. They worked well in many applications but they are not recommended for vehicles produced after the early 1980s. The use of aluminum cylinder heads mounted on cast iron blocks produced some engineering problems on sealing this area. Aluminum expands twice as quickly as cast iron. Also, current engines and heads are lighter and more flexible. Head gaskets must expand and flex with the materials above and below them. To correct this, the manufacturers redesigned the components, selected new materials, and created new manufacturing processes to produce workable gaskets.

The modern head gasket has a steel core made of solid or clinched steel plate. A dense *composite* facing is placed on each side of the core and covered with a low- or anti-friction material. Holes are punched though the gasket to line up with the engine cylinders and coolant and oil passages. The rim around each cylinder hole has a stainless steel fire ring to help protect the gasket material from burning air and fuel in the combustion chamber. The anti-friction coating is usually Teflon or a silicone-based material. Some types of head gaskets have a silicone bead placed around the oil and coolant holes to prevent liquid leakage (Figure 7-19).

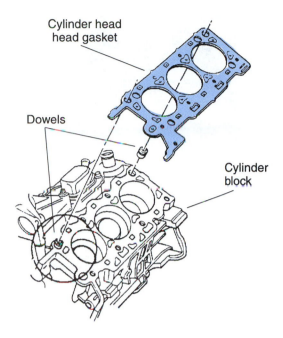

Cylinder head
head gasket

Dowels

Cylinder
block

Figure 7-17 A head gasket must withstand high temperatures, stress, and flexing while still maintaining an excellent seal between the cylinder head and engine block. (Reprinted with the permission of Ford Motor Company)

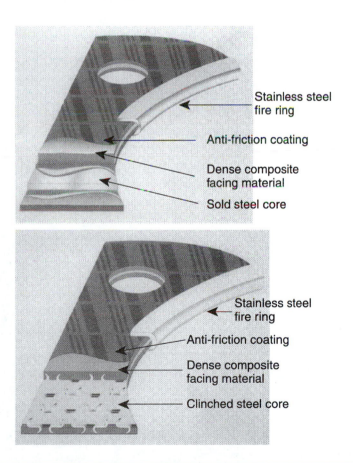

Stainless steel
fire ring

Anti-friction coating

Dense composite
facing material

Sold steel core

Stainless steel
fire ring

Anti-friction coating

Dense composite
facing material

Clinched steel core

Figure 7-18 The anti-friction coating allows the sealed components to slide over the gasket without damaging it. This allows expansion and contraction during temperature changes. (Courtesy of Fel-Pro Incorporated)

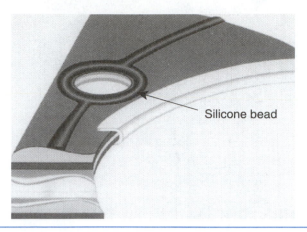

Figure 7-19 A silicone bead can be laid on the head gasket to seal coolant and oil crossovers between the block and head. (Courtesy of Fel-Pro Incorporated)

Similar to head gaskets are *intake and exhaust gaskets* (Figure 7-20). They differ in construction and material and must perform well after being exposed to high temperatures and chemicals. Sometimes they are placed between two different metals and must be able to expand properly with both.

Other Gaskets

Cork gaskets are used to seal low-pressure areas like the valve covers and oil pans. When heated, cork will become brittle and crack if over tightened. Rubberized cork gaskets are often used as valve cover seals. This type of cork gasket is treated to a reduce-shrinking process and maintains resiliency (the ability to return to its original shape) when clamped and released. Cork gaskets may also be found on older engine and transaxle oils pans.

Synthetic rubber gaskets are made from man-made materials. The material is usually *neoprene* and the gasket is reusable. For this reason, they are often used as valve cover gaskets because valve covers require removal for routine maintenance (Figure 7-21). They may also be found as oil pan gaskets and in other areas.

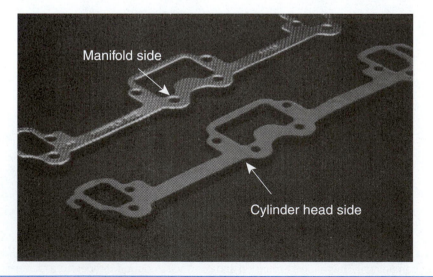

Figure 7-20 This type of gasket is used to seal two components that create high temperatures. (Courtesy of Fel-Pro Incorporated)

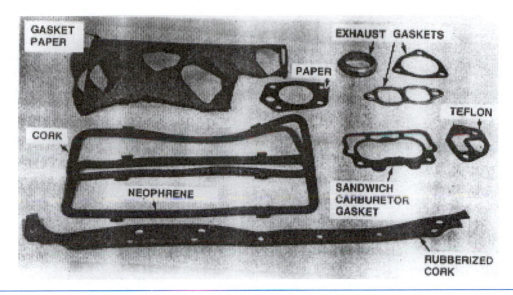

Figure 7-21 Neoprene gaskets can be reused if handled carefully during removal and installation.

Paper gaskets are made of treated paper that retains low-pressure, low-temperature areas between two stationary components. A typical application is between the different housings and covers on a manual driveline or on some water pumps. Paper gaskets may be relatively thick for some applications, like the water pump, or they may be as thin as a regular piece of paper.

Chemical gaskets are used on many new cars. They are installed by robotic equipment capable of laying an almost perfect $\frac{1}{8}$-inch bead around the entire sealing area, thereby making a formed-in-place gasket. Because of this, they are difficult to use during repairs since a poorly-laid bead will leak. Aftermarket part vendors may use cork or neoprene gaskets for repairs.

Gasket sealers are sometimes used to hold the gasket in place and help it seal. Care should be taken when using gasket sealers. Instructions should be followed, as many gasket manufacturers do not want any sealer on their gasket or placed in certain amounts at specific places. The sealer may be applied with a brush or pressed from a tube. Gasket sealers are *anaerobic* chemicals, meaning they will cure only in the absence of air. The two components have to be fitted and torqued before the sealer will cure.

A gasket sealer may be required on some fasteners that extend into the cooling passages. The sealer is spread lightly over the fastener's threads to seal the area between the threads.

Chemical Sealants

Chemical sealants include the sealing materials mentioned above and a popular sealer known as **room temperature vulcanizing (RTV)** (Figure 7-22). RTV is an *aerobic* sealant and may be known as "silicone rubber." Aerobic means RTV cures when exposed to air. It can be used to seal two stationary components like an oil pan and an engine block. It cannot be used as a head gasket or an exhaust manifold (high heat and pressure). It cannot be used on fuel systems either because fuel deteriorates RTV. RTV will set in about 15 minutes in warm temperatures and high humidity. It cures in about 24 hours. It will not set or bond to a gasket. RTV comes in different colors.

Shop Manual
page 170

Chemical sealants and sealers should be checked for their use with electronic engine controls. While all products of this type are made for the new engine, it is always best to ensure the product is correct for the application.

Room temperature vulcanizing (RTV) is an aerobic sealant also known as "silicone rubber." It is used to seal two stationary components.

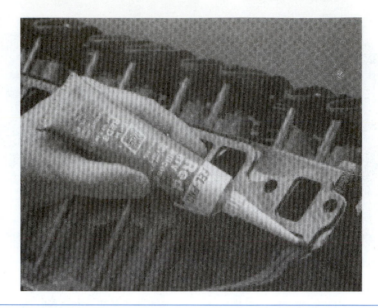

Figure 7-22 Use silicone only where directed by the gasket manufacturer. Silicone applied improperly can prevent the gasket from sealing. (Courtesy of Fel-Pro Incorporated)

Shop Manual
page 169

Seals are used between moving components and their housings to retain liquids, especially in high-pressure and high-temperature areas.

A *lip seal* is installed with the lip opening facing the lubricant such that pressure created by the oil against inner side of the lip forces its edges tighter around the moving part.

A *garter spring* is a small coil spring that is made similar to a bracelet.

Seals

Seals are normally used between moving components and their housings to retain liquids. Seals can be used in high-pressure and high-temperature areas and are usually make of butyl rubber or neoprene within a metal ring. The ring provides the connection to the stationary component.

A **lip seal** is the type most commonly used on vehicles (Figure 7-23). The seal has a lip that is installed facing the lubricant. In this way, the pressure of the oil against the inner side of the lip forces the lip edge tightly around the moving part. The lip may have a **garter spring** behind it so it is held to the shaft better (Figure 7-24). Some seals have a smaller lip called a *dust lip* that faces away from the lubricant. It does as its name implies, keeping dirt from entering the component. Seals are found at the front and rear of the engine crankshaft, front and rear of a transaxle, and around drive axles.

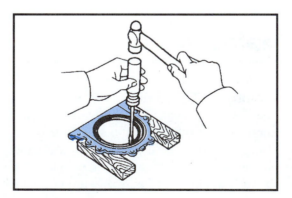

Figure 7-23 This lip seal fits into a cover plate. (Reprinted with permission)

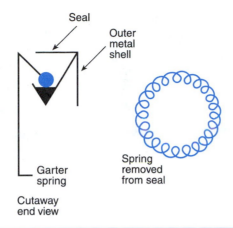

Figure 7-24 The garter spring is a small bracelet that fits behind the lid of a seal.

O-Rings

An **O-ring** is so named because of its shape and the manner in which it is used. The O-ring is shaped like a rubber bracelet without a latch (Figure 7-25). It fits into a groove, either in the moving component or in the stationary component. The ring extends above the groove and is pressed against the second component (Figure 7-26). O-rings are commonly used on high-pressure, hydraulic fittings like power steering, brakes, and air-conditioning systems. A specially-cut type called a *square-cut O-ring* is used on disc brake systems (Figure 7-27).

A third type of seal does not necessarily keep lubricants in, but keeps dirt and water from entering the component. In Shop Manual, Chapter 4, "Diagnosing by Theories and Inspection," we found two rubber boots on the front drive axles that protect the components and retain the grease. Other rubber boots are used in steering, suspension, and brake system to protect the various components and retain lubricants.

An *O-ring* is shaped like a rubber bracelet and fits into a groove in either the stationary or moving part. O-rings are used in high-pressure, hydraulic fittings.

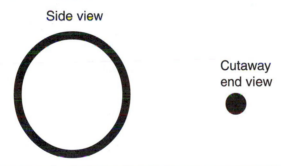

Figure 7-25 An O-ring.

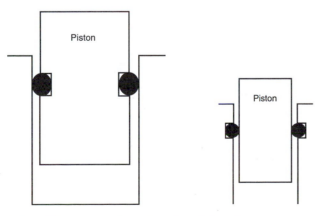

Figure 7-26 The O-ring may move with the piston or be installed in the wall so the piston moves over the ring.

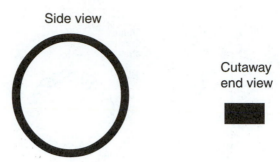

Side view

Cutaway
end view

Figure 7-27 This type of O-ring is square cut. When bent and released, it tends to return to its original shape.

Summary

- ❑ Automotive sealants include gaskets, chemicals, or seals.
- ❑ Automotive bearings may be inserts, rollers, or the ball type.
- ❑ Bearings must support axial and radial loads, but they may be designed for one type of load specifically.
- ❑ Generally, automotive grease is classified as either L or G.
- ❑ L grease is for suspension and A is for wheel bearings.
- ❑ Gaskets are used between two stationary parts.
- ❑ Gaskets can be made to withstand high temperature and pressure.
- ❑ Gaskets can be very thin and made from paper.
- ❑ Gaskets can be made from a bead of chemical compound.
- ❑ RTV is a chemical sealant used only on stationary components.
- ❑ Seals are usually used between a moving component and a stationary component.
- ❑ Seals are installed with the lip facing the lubricant.
- ❑ O-rings are special-purpose seals.
- ❑ A rubber boot may seal the component from dirt and water.

Terms to Know

Aerobic

Anaerobic

Ball bearing

Bearing

Bearing insert

Bearing journal

Bearing load

Bearing races

Bushing

Camshaft

Camshaft bearing

Composite

Gasket

Grease

Lip seal

Main bearing

Needle bearing

Neoprene

O-ring

Rod bearing

Roller bearing

Room temperature vulcanizing (RTV)

Seal

Tapered roller bearing

Thrust bearing

Review Questions

Short Answer Essays

1. Explain the construction of a typical head gasket for newer engines.

2. Describe the construction and use of a lip seal.

3. List the areas where RTV may be used.

4. Explain how head gasket material is protected from flame and chemicals.

5. List and describe the uses of the two major classification of automotive grease.

6. Describe the construction of a tapered roller bearing.

7. List the components of a roller bearing.

8. Describe how the load contact areas of a ball bearing and roller bearing differ.

9. Describe bearing inserts.

10. Explain how a rod bearing is installed.

Fill-in-the-Blanks

1. A(n) _____ _____ is used to support the thrust load of a crankshaft.

2. A type of _____ _____ is used to support the wheel assembly.

3. L type grease may fail if used on _____ _____.

4. The oil pan may be sealed by a(n) _____ or _____ _____.

5. High pressure within the oil pan may cause the crankshaft _____ _____ to leak.

6. The head gasket has holes punched through it to allow passage of _____ and _____.

7. Gaskets may use a(n) _____ to hold it in place.

8. A seal is installed with the dust lip _____ from the lubricant.

9. An O-ring is installed in a(n) _____ in the moving component.

10. Air cures a(n) _____ sealant.

ASE Style Review Questions

1. Seals are being discussed. *Technician A* says the lip faces the lubricant.
   Technician B says a garter spring may be used on some seals. Who is correct?
   A. A only
   B. B only
   C. Both A and B
   D. Neither A nor B

2. The head gasket is being discussed. *Technician A* says the steel ring is placed around the cooling passage.
   Technician B says silicone may be used to seal the oil passages. Who is correct?
   A. A only
   B. B only
   C. Both A and B
   D. Neither A nor B

3. *Technician A* says a gasket can be used to replace a seal.
   Technician B says a chemical can be used to replace a gasket. Who is correct?
   A. A only
   B. B only
   C. Both A and B
   D. Neither A nor B

4. *Technician A* says lubricant pressure can be used to better seal the lip to the shaft.
   Technician B says a garter spring is used to better seal the dust lip. Who is correct?
   A. A only
   B. B only
   C. Both A and B
   D. Neither A nor B

5. RTV is being discussed. *Technician A* says RTV is an aerobic sealant.
   Technician B says the absence of air cures RTV. Who is correct?
   A. A only
   B. B only
   C. Both A and B
   D. Neither A nor B

6. Gasket sealers are being discussed. *Technician A* says a sealer can be used to replace the gaskets.
   Technician B says sealers are cured by the absence of air. Who is correct?
   A. A only
   B. B only
   C. Both A and B
   D. Neither A nor B

7. *Technician A* says anaerobic chemicals cure when the parts are assembled and torqued.
   Technician B says aerobic chemicals cured completely in about 24 hours. Who is correct?
   A. A only
   B. B only
   C. Both A and B
   D. Neither A nor B

8. The use of O-rings is being discussed. *Technician A* says O-rings should be used in low-pressure systems.
   Technician B says a square cut O-ring is used on drum brakes. Who is correct?
   A. A only
   B. B only
   C. Both A and B
   D. Neither A nor B

9. Bearings are being discussed. *Technician A* says needle bearings are a type of roller bearing.
   Technician B says flat roller bearings are used to support a wheel. Who is correct?
   A. A only
   B. B only
   C. Both A and B
   D. Neither A nor B

10. The use of bearings is being discussed. *Technician A* says a roller bearing may be used in alternator.
    Technician B says a bushing is used to support the camshaft. Who is correct?
    A. A only
    B. B only
    C. Both A and B
    D. Neither A nor B

Automotive Systems and Engine Operations

Upon completion and review of this chapter, you should be able to:

❏ List the major systems of a vehicle and their purposes.

❏ Describe the design of an internal combustion engine and its general operation.

❏ Discuss the starting and charging systems.

❏ Discuss the lubrication system and engine oils.

❏ Discuss the cooling system and coolant.

❏ Discuss the fuel and air intake and exhaust systems.

❏ Discuss the ignition system.

❏ Describe the operation of a four-stroke gasoline engine.

❏ List and discuss the six major emission systems.

Introduction

The previous chapters have presented some information on theories, shop management, and what holds the vehicle together. Starting with this chapter, we will discuss an overview of the various systems, beginning with the first major system: the engine.

Automotive Systems

Electrical Systems

The electrical system was once used for starting and running the engine and lights. Today's automobile does not have a single system that does not rely on electricity or can be engineered to work with electricity and electronics. Even tire pressure can now be measured and transmitted to the dash for the operator's information by electricity and electronics. The electrical system is composed of several subsystems.

The starting system is used to **crank and start the engine.** The major components are the *starter motor* and the *starter control circuits* (Figure 8-1). The starter motor is a direct current (DC) motor that drives the engine's flywheel through a gear mechanism. The motor receives power directly from the battery. Current flow from the battery is turned on or off using contacts within the ignition switch. When the ignition key is rotated to the start position, current is sent to a **relay,** which closes a larger switch and allows full battery current to flow to the starter. A system of magnets and contacts (brushes) cause an armature to turn the gear mechanism.

Shop Manual pages 190–195, 201–203

Cranking the engine means to turn the crankshaft to draw fuel and air into the cylinders. *Starting* is the firing of the fuel/air mixture to power the engine.

The starter *relay* can be mounted on the starter or on the inner side of the fender.

> ⚠ **WARNING:** Always disconnect the negative cable first before working on the vehicle's electrical systems. Always connect the negative cable last. Damage could result to electronic circuits if the positive cable is removed first or replaced last.

> ■ **CAUTION:** Ventilation is required when recharging a battery. The area around the battery must be cleared of flame-producing devices. A recharging battery produces hydrogen gas. Serious injury could result if the gas is exposed to a flame.

> ⚠ **WARNING:** Never operate the charging system without the battery being connected. Damage to the charging system and other systems could result.

M/T MODELS

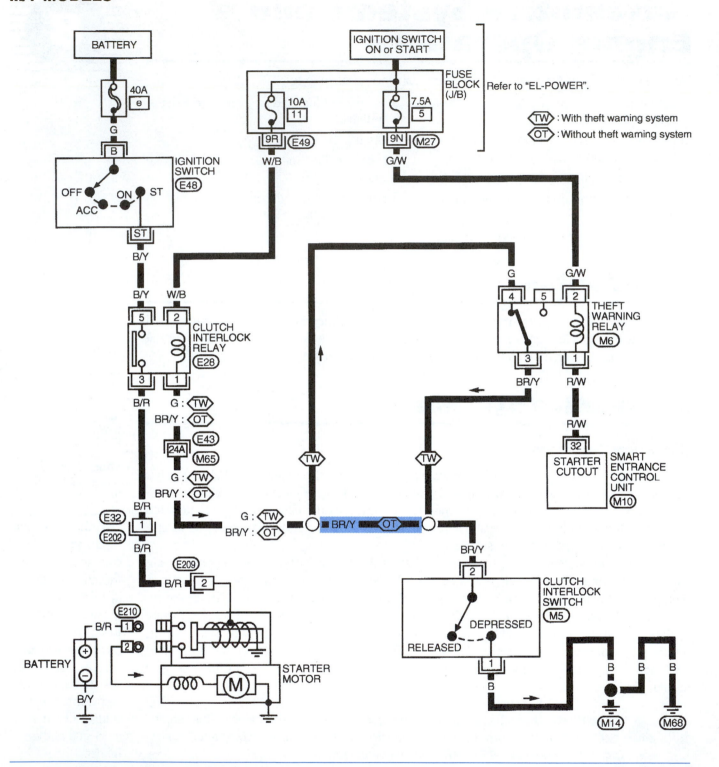

Figure 8-1 A typical starter circuit. (Courtesy of DaimlerChrysler Corporation)

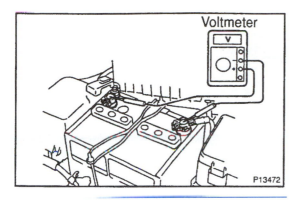

Figure 8-2 The battery provides a storage area for direct current. (Courtesy of DaimlerChrysler Corporation)

Figure 8-3 This generator or alternator produces alternating current.

The battery is an electrical storage device that must be recharged (Figure 8-2). The charging system recharges the battery and powers the vehicle system when the engine is running. The major components of the charging system include controls, the **alternating current (AC)** generator or alternator, the battery, and the alternator drive mechanism (Figure 8-3).

Before we continue, we need to discuss types of current as they apply to the vehicle. An alternator can produce AC easier than **direct current (DC)** at lower engine speeds. However, alternating current cannot be stored. Today's alternators are wired to generate alternating current and send it through a control that changes it to direct current. The direct current is stored or used by the vehicle.

The battery stores current by chemical reaction. Lead plates are surrounded by **electrolyte**, a mixture of water and sulfuric acid, 64 percent and 36 percent respectively. Electrolyte is conductive (carries current) and reactive (reacts with other elements). The electrolyte reacts with the lead to release electrons from atoms (Figure 8-4). When a circuit is closed or turned on, the free electrons are released into the conductors and electricity flows.

The electrons in *direct current (DC)* flow in one direction. In *alternating current (AC)*, they flow from positive to negative polarity.

Electrolyte is a mixture of water and sulfuric acid. It reacts with lead to release electrons from atoms and causes electricity to flow when a circuit is closed or turned on.

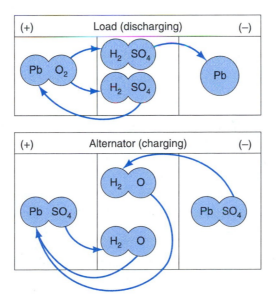

Figure 8-4 The electrolyte acts upon the lead plate to store direct current. (Courtesy of Davis)

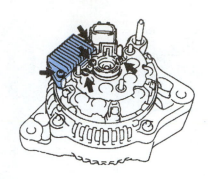

Figure 8-5 The regulator controls alternator output by sensing battery charge and electrical system load. (Courtesy of Nissan North America)

A *regulator* in the charging circuit senses battery charge and electrical load and turns the AC generator on and off.

The *PCM* is the main computer in the vehicle. It receives data from engine and transmission sensors, calculate changes, and activates various actuators.

Vehicle *sensors* measure electrical actions and transmit the data to the vehicle's computer.

Electrical actuators are working devices turned on or off by the computer based on sensor data.

On Board Diagnostics II (OBD II) monitors emission devices on an engine and is required on passenger cars and some light trucks. *Diagnostic Trouble Codes (DTCs)* are data retrieved by using a scan tool that assist in repairing a vehicle.

Alternator controls consist primarily of an electrical **regulator** (Figure 8-5). The regulator monitors the battery charge and electrical load throughout the vehicle. When the charge is low or the load is high, the regulator supplies additional current to the magnetic fields inside the generator and produces more output.

Computerized Controls and Actuators

A computer commonly known as the **powertrain control module (PCM)** controls most of the engine and transmission operations in modern vehicles. Almost every system on the vehicle is or can be designed to use electronics. Computers can analyze data from the **sensors,** compare it to programmed instructions, and transmit a command for a task to be performed. In most cases, a mechanical or **electromechanical actuator** performs the necessary actions.

A vehicle's computer system consists of sensors, actuators, and the computer. *Sensors* tell the computer what is happening and *actuators* perform the work. The *computer* collects the data and compares it to a preloaded program. A certain signal will require a certain action. If the signal is outside the computer's program limits, it may ignore the signal or take actions to notify the driver that a problem exists.

A typical example is the driver switching on the air conditioner. The operator presses a switch, signaling the computer that air conditioning is requested. The computer analyzes the engine temperature, speed, and the length of time the engine has been operating. It also looks at information on the throttle position and any other data that the program requires before the air conditioning can be activated. If all conditions are satisfactory, the computer switches on the air conditioner compressor and cooling fan relays. The compressor engages and the fan starts. All of this is done by the time the driver has removed her or his finger from the button and placed it back on the radio dial.

In this case, a mechanical switch (sensor) creates an electrical signal. This signal and other signals from sensors on and in the engine or elsewhere on the vehicle are analyzed and an electrical command is issued. The command closes two electromechanical switches (actuator). Current is supplied to the electric cooling fan motor and the electromagnetic clutch (actuator) on the compressor. The compressor is basically a mechanical, high-pressure hydraulic pump or the final actuator in this series of actions.

A federally mandated computer system known as **On Board Diagnosis II (OBD II)** is required on passenger cars and many types of light trucks. A companion system is being installed on heavy vehicles. OBD II is primarily required to monitor the emission devices on the engine. Any technician with the proper scan tool can retrieve certain mandated **Diagnostic Trouble Codes (DTCs)** that assist him or her in repairing the vehicle. The older system allowed the manufacturers to limit the amount of data accessible to non-dealership technicians.

Most of this inaccessibility was due to (a) the manufacturers' desire to protect engineering designs from competitors, and (b) the capability of the computer system in use. Since typical vehicles burn gasoline in the same type of engines, the options available to the manufacturers were limited to refining the same type of emission control devices. OBD II allows the manufacturer to protect some information, but allows repair technicians to access any data pertaining to the vehicle's emission control devices.

OBD II not only informs the operator and technician of a broken component, it monitors the emission control devices and will detect a component showing signs of a pending failure. In this manner, the repair can be accomplished before an actual failure. OBD II not only monitors dedicated emission control devices on the vehicle but it also monitors systems like the ignition to ensure that they are performing at their designed level.

Although the diagnostic system required by the EPA is for emission control, most vehicle manufacturers have utilized the more powerful PCM to monitor many actions and operations of other components of the engine and transmission. The PCM is usually linked to a communication system with other computers on the vehicle. The other computers control the electric windows and door locks, the anti-theft devices, climate control, memory seats, and other operational and passenger comfort features. This system of computers shares data from some of the same sensors and uses the information in conjunction with individual programs to determine what, if any, action should be commanded. The vehicles of today have a computer network system similar to that of a small business.

Signals, commands, and actions occur up to a million times a second on the newest computer systems. They range from operating the engine, shifting the transmission, locking the doors, and alerting the neighborhood to vehicle theft or tampering. Each model year brings more electronic devices and almost every device is controlled or monitored by a computer.

Drivelines

The driveline delivers the engine's output to the wheels. There are several major, expensive components in this system. In a rear-wheel-drive (RWD) vehicle, they include a transmission, differential, axles, drive shafts, and all of the mountings and housings. In the front-wheel drive (FWD), in the same order, the components are the transaxle, final drive, drive axles, mountings, and the housing. There is no drive shaft in a FWD. Most of the driveline is located in the engine compartment or under the vehicle. This system is covered in detail in Chapter 10, "Suspension and Steering."

The transmission/transaxle provides a means of selecting a gear size that smoothly marries engine output to vehicle load (Figure 8-6). It may be an **automatic** or a **manual** where the

An automatic transmission changes forward gears by using valves and hydraulic pressure. A *manual transmission* requires the driver to select a gear using a shift lever.

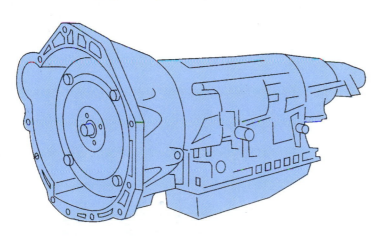

Figure 8-6 Shown is an automatic transmission for a Nissan truck. (Courtesy of Nissan North America)

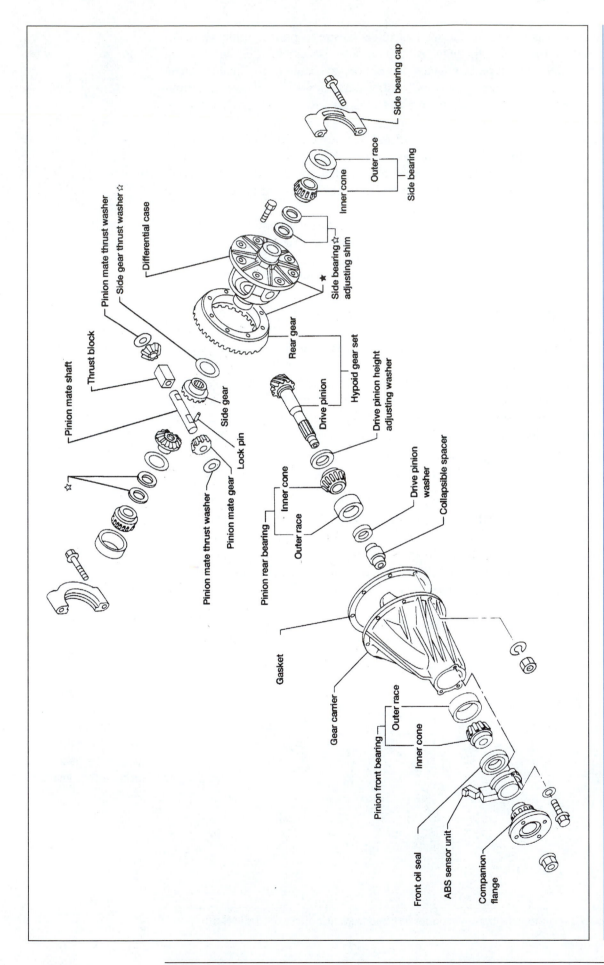

Figure 8-7 The differential allows the rear-drive wheels to turn at different speeds during turns. The final drive does the same for front-wheel-drive vehicles. (Courtesy of Nissan North America)

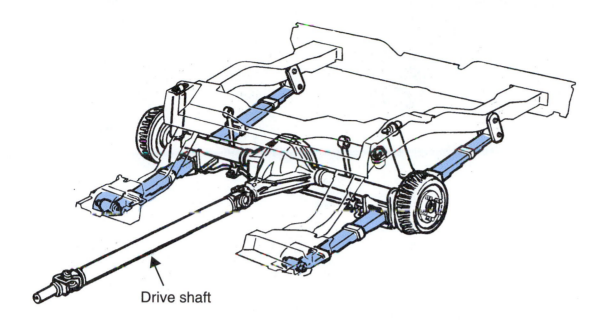

Drive shaft

Figure 8-8 The drive shaft delivers transmission output to the differential. (Courtesy of Chevrolet Motor Division, General Motors Corporation)

driver selects forward gears through lever and linkages. The differential/final drive allows the drive wheels to turn at different speeds when the vehicle is turning a corner or curve (Figure 8-7). Drive shafts provide the link between the RWD transmission and differential (Figure 8-8). Drive axles deliver the power from the transaxle final drive to the wheels on a front-wheel-drive vehicle.

Some vehicles have four- or all-wheel drive. Four-wheel drives allow the driver to use the front wheels to pull the vehicle when the rear wheels lose traction. All-wheel systems drive all four wheels at all times.

Steering and Suspension Systems

Steering, as the name implies, allows the driver to steer the vehicle in the direction desired. The steering system provides a system of linkages that allows the driver to turn the front wheels (Figure 8-9). The steering wheel is connected through a shaft to the steering gear unit. The gear multiplies the input force to make steering easier.

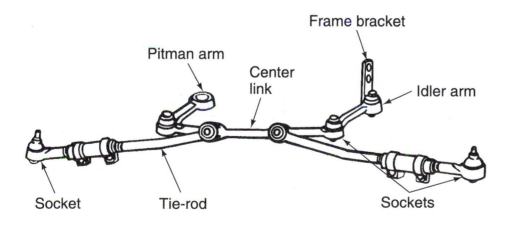

Figure 8-9 This type of steering system is still found on many trucks but few passenger cars. (Courtesy of Moog Automotive, Inc.)

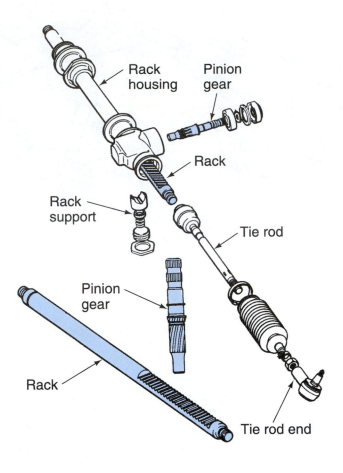

Figure 8-10 Rack and pinion steering is found on almost all newer passenger cars and some light trucks. (Courtesy of DaimlerChrysler Corporation)

Most systems have power-assisted steering. Valves in the gear unit direct pressurized, hydraulic fluid against a piston. This reduces driver effort. The major components are the steering column with the steering wheel, linkage, steering knuckle, and gearing. Newer automobiles have rack and pinion systems where the gear unit has been replaced (Figure 8-10). The rack system takes less room and connects the steering column directly to the wheels.

Suspensions

The suspension supports the vehicle and helps provide a smoother ride. Major components include springs and shock absorbers. Generally, each wheel has a spring and shock absorber. Any vehicle component mounted above the springs is supported by the springs (Figure 8-11). When a wheel strikes a bump or pothole, the springs allows the wheel assembly to move away from the bump, thereby preventing some of the impact being delivered to the passenger compartment. Springs have a tendency to keep bounding and rebounding even after the bump has been passed. Shock absorbers are designed to slow and stop the movement of the spring (Figure 8-12). They also help prevent some of the impact being transmitted to the passengers. Shock absorbers do not support the weight of the vehicle. Suspension and steering are covered in Chapter 11, "Brakes."

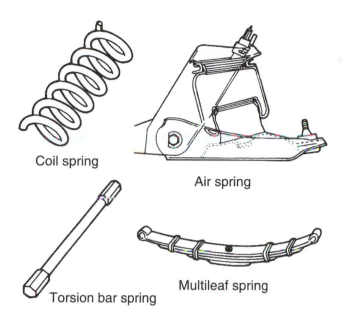

Coil spring

Air spring

Torsion bar spring

Multileaf spring

Figure 8-11 Four basic types of springs are shown.

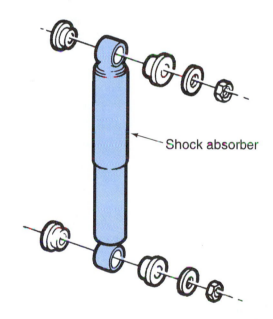

Shock absorber

Figure 8-12 The shock controls the bouncing action of the springs. (Courtesy of American Motors)

Brakes

There has to be a means for a controlled stop of the vehicle once it is moving. The brake system connects the driver to the wheel through linkage and hydraulic circuits (Figure 8-13). The brake system on a vehicle is hydraulically activated by a mechanical linkage in the passenger compartment. The applied forced is transmitted to the brakes at the front and rear axles. Mechanical devices apply the force to the braking components at the wheels.

Many new systems use an antilock system to prevent brake lockup during emergency stops. Antilock systems rely upon electronic devices and computers to control braking. Brakes are discussed in detail in Chapter 12, "Auxiliary Systems and Climate Control."

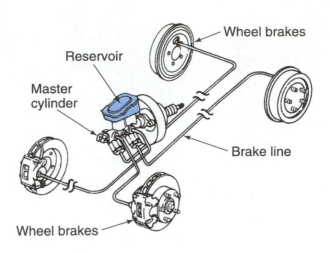

Figure 8-13 The brake system uses hydraulic pressure generated by the master cylinder to apply the braking pads and shoes. (Courtesy of Delco-Remy, Division of GMC)

Frames and Bodies

Much as bones support the body, the frame provides support for all components of the vehicle. There are various frame designs (Figure 8-14), but newer systems use a *unibody design* where the body itself provides some of the support (Figure 8-15). This makes the vehicle lighter and more fuel efficient.

The frame has attachment points for suspension and steering components, bumpers, and the body. A *full frame* has rigid parallel channels with crossmembers, which help keep the parallel channels in place and provide mounting points for items like the engine and transmission (Figure 8-16). Full frames are heavy, but can support heavy loads. Most passenger cars

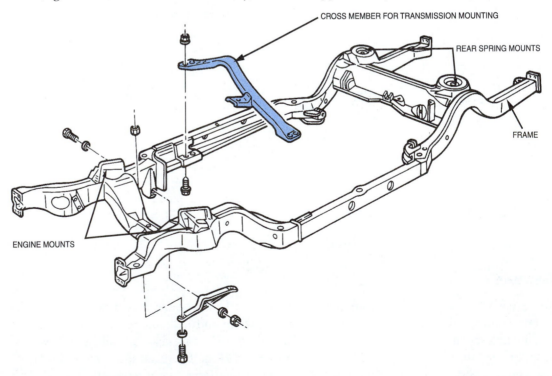

Figure 8-14 This type of frame is used on older cars and is still used on many trucks. (Courtesy of Chevrolet Motor Division, General Motors Corporation)

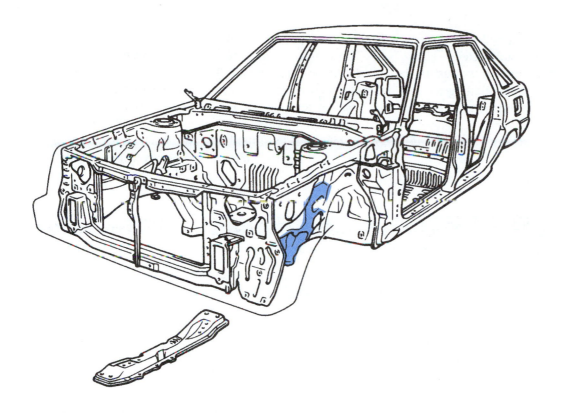

Figure 8-15 The unitized body directly supports almost all of the vehicle's components except for the engine and transaxle. (Reprinted with permission)

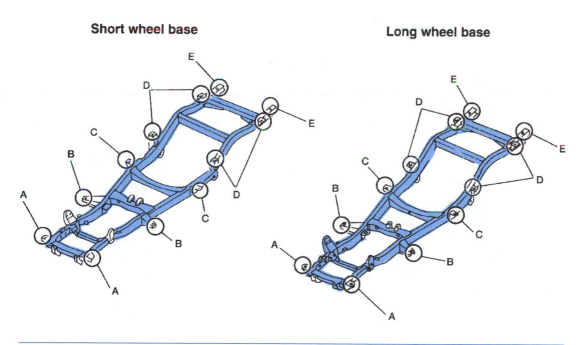

Short wheel base

Long wheel base

Figure 8-16 The connection points on this frame are for mounting the body and other vehicle components. (Courtesy of Nissan North America)

now use the unibody, while trucks still use the full frame. Some rear-wheel-drive cars still use a version of the full frame.

Unibody construction relies on the body to support the passengers and interior components. However, the unibody by itself cannot support the torque actions and movement of the powertrain. The steering and suspension systems also stress a unibody. To counter this problem, sub-frames are used to support the powertrain and at least some portion of the steering and suspension systems (Figure 8-17). The sub-frames are constructed to support various loads without stressing the rest of the vehicle. Some vehicles only have a sub-frame at the front of the vehicle supporting the engine, transaxle, steering, and the front suspension. The rear portion has strengthened areas for attaching the rear wheels and suspension. On vehicles with two true sub-frames, the body becomes the connection between the two units. This reduces weight greatly, but a small "parking lot" accident can twist the body and sub-frames out of alignment on older designs. Newer technology has made the unibody stronger with more protection for the passengers as discussed in Chapter 4, "Automobile Theories of Operation."

A BIT OF HISTORY

The first car enclosed with a glass windshield was introduced in 1903.

The *body* is the shell that surrounds most of the vehicle's components and passenger compartment (Figure 8-18). The body also provides a mount for several automotive subsystems. Almost all lights, interior and exterior, are mounted to body panels. The doors, hood, and trunk lid become supporting members of the body when closed. Door posts or pillars provide latching and locking devices for the door and help support the roof. Front and rear windshields and other glass components are mounted to the body. Other devices mounted directly to the body are windshield wipers, instrument panels, seats, steering columns, and rearview mirrors. Further information on the frame and body may be found in Delmar Publisher's publication *Motor Auto Body Repair, 3E* by Robert Scharff and James E. Duffy.

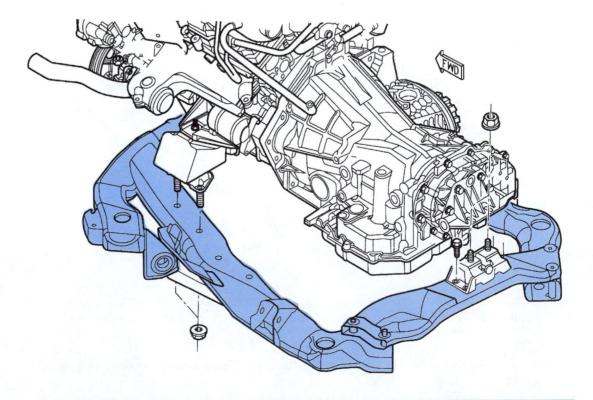

Figure 8-17 This is a typical sub-frame for supporting the engine, transaxle, and some portions of the steering and suspension systems. (Courtesy of DaimlerChrysler Corporation)

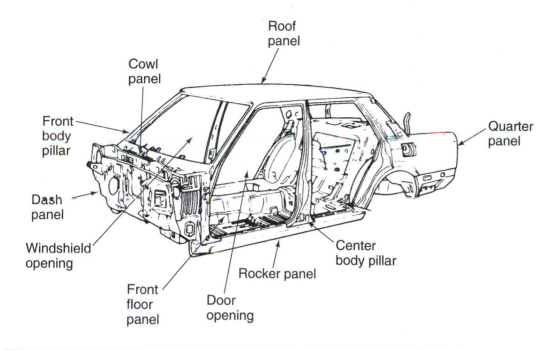

Figure 8-18 The body provides protection for the passengers and mounting areas for various components like the glass and doors. (Reprinted with permission)

Climate Control Systems

Climate control includes the heating and air conditioning systems. In many models, the two systems have been combined into a single system that may be controlled manually or automatically (Figure 8-19). Either system can keep the passengers comfortable by using liquids to transfer heat.

> **CAUTION:** Do not attempt to open a hot engine cooling systems. Serious burn injury could result.

> **CAUTION:** Do not work on an air conditioning system without protective clothing and equipment. Serious injury could result.

> **CAUTION:** Do not open a charged air conditioning system. The refrigerant can freeze skin, and releasing refrigerant into the atmosphere is illegal.

The heater system blows air over hot engine coolant flowing through a small radiator-type device. This warms the air and the passenger compartment. Air conditioning works in the reverse,

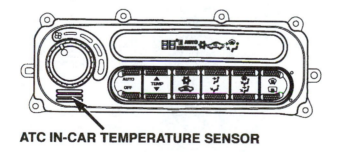

ATC IN-CAR TEMPERATURE SENSOR

Figure 8-19 The climate control system can be set at a comfortable temperature and the driver does not need to select heating or cooling. It operates the correct system as needed. (Courtesy of DaimlerChrysler Corporation)

The two *refrigerants* most commonly used today are R12 and R134a. The newest and most environmentally safe is the R134a.

A *heads-up display* projects instrument or dash gauge readings onto the windshield in the driver's line of sight.

but uses a special liquid referred to as **refrigerant** or freon and varying pressures within the systems. Both systems can affect the operation of the engine and both should be serviced or supervised by experienced technician. In fact, federal, state, and local laws require an air conditioning technician to be certified through a special written test. Climate control is covered in Chapter 12, "Auxiliary Systems and Climate Control."

Accessories

At one point, accessories included items like roll-up windows and windshield wipers. Then, heaters, radios, and air conditioning were added over the years. Now, there are all kinds of accessories available, including automatic temperature control, power seats and locks, heated seats, and all types of passenger comfort and care devices. In addition, there are **heads-up displays (HUDs),** digital gauges, memory seats, automatic-dimming rear view mirrors, and electric remote-controlled outside mirrors. Almost every day, automotive designers come up with a new accessory or a more refined version of an old one. Satellite position sensors and route mapping will soon be available in most cars. Accessory systems provide a great deal of work for technicians versed in electronics.

Engine Design

The engine has become the most complex system in the vehicle. Its orginal design and purpose is still intact. However, a quick glance under the hood would cause most reasonable shade-tree mechanics to find a telephone and ask for help. The internal combustion engine operation has been refined internally and electronically. Working on today's automotive engine is no longer a Saturday morning job.

The *mechanical linkages* are the piston, connecting rod, crankshaft, and the supporting bearings and rings.

An internal combustion engine is nothing more than an air pump. When the starter is engaged, air is sucked into the cylinders. Power is possible when fuel is mixed into the airflow and lit with an electrical spark. The expanding gases (thermodynamic theory) supply the necessary power that pushes on the **mechanical linkages** in the engine (Figure 8-20). An engine is classified using three criteria: *block configuration, displacement,* and the *valvetrain* designs. Its operation is classified as a *two-stroke, four-stroke,* or *rotary*. When an engine is referred to as "gasoline" or "diesel," it is a reference to the fuel used, not the type of engine.

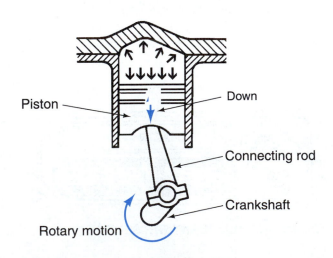

Figure 8-20 The ignition spark lights the air/fuel and the piston is forced downward, thereby turning the crankshaft.

Engine Block Design

The engine **block** starts as a large block of treated cast iron or aluminum. It is bored, drilled, machined, and polished to form the basic support for all the engine's operating systems, internally and externally.

The block may be shaped in various configurations. An *in-line engine* has all of the cylinders aligned in a series of vertical bores. The most common in-line blocks are a four and six-cylinder design (Figure 8-21). More cylinders involve a longer and heavier block. Four-cylinder engines are available in almost all models of cars and light trucks. Six-cylinder, in-line, gasoline engines are used less frequently in modern vehicles because of the block's weight and size. They are most commonly used in light trucks. Many over-the-road trucks still use the in line six and in some cases, in-line eight cylinders.

Another block is known as the *V-block*. The numbers of cylinders are divided into two rows set at an angle to each other (Figure 8-22). The angle between the rows of cylinders is 90 degrees or 60 degrees. The V-block allows for a lower hood line and for more cylinders to be added in a shorter, linear space. Then, the engine could be more powerful and contained in a smaller block. V-block engines also tend to create less vibration than a comparably sized in-line.

Volkswagen made the *horizontally-opposed engine* famous during its use in the orginal **VW Beetle.** The rows of cylinders are laid out 180 degrees apart (Figure 8-23). This provided a very low profile and air cooling. The engine was mounted in the rear of the vehicle. Subaru still uses a front-mounted, horizontally-opposed engine in some of its models. Another thing about VWs and Subarus is the location of their spare tires. Both are mounted in the *front* of the car. The VW Beetle has storage space under the front hood along with the tire stored there. The spare tire in a Subaru is laid across the top of the engine.

Another engine design was used almost exclusively by Chrysler. It is an in-line six-cylinder that leans or slants to the right. It became known as the *slant 6*. This engine is still operating in many types of industrial equipment.

> The machined portions of a basic *block* are the cylinders, mounting deck for the cylinder head, lifter bores, and the saddles or bores for the bearings.

> The new *VW Beetle* uses an inline engine.

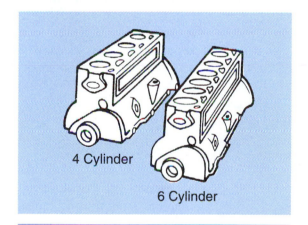

Figure 8-21 In-line engines have all cylinders in a straight line.

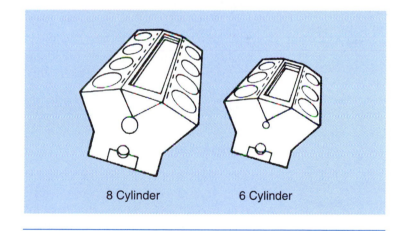

Figure 8-22 V-blocks have the cylinders in two equal rows.

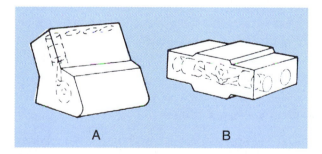

Figure 8-23 The horizontally opposed (B) and slant 6 (A) engines were only used in certain models.

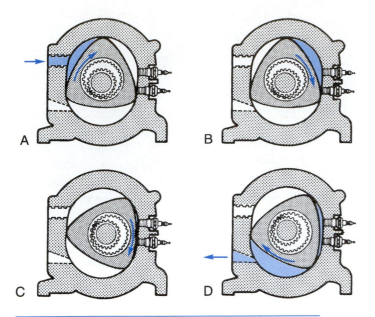

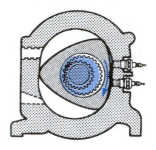

Figure 8-24 The rotary's oval chamber is shaped to create the same internal combustion conditions found in a piston-driven engine.

Figure 8-25 Planetary gear sets are also found in transmissions and transaxles. They are used to multiply torque.

A simple *planetary gear set* has a sun gear that fits inside a planetary carrier. The carrier, in turn, fits inside a ring gear. This allows torque changing in a small amount of space.

Mazda uses another engine type for its RX7 model that does not use pistons or cylinders to generate and transfer the combustion power to the crankshaft. Instead, two triangular rotors within matching oblong chambers intake, compress, and exhaust the air/fuel mixture (Figure 8-24). The rotors are attached to the crankshaft through a **planetary gear set,** which increases the force delivered by the rotors (Figure 8-25). This design is known as a *rotary engine.*

Fuel economy has caused some types of engines to be discontinued in many car models. The most common type put into today's car is either a four-cylinder in-line or the V-6. During the 1998 and 1999 model years, some V-8 engines were re-introduced in high-performance and luxury model vehicles. The V-8 engine produces more horsepower, and new electronics and better engineering keep the fuel mileage and emissions within mandated limits. Some engines are built using aluminum for the blocks and cylinder heads. A few research engines have been made of ceramic in an attempt to reduce weight and durability. Ceramic material is expensive, but the research provides extensive data on how to use other materials.

Displacement

Engine *displacement* may be measured in cubic inches displacement (cid), liter (L), or cubic centimeters (cc).

The amount of air an engine can move in one cycle is its **displacement**. Displacement is the total volume of the cylinders calculated by using the bore, stroke, and number of cylinders (Figure 8-26). The *bore* is the diameter of the cylinder. The *stroke* is measured from the top of the cylinder to the top of a piston that is resting at bottom dead center (BDC). The *volume* of a cylinder is equal to the square of the bore's radius multiplied by 3.1414 and then multiplied by the stroke. The volume is multiplied by the *number of cylinders* to find total engine displacement. A typical displacement of late 1990's engines is 2.0, 3.0, 3.8, 4.0, and 5.0 liters (L). There are others that vary in displacement, but they generally fall within the ranges mentioned here. As a point of reference, an engine with 318 cubic inch displacement (cid) is listed as a 5.2 L while a 360 is a 5.9 L.

A Comparison—Old versus New

During the 1950s and 1960s, high-performance, high-horsepower engines ranged from 350 cid (5.7 L) to 440 cid (7.2 L). Stock engines were capable of producing up to 400 horsepower and high vehicle speeds, but they were fuel-hungry and produced enormous amounts of harmful

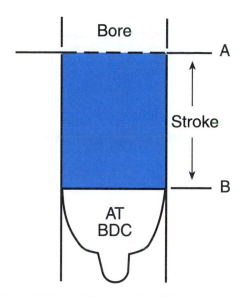

Figure 8-26 Displacement is the volume between the top of the cylinder (A) and the piston at BDC (B).

emissions. Late 1990's engines can produce 300 horsepower with high fuel economy, high ve-hicle speeds, and very low levels of harmful emissions, all with a smaller displacement.

Valve Trains

There are two basic **valve trains** in use today. The older version, **overhead valve (OHV),** has been used in all types of engine block designs. The overhead valve (OHV) system uses a crankshaft-driven camshaft in the block to move the valve mechanism. The camshaft turns at half the crankshaft speed and must be timed to open and close valves in time with piston position (Figure 8-27). The camshaft pushes upward on a solid or hydraulic **lifter** moving a push rod. The push rod applies upward force to one end of a **rocker arm**. The rocker arm is pivot-mounted and the opposite end applies downward force to the valve (Figure 8-28). With the valve open, air and fuel can enter the cylinder or exhaust can exit. As the camshaft rotates, force is removed from the valve train. A spring mounted around the valve stem forces the valve upward and closed. The valve spring also applies force throughout the mechanism to

The *valve train* consists of valves, valve springs, a camshaft, valve retaining devices, and may have valve *lifters*, pushrods, and *rocker arms*. The older version of the valve train is the *overhead valve (OHV)* which uses a crankshaft-driven camshaft to move the valve mechanism.

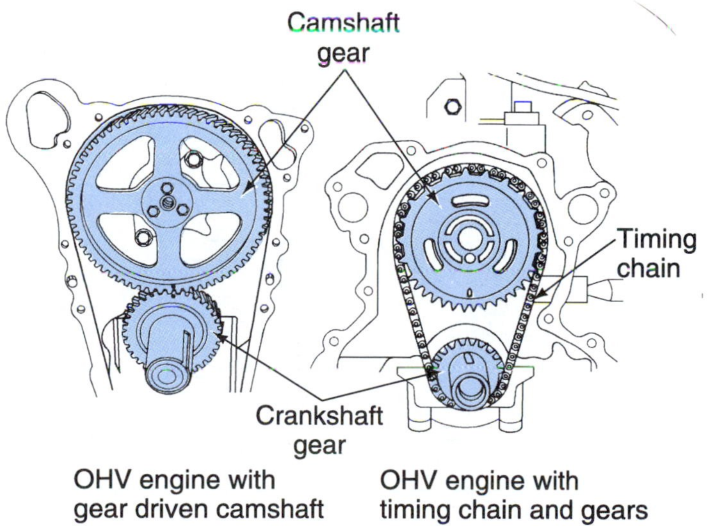

Camshaft gear

Timing chain

Crankshaft gear

OHV engine with gear driven camshaft

OHV engine with timing chain and gears

Figure 8-27 The camshaft and lifters are located in the block for an OHV engine.

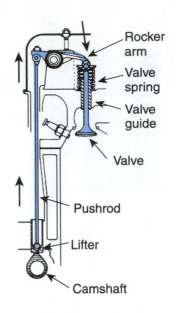

Figure 8-28 The rocker arm makes the connection between the pushrod and the valve stem.

An *overhead cam (OHC)* engine is used on many newer vehicles. The camshaft is installed in the cylinder head with the valves.

maintain constant contact between the various components (Figure 8-29). There is usually one intake and one exhaust valve per cylinder. Newer, more powerful and fuel-efficient engines may have two intake and two exhaust valves. This provides better flow characteristics for the air/fuel mixture and exhaust gases.

The system being used on many newer engines has the camshaft installed in the cylinder head with the valves. It is known as an **overhead cam (OHC)** engine. The camshaft is mounted directly over the valve and exerts downward force directly onto the valve stem (Figure 8-30). This system eliminates the lifter, push rods, and rocker arms that provide better durability, less maintenance, and less noise. Newer versions of the OHC system may have small, hydraulic lifter-type mechanisms that absorb the vibration and shock of the valve opening and closing. Other versions act directly on the rocker arms to operate the valves. There may be a roller bearing between the camshaft and rocker arm. (Figure 8-31).

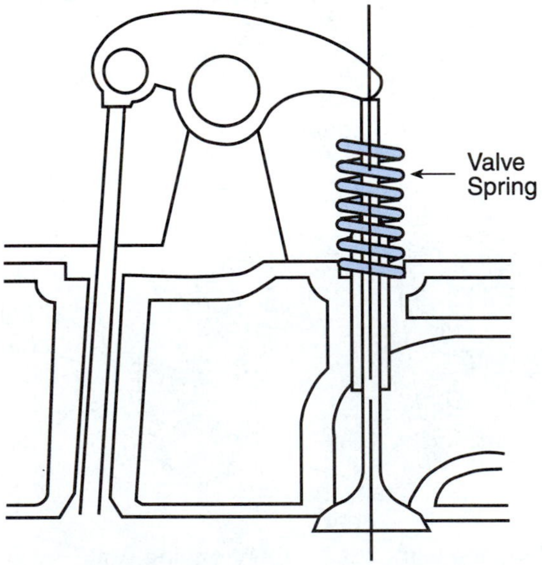

Figure 8-29 The valve spring applies continuous force on all components to prevent sloppy or noisy movement and returns the valve to the closed position.

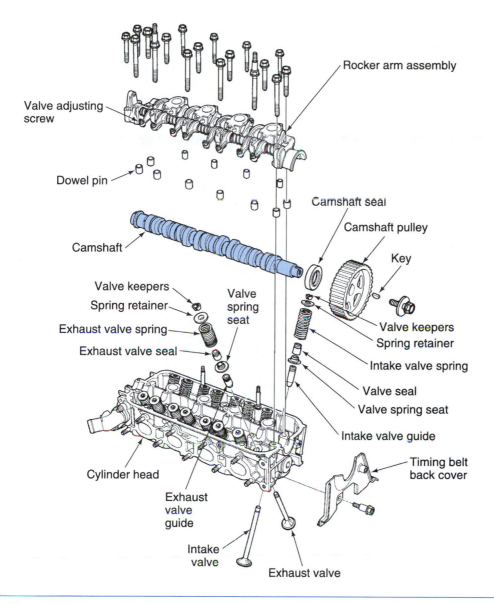

Figure 8-30 The OHC design eliminated the lifter and pushrod. (Courtesy of Honda Motor Co. Ltd.)

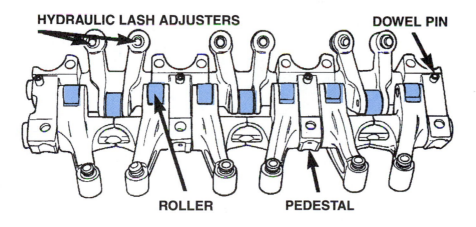

Figure 8-31 The roller bearing provides quicker, smoother operation and a longer component life. (Courtesy of DaimlerChrysler Corporation)

Some *dual overhead cam (DOHC)* engines use a single belt to drive all camshafts. Others use a belt to drive one camshaft on each head with the second camshaft gear driven by the first.

More powerful OHC engines use two camshafts per cylinder head. Each camshaft operates either the intake or the exhaust valves. The **dual overhead cam (DOHC)** reduces stress on the camshaft and allows for multiple valves per cylinder (Figure 8-32). Each camshaft is timed to another and the crankshaft.

The camshafts on an OHV engine are driven through meshed gears or a chain connecting the crankshaft and camshaft gears. When worn, both systems must be lubricated and replaced as a set. The gears or chain are not readily visible, and noise or a breakdown is the first sign of a trouble. A chain and gear may drive the camshafts on an OHC system, but many use a flexible belt. The belt does not require lubrication, is quieter, and is replaced based on mileage or age (Figure 8-33). If the belt is replaced at the recommended intervals, breakdowns because of belt failure become very uncommon. Failure to replace the belt as scheduled can result in serious damage to the engine.

If the valve is open as the piston reaches the top of its travel, the valve, piston, and the cylinder head may be damaged beyond repair. This is known as an *interference* engine because the open valve extends below the cylinder head and into the cylinder bore. *Non-*

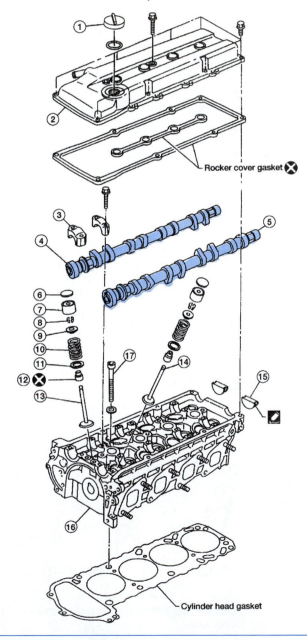

Figure 8-32 Multi-valves per cylinder and increased horsepower/torque are possible with dual overhead cams (DOHC). (Courtesy of Nissan North America)

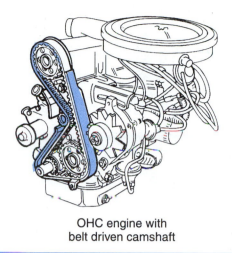

OHC engine with
belt driven camshaft

Figure 8-33 Timing belts are quicker and easier to maintain than chains.

interference or freewheeling engine valves open into the combustion chamber but do not extend into the cylinder bore (Figure 8-34).

Based on the above discussion, we can determine the size and possibly the application of the engine being repaired or researched. A typical engine would be a 2.0 L OHC in-line 4. This indicates that an engine of this description is an in-line four-cylinder with two liters of displacement and an overhead cam valvetrain. Further research could indicate that this engine was used in many Chevrolet Cavalier models in the late 1980s and the 1990s and is still being used in many domestic and import vehicles.

Engine Components

A typical engine is an assembly of four major sub-assemblies or components. They are the block assembly, cylinder head, intake manifold, and exhaust manifold (Figure 8-35). The block assembly houses the piston assemblies, crankshaft, and the camshaft and valve lifters on an OHV-type valvetrain. An oil pump is mounted either into the block or suspended from the bottom of the block (Figure 8-36). External areas for mounting brackets of external components are machined on the block.

On top of the block is the cylinder head which houses the valve and valve mechanism. On an OHC engine, the camshaft is mounted in the head (Figure 8-37). In the cylinder head and block, passageways for oil and coolant flow exist within the component and crossover into the other components. There are external ports in the cylinder head(s) that match the ports on the intake and exhaust manifolds.

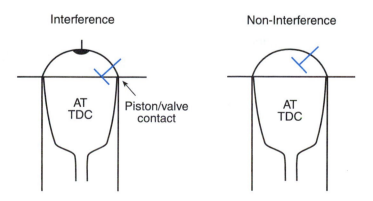

Figure 8-34 Non-interference engines keep the valve within the combustion chamber.

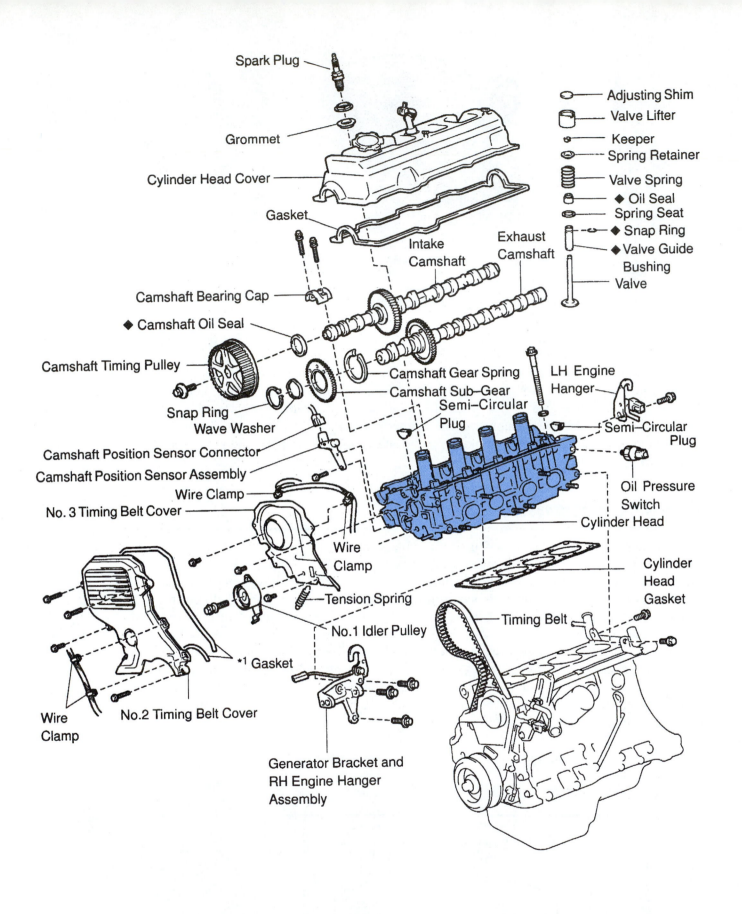

Figure 8-35 The engine is an assembly of hundreds of moving and stationary components. (Reprinted with permission)

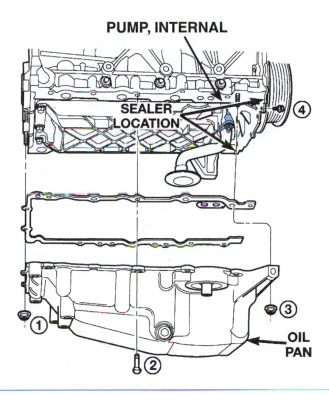

PUMP, INTERNAL

SEALER LOCATION

OIL PAN

Figure 8-36 The oil pump on many newer vehicles is mounted internally on the crankshaft just behind the large crankshaft pulley. (Courtesy of DaimlerChrysler Corporation)

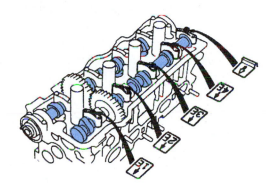

Figure 8-37 An overhead cam is mounted above or right beside the line of valves. (Reprinted with permission)

The two manifolds are mounted to the outside of the cylinder heads (Figure 8-38). An in-line engine has one of each. A V-block has two exhaust manifolds with one intake manifold placed in the V between the two cylinder heads. The exhaust manifold has no moving parts, but may have threaded holes for mounting emission control devices. The intake also has no moving parts, but has external mounting areas for attaching items like an upper intake manifold, fuel delivery, and a throttle valve. Intake manifolds have runners that connect the air entrance to each of the cylinders. The runners are designed to deliver the same volume of air to each cylinder. Exhaust runners are provided for each cylinder, but they are joined at a point away from the cylinder head. This junction is the point where the exhaust pipe is connected. It should be noted that the number and type of attachments on each of the above components depends on the engine design and passenger comfort available on a particular vehicle model. It could be a very simple system with the engine assembly, a water pump, and alternator or a system that is so loaded with extras that the engine assembly is not visible at first glance under the hood.

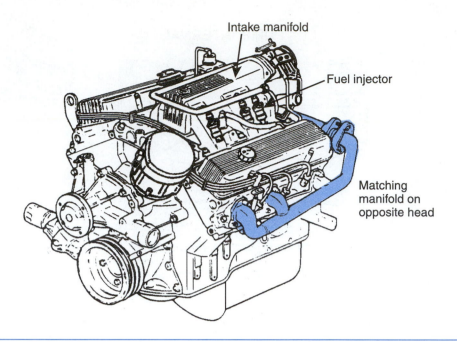

Intake manifold

Fuel injector

Matching manifold on opposite head

Figure 8-38 The exhaust manifolds are mounted to the outside of the cylinder heads for exhausting the cylinders. (Courtesy of Chevrolet Motor Division, General Motors Corporation)

Engine Operation

Most small engines except for marine are *four-stroke* types, including most one-cylinder lawn mowers.

A typical gasoline-fueled engine uses a **four-stroke** cycle operation. Other types include the two-stroke and the rotary. Each type uses similar systems for fueling, lubrication, cooling, and the other components needed for operation. Two-stroke engines are used primarily in light equipment. However, they are generally less efficient then four-strokes and some require the lubrication to be mixed with the fuel. They also tend to emit excessive smoke and pollutants if not properly tuned.

Two-Stroke Operation

The two-stroke engine intakes and compresses the fuel/air mixture during the upward movement of the piston. The downward movement of the piston during the power stroke rotates the crankshaft and opens the exhaust valve. The lower portion of the piston's next upward movement starts clearing the cylinder of exhaust but also begins the intake of fresh air/fuel. The operation performed on this stroke is known as *scavenging* because of the method of clearing and filling the cylinder. Intake and compression are completed on one stroke while power and exhaust are accomplished on the next (Figure 8-39). Intake and exhaust valves are used to open or seal the cylinder. Power is produced on each revolution of the crankshaft.

Rotary Engines

Rotary engines usually are not rebuilt by a shop, but are purchased from Mazda. This is because of the type of machining required.

In a **rotary engine,** the air/fuel mixture is compressed using a triangular rotor inside an oblong chamber (Figure 8-40). At each apex (corner) of the rotor, a seal keeps contact with the chamber wall and pushes the air/fuel ahead of it or is pushed by expanding gases (Figure 8-41).

As the rotor turns, a seal forces the air/fuel mixture into a small area of the chamber. This reduced chamber volume compresses the mixture until ignition occurs. The expanding gases are directed against the seal ahead of the compression seal. This power is transmitted through the rotor to the crankshaft. As the rotor continues to turn, the gases enter a larger area where the spent gases are exhausted. As this area widens, a low-pressure area is formed under

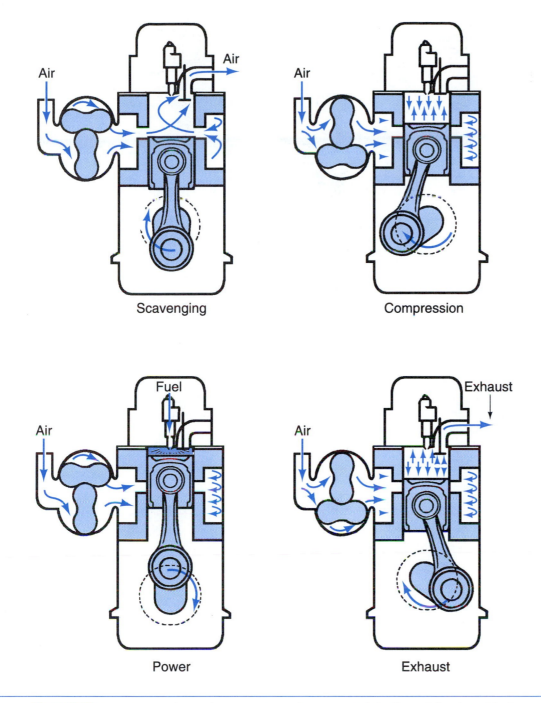

Figure 8-39 Some two-stroke engines use rotary valves to control the flow of the air and fuel.

the intake valve. A new mixture is drawn in and the cycle is repeated. It should be noted that each side of the rotor and each seal act as the power, intake, compression, and exhaust component during each rotation of the rotor. Since the rotor always turns in one direction without the stop/start motion of a piston, the rotary engine could produce more horsepower with less waste energy than a conventional gasoline engine. However, the rotary engine does not produce the same torque as a piston engine of the same displacement and is considered less efficient. Rotary engines have the same problem with harmful emissions as piston engines.

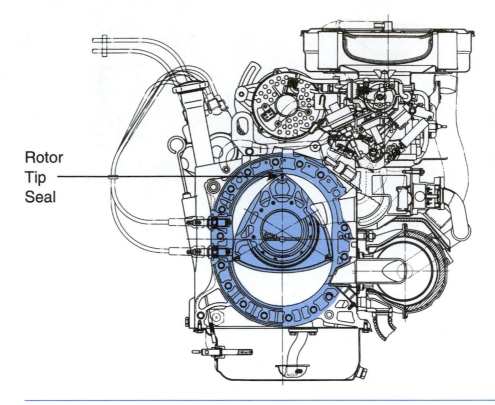

Rotor Tip Seal

Figure 8-40 There are two chambers and two rotors in a Mazda RX7 engine. (Courtesy of Mazda Motor of America, Inc.)

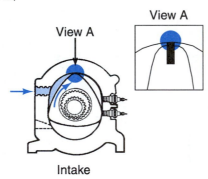

Figure 8-41 The seal slides along the inside of the chamber to capture and move the air/fuel mixture.

Four-Stroke Operation

The four-stroke engine is used in nearly every car and truck currently being produced. A *power stroke* is produced every two revolutions of the crankshaft. This engine requires four movements of the piston to complete a cycle.

The first stroke is the *intake stroke*. As the piston moves from **top dead center (TDC)** downward, the camshaft opens the intake valve. Reduced pressure in the cylinder, combined with the atmospheric pressure in the intake manifold, force the air into the bore (Figure 8-42). Depending on the fuel delivery system, fuel may be drawn or injected into the air stream at the entrance of the intake manifold, into the air stream near the end of the intake runners, or directly into the combustion chamber.

The intake valve remains open until the piston nears **bottom dead center (BDC)**. During the upper movement or *compression stroke*, the air/fuel mixture is compressed into the combustion chamber (Figure 8-43). Both valves are closed to seal the chamber during this stroke

Top dead center (TDC) is when the piston is at the very top of its travel in the cylinder. Bottom dead center (BDC) is just the opposite.

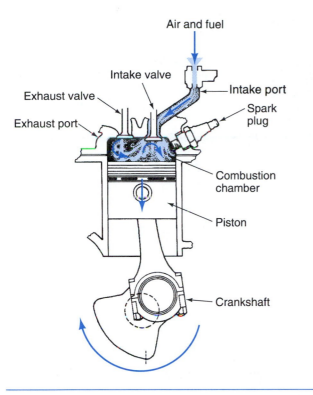

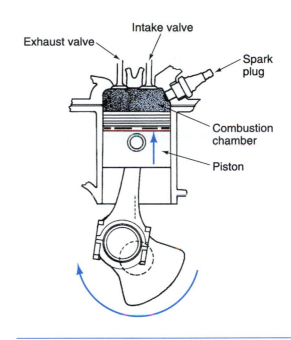

Figure 8-43 The piston compresses the air/fuel mixture into the compression chamber. (Courtesy of Breton Publishers)

Figure 8-42 The air is drawn into the cylinder as the piston moves downward during the intake stroke. (Courtesy of Breton Publishers)

(Figure 8-44). At a specific point in piston travel, an electrical spark is used to ignite the mixture. On multiple fuel injection (MFI), the fuel is injected over the intake valve or into the chamber just before the spark. The expanding gases force the piston downward in a power stroke, which turns the crankshaft (Figure 8-45). Note that the valves are closed during both the compression and power stroke.

Near BDC the exhaust valve opens and the upward movement or *exhaust stroke* of the piston exhausts the spent gases. As the piston nears TDC, the intake valve begins to open just

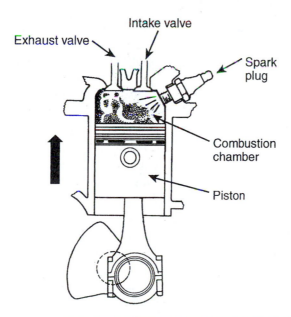

Figure 8-44 The spark plug fires about 10 degrees before the piston reaches top dead center on the compression stroke. (Courtesy of Breton Publishers)

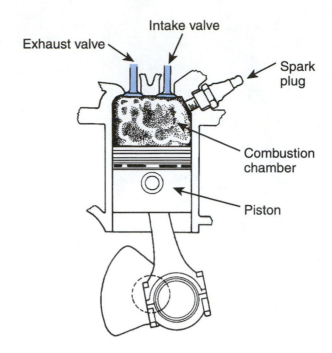

Figure 8-45 The expanding gases of the burning air/fuel mixture create the energy to drive the piston. (Courtesy of Breton Publishers)

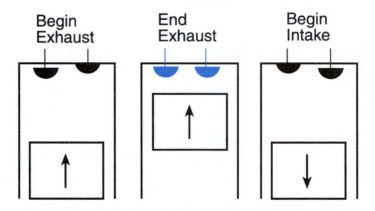

Figure 8-46 Valve overlap occurs at the end of the exhaust stroke and the beginning of the intake.

as the exhaust begins to close. The resulting *valve overlap* helps exhaust the spent gases and assists in starting the new mixture into the cylinder (Figure 8-46).

The **timing of the valves'** opening and closing, the ignition timing, and the air/fuel mixture are critical to smooth engine operation. If any of the three is worn, broken, or out of adjustment, the engine will perform poorly and may be seriously damaged. Other operating systems of the engine can cause similar problems.

Newer technologies allow the engineers to design an engine that can change valve timing and valve opening. The PCM uses sensor data to change intake and exhaust valve timing and amount of opening. This helps reduce harmful emissions and increases fuel mileage. One system eliminates the entire valve drive mechanism with a solenoid, which opens and closes the valve on command from the PCM.

The four-stroke engine operation is fairly simple if only one piston is being considered and the engine speed is low. However, when multiple pistons, piston weight, piston speed, and other stress factors are considered, the operation is no longer simple. A four-cylinder engine running at 2,000 rpms will have each of the four pistons stop and start about 4,000 times in one minute.

Valve timing is set by matching the crankshaft and camshaft gears to timing marks. The camshaft can be timed-advanced or retarded for performance engines.

Lubrication

All components that come in contact with each other need some type of lubrication. The engine with its close fitting parts, high speed, and high temperatures is no exception. Poor or no lubrication will be apparent within several miles or minutes of operation. Most damage from poor lubrication requires expensive repair or replacement of the complete engine.

The lubrication used today is a high-tech combination of lubricants and cleaning agents designed to reduce friction and improve fuel mileage. There are many different brands and combinations on the market today. Engine oil is rated by its thickness and resistance to breakdown by the Society of Automotive Engineers (SAE) and the American Petroleum Institute (API). The oil is rated by numbers, which refer to its thickness or weight (Figure 8-47). Common engine oil used in modern engines is rated as 5W30. The multiple designations, 5 and 30, indicates that this oil can be used in cold or hot weather. The number 5 indicates thin oil while 30 is thicker oil. The newest engines have lower tolerances between moving parts and require thinner oil for lubrication. Older engines typically used thicker oil than today's engine.

Also available on the market are synthetic engine oils (Figure 8-48). Synthetic oil is made from chemicals, does not use natural petroleum products, and is rated by thickness in the same

Figure 8-47 Engine oil comes in several ratings, but only use the one specified by the manufacturer.

Figure 8-48 Synthetic oils are not made from natural products but are rated the same as natural petroleum oil.

manner as natural oils. In some cases, synthetic oils are better than natural oils, however, they are usually much more expensive then regular petroleum products. Synthetic oils are rated the same as the naturally-formulated oils. There are also mixtures of synthetics and natural oils available known as *synthetic blends*. The blends are a little cheaper than full synthetics, but more than naturals.

One of the problems with lubricating an engine is the time during shutdown. The oil tends to run or drain back from engine components, resulting in a dry or nearly dry startup where the moving components have little, if any, lubrication. This causes metal-to-metal contact and wear. Some synthetic brands, Castrol for one, offer a lubricant with magnetic tendencies. The oil has synthetic ester molecules that attach to the components and prevent drain back. In this manner, the moving components are lubed sufficiently even after an overnight shutdown.

The main point to remember is to check the service manual for the type of oil to be used in the various systems of the vehicle. In some engines, the use of synthetic or blends may be a total waste of time and money. Other engines can have longer life, better fuel economy, and a quieter run with synthetics or blended lubricants. There are synthetic and blend lubricants available for automatic transmission/transaxles as well. Always consult the service manual.

The engine's oil is stored in an oil sump attached to the bottom of the block. A crankshaft or camshaft-driven oil pump is mounted directly over the oil with a pickup tube extending into the oil. The tube has a coarse screen pickup to filter out larger trash particles. The pump draws oil from the sump, pressurizes it, and forces it into the oil filter before it enters the engine's oil galleries. A **check valve** is installed in most pumps to prevent oil drainback during engine shut down.

An oil filter removes most of the harmful particles floating in the oil. The particles are metal and carbon that result during normal engine operation. The filter adapter has a *bypass* or *pressure relief* valve that opens when oil pressure within the filter becomes too high (Figure 8-49). This allows oil to flow past the filter and into the engine. The valve is primarily designed to allow oil flow when the engine and oil temperatures are cold. However, if the filter is clogged or the oil is extremely dirty, the valve will open. In both cases, unfiltered oil is pumped into the engine. Damage could result.

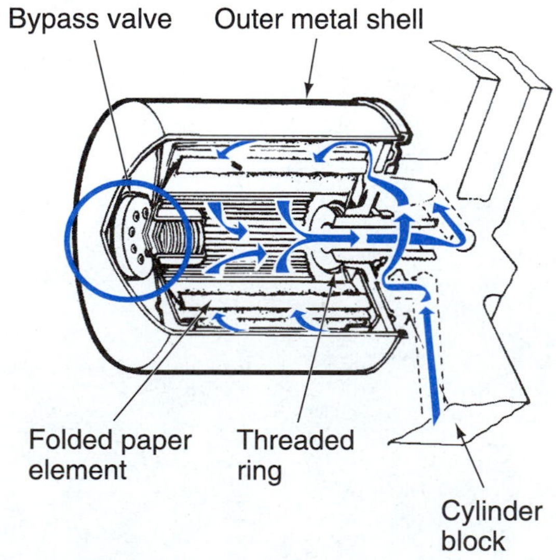

If the oil pump *check valve* is stuck in the open position, the engine would receive little or no lubrication during the first few moments of startup. This increases bearing wear tremendously.

Figure 8-49 The bypass valve cools oil to flow past the filter, thereby preventing damage to the engine. (Courtesy of DaimlerChrysler Corporation)

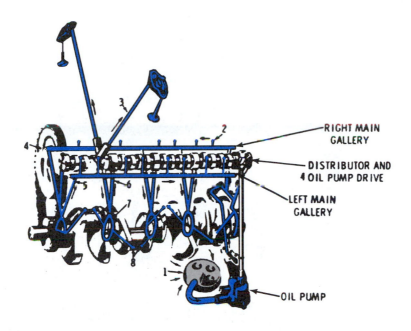

Figure 8-50 All mated, moving components must have pressurized lubricant between them to reduce friction and heat. (Courtesy of General Motors Corporation, Service Operations)

The oil moves under pressure into the block's oil galleries. Some engine components are *pressure lubricated* while others are *splash lubricated*. Pressure lubricated components include the crankshaft bearings, connecting rod bearings, camshaft bearings, and some rocker arms (Figure 8-50). All of the bearings operate under extreme conditions and the pressurized oil is required to maintain a **clearance** between the bearing and the supported component. Splash lubrication is accomplished when the pump supplies a large pool of oil for the component to move through. The oil is splashed over the components before returning to the sump though large passageways or galleries. Some components using splash lubrication are camshaft drive gears and chains, valve train mechanisms, and cylinder walls (Figure 8-51).

Some people believe that the crankshaft rotates through the oil in sump. If this did occur, **cavitation** or foaming would cause the oil pump to suck air and the engine would be without lubrication. Too much oil can do as much damage as too little oil. The type and amount of lubricant specified by the manufacturer must be followed.

Starting System Safety Devices

Different circuits control the starting and charging systems. The starter controls are the ignition switch, a park/neutral or clutch switch, and the relay. Anti-theft circuits may also be in series with or controlling power to the starter relay. The *park/neutral switch* is in series with the ignition switch, starter relay, and the PCM. If the automatic transmission is not in park or neutral, the switch is opened and current cannot flow to the starter relay or PCM. A *clutch switch* performs the same function on a manual transmission vehicle. If the clutch pedal is not fully depressed, the starter and PCM will not function. Some vehicles also have an anti-theft system that will disable the starter and/or ignition system if not disarmed by the correct security code.

The *clearance* between a bearing and its journal is called "oil clearance" and is usually about 0.001 inch to 0.003 inch.

Cavitation is the introduction of air into a fluid, in this case, air forced into the oil by the rotating crankshaft and connecting rod.

Newer vehicles have a key-transmission interlock which requires the transmission to be placed a certain gear before the key can be removed from the switch.

Newer vehicles may also have a brake-transmission interlock, which requires the operator to apply the brake before the transmission can be shifted from PARK.

209

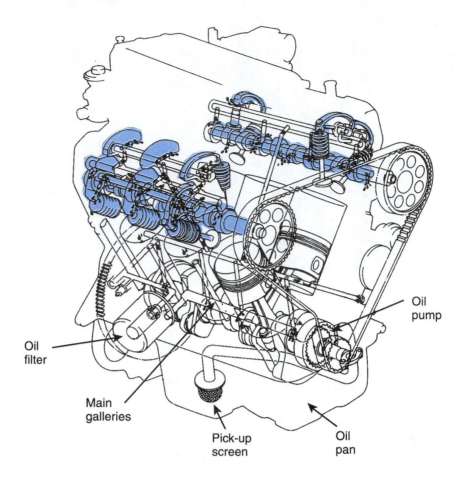

Oil
pump

Oil
filter

Main
galleries

Pick-up
screen

Oil
pan

Figure 8-51 The highlights show some of the splash lubricated areas. (Courtesy of Nissan North America)

Shop Manual
pages 185–190

Fuel and Air Systems

The fuel and air systems are designed to deliver fuel and air in given proportions at any engine speed or load. Improper fuel/air mixtures will cause a no-start or low-performance conditions. Fuel mileage will be lowered and harmful exhaust emissions will increase.

Mechanical Fuel Delivery Systems

Fuel is delivered to the mixing unit by a fuel pump. A mechanical pump is used on most *carburetor* systems. The pump operates from a camshaft-driven push rod. Its fuel delivery pulses on and off as the camshaft turns. The carburetor is the mixing unit, which has a reservoir or bowl to hold sufficient fuel. The bowl is required due to pump pulses and the pump's inconstant pressure and volume. The mechanical pump delivers fuel to the carburetor on each pump stroke.

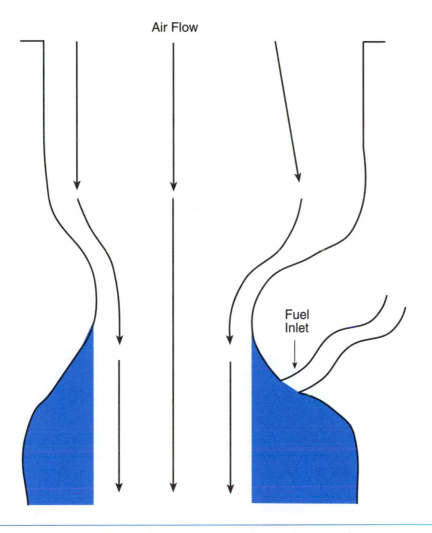

Air Flow

Fuel
Inlet

Figure 8-52 Low pressure is created by air moving rapidly past the recessed area.

The carburetor has various internal passageways to deliver fuel into the airflow moving through its **venturi** (Figure 8-52). The venturi is a large, vertical air passage near the center of the carburetor body. It opens directly into the top of the intake manifold. As the air flows past a narrow portion of the venturi, a low-pressure area is established at the point where the venturi widens again. In the bowl, atmospheric pressure pushes the fuel into the low-pressure area to mix with the air. The fuel passages are sized to deliver a certain amount of fuel depending on the volume of air moving through the carburetor.

Mechanical fuel systems are almost non-existence in modern vehicles. However, there are vehicles and equipment in use today that rely on this type of fuel system.

Electronic Fuel Injection (EFI)

The electronic **fuel injection** (EFI) system requires an electric fuel pump (Figure 8-53). This type of pump delivers fuel at a constant pressure and volume. The fuel may be delivered to injectors in a throttle body unit similar to the carburetor or individual injectors at each cylinder. Throttle body injection has some of the same problems as the carburetor system. However, the pressure to the injectors can be regulated with an engine load-sensing fuel regulator to ensure a nearly constant air/fuel ratio at all engine speeds and loads.

Some vehicles, most notably Ford, use a low-pressure supply pump in the tank and a high-pressure pump mounted on the frame rail. Most vehicles have one electric pump mounted

During the 1980s and early 1990s, many carburetors were equipped with electronic devices to better control the air/fuel mixture. A *venturi* is a vertical air passage near the center of the carburetor's body. As air passes through, a low-pressure area is established which helps mix fuel and air.

Some *fuel injections* are known as "multi-point fuel injection (MFI)."

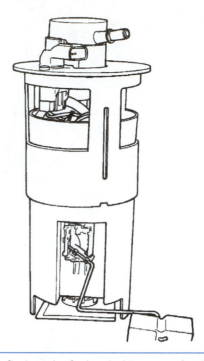

Figure 8-53 This fuel pump fits into the fuel tank. (Courtesy of DaimlerChrysler Corporation)

inside the tank. There are two critical concerns with electronic fuel delivery maintenance. This type of system depends on a constant pressure and the volume of fuel. A clogged or restricted filter can cause numerous problems with engine operations. Sometimes, the filter is forgotten since some vehicle manufacturers do not require a typical service for 100,000 miles. The technician should keep this in mind when the vehicle comes in for an oil change. If the fuel filter is factory-installed and the vehicle mileage is over 30,000 miles, it may be best to question the owner on the history of the fuel filter.

A second and potentially more expensive problem is running the fuel tank dry. An in-tank electric pump is cooled and lubricated to some extent by the fuel passing through it. A pump running without fuel in the tank can be damaged, but this may not become obvious until a later date when it fails. If a fuel pressure problem develops and the filter is good, ask the owner if he or she knows whether or not the vehicle ran out of gas recently. Sometimes, a simple question and answer can resolve a complex problem, and the owner is usually the best source of vehicle history.

Most modern EFI systems have individual injectors for each cylinder. The injectors are mounted in the intake runners at the intake valve or into the cylinder head (Figure 8-54). The injectors are opened on command from the PCM for a specific period of time depending on engine speed, load, temperature, and other factors. The constant fuel pressure and volume ensure that the air/fuel ratio meets the operating condition of the engine.

The fuel is injected over the open intake valve on most vehicles. However, a new system called *direct injection* has the fuel sprayed (injected) directly into the combustion chamber near the intake valve. This provides a better burn and cleaner emissions since the fuel is injected into the air flowing past and around the intake valve. Some direct injection systems use an engine-mounted injector pump to pressurized the fuel. Direct injection with an injector pump is the fuel system used on diesel-fueled engines.

Air for an EFI system is drawn into the engine in a manner similar to the carburetor. However, dirty air and fuel filters affect EFI systems more than a carburetor system.

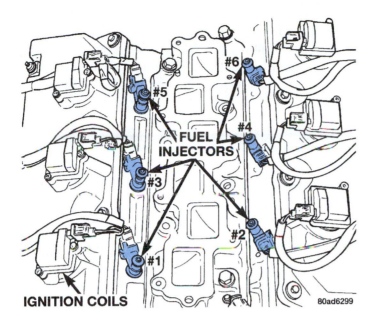

Figure 8-54 The injectors shown are for a multi-port fuel injection system. There is one injector per cylinder. (Courtesy of DaimlerChrysler Corporation)

Controls for the electrical fuel pump are the ignition switch, pump relay, and the PCM (Figure 8-55). Some vehicles use a separate switch in series with the relay and pump. This switch may be referred to as a *rollover* or *inertia* switch. Its purpose is to cut power to the pump if the vehicle is involved in an accident. Other systems use an oil pressure switch or engine speed sensor to signal the PCM that the engine is not running or oil pressure has been lost. General Motors uses an oil pressure sensor to stop the pump if oil pressure is lost. A bypass circuit in the fuel pump relay will supply power to the pump during starting. Like the starter circuit, an anti-theft circuit may be incorporated into the fuel pump circuit.

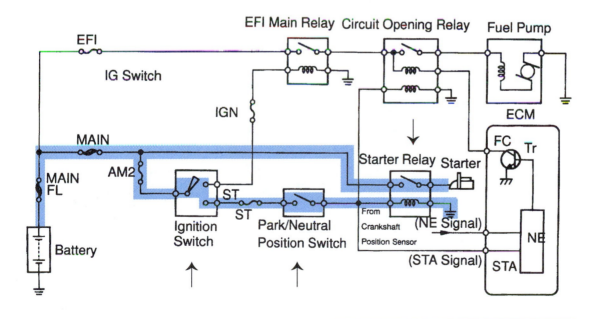

Figure 8-55 The arrows indicate the starter circuit control devices. The highlight shows the running current. (Reprinted with permission)

The best air/fuel mixture with current engine technology is 14.7 parts air to 1 part fuel, or a 14.7:1 ratio. A cold engine will need a rich mixture of about 13:1 for starting. A warm engine can operate on a leaner mixture of about 15:1 ratio or higher. The volume or amount of intake air depends on engine displacement, the fuel system, and the mechanical condition of the engine. A dirty air filter can enrich the fuel mixture to the point of flooding the engine by restricting the air intake.

Ignition Systems

The ignition system provides the spark to ignite the air/fuel mixture. The system consists of controls, an ignition coil, spark plugs, and spark plug wires. The heavy current circuit for the spark plugs is called the *secondary circuit*. A mechanical *distributor* or the PCM controls the system. The distributor is geared to the camshaft. The distributor shaft turns a rotor inside the cap (Figure 8-56). The cap has an internal ignition coil terminal and a terminal for each spark plug. The rotor points at each spark plug terminal once in each revolution. If the coil is discharging, the current moves through the cap's coil terminal, travels along the rotor's conductor, and bridges the gap between the rotor tip and the nearest spark plug terminal. The distributor has an advanced mechanism that adjusts timing of the spark based on engine speed and load.

Distributor systems have gone the way of carburetors. They have been replaced with electronic systems that are faster and more accurate in controlling ignition spark. But, like the carburetor fuel system, there are many vehicles and equipment still in use with the distributor ignition system.

The newer systems employ electronics to control the ignition spark. The distributor has been replaced completely with an electronic package that determines crankshaft and camshaft position, engine load, engine temperature, and **engine speed** to decide the point of ignition within each individual cylinder (Figure 8-57). This same data is also used to fire the fuel injectors and vary the time of injector opening.

Shop Manual
pages 195–199

Diesel fueled engines use the heat of the compressed air to ignite the fuel. Diesel engines do not have an electric ignition system.

The ignition timing must be advanced because the speed of the pistons increases as the *engine speed* increases.

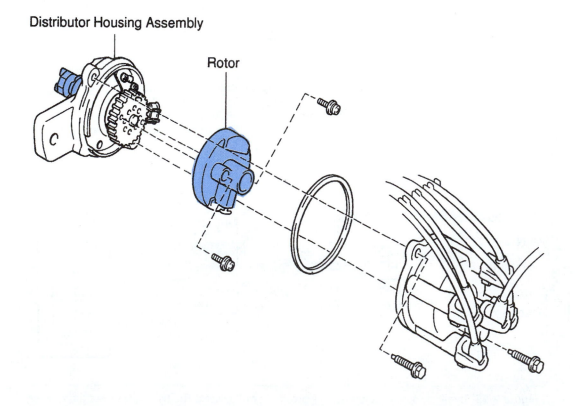

Distributor Housing Assembly

Rotor

Figure 8-56 The distributor shaft is driven by a camshaft (left). The rotor (right) directs current to the proper spark plug. (Reprinted with permission)

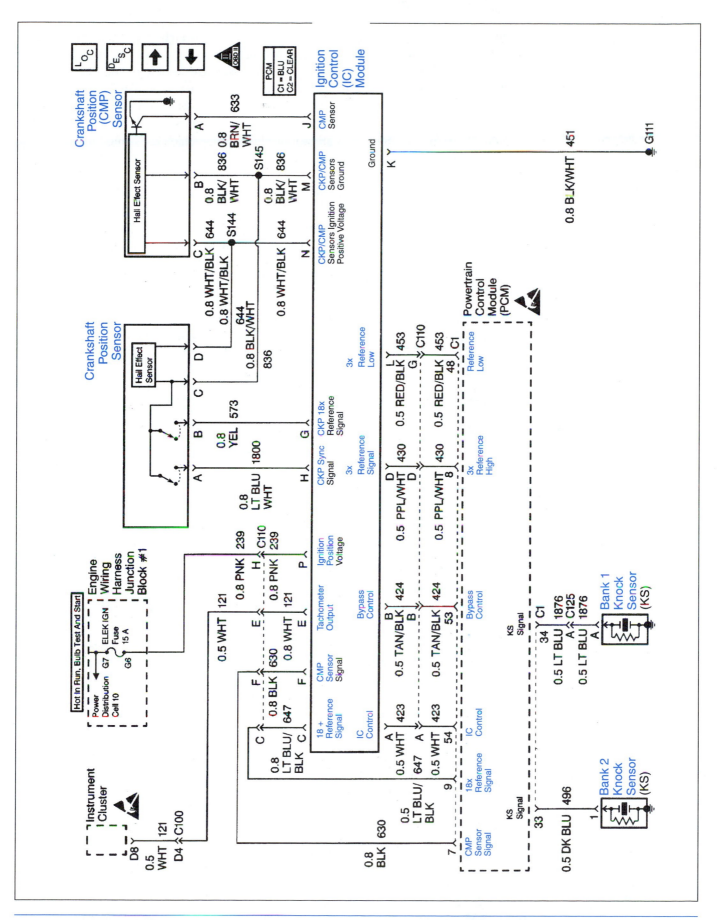

Figure 8-57 Note the sensors and control circuits in this ignition system. (Courtesy of Chevrolet Motor Division, General Motors Corporation)

The PCM performs all the calculations necessary based on engine and transmission sensor input. This provides maximum fuel mileage, cleaner emissions, and more available power for any given engine application. Some of the vehicle **sensors** and actuators are shown in Figure 8-58. Some of the terms and explanations listed there may be unfamiliar, but they are only offered at this point as an indication of the electronics used on an engine. Also, this figure does not list all of the electronic devices, and some manufacturers use different terms for the same type device. Some of the devices are not used on some vehicles. Detailed information is available in the *Today's Technician* series, specifically *Automotive Engine Performance* and *Automotive Computer Systems*.

Delivering current to the spark plug is similar to the mechanical ignition system. The major mechanical difference is a separate **coil pack** for each pair of spark plugs or a coil for each spark plug (Figure 8-59). On the paired system, each time a coil pack fires, it ignites two separate spark plugs. One of the spark plugs will fire a compression stroke and generate the controlled burn to power the piston. The second plug is on the exhaust stroke and acts as a waste spark. There may be very little fuel left in that cylinder, but, primarily, the spark is used to

Sensor/Actuator	Information	Determines
Throttle position (TPS)	Throttle opening	Driver's speed intention
Oxygen sensor (O$_2$)	Oxygen in exhaust gases	Air/fuel mixture ratio
Manifold absolute pressure (MAP)	Pressure in intake manifold	Engine load
Mass air flow (MAF)	Air flow in grams per minute	Engine load (may be used with or in place of MAP)
Engine coolant temperature (ECT)	Engine temperature	Engine temperature
Air charge temperature (ACT)	Intake air temperature	Intake air temperature
Vehicle speed	Vehicle speed	Vehicle load (may be used to send data to speedometer)
Overdrive switch	Engages/disengages overdrive	Provide better fuel mileage
Torque converter clutch (TCC)	Engages/disengages lockup clutch	Locks/unlocks engine output directly to driveline
P/N; clutch switch	Gear or clutch position	Allows engine to be started
Brake light switch	Brake on or off	Turns off cruise control, activates antilock brakes, vehicle is to be slowed
Injector	Fuel delivery	Injects fuel into intake or cylinder
Idle air bypass (IAC)	Air control during idle speed	Controls engine idle speed by adjusting air flow
Cooling fan	Cools coolant	Draws air over radiator fins
Air condition clutch	Drives air conditioning compressor	Cool air to passenger compartment

Figure 8-58 The sensors listed here are standard in most vehicles, but they constitute only a small portion of the electronic devices used on today's vehicles.

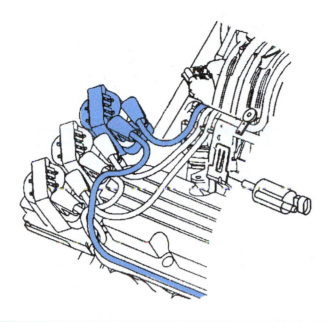

Figure 8-59 This set of coil packs is for a V-6 engine. Note that there are two spark plug wires per coil. (Courtesy of Chevrolet Motor Division, General Motors Corporation)

discharge the coil pack. The very newest systems use a coil per spark plug. The coil pack is mounted where the spark plug wire would normally be attached to the plug (Figure 8-60). This system eliminates the distributor and spark plug wires. It also allows very precise timing of the ignition spark.

Diagnosing an electronic ignition or fuel system can be done accurately as long as the technician understands how the system works. With the systems produced in the mid-1990s

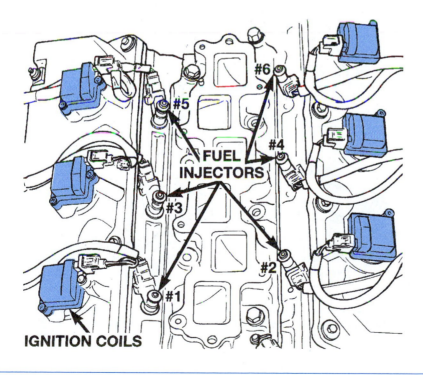

Figure 8-60 This type of coil-on-plug ignition system is replacing the coil pack. (Courtesy of DaimlerChrysler Corporation)

and later, electronic diagnostic equipment can tell the technician almost everything that is happening in the two systems by connecting one cable between the equipment and the vehicle. However, if the technician does not understand the operation of the systems, all of that data is worthless.

Cooling Systems

Horsepower is produced from the expanding gases in the combustion chamber. High temperature is created and must be dissipated. Some of the heat is absorbed by the engine's lubricant and carried to the oil sump. However, most of the engine heat is absorbed by the cooling system.

A crankshaft-driven **water pump** assists in moving cool coolant into the engine block. The coolant circulates through passageways in the block and cylinder head. Heat from combustion and friction is transferred to the coolant. The coolant leaves through a thermostat housing near the top of the engine. The thermostat is a temperature-sensitive valve mounted at or near the highest and hottest part of the engine (Figure 8-61). Some vehicles have the thermostat mounted at the top entrance into the radiator. During cold and warm-up conditions, the thermostat remains closed, forcing the coolant to circulate back into the engine block. By not cooling the coolant, the engine heats quicker. Most thermostats are fully open at about 190 degrees Fahrenheit (88 centigrade). As the temperature rises, coolant is allowed to flow through the upper hose into the top of the radiator.

Shop Manual
pages 199–201

The *water pump* may be mounted behind the timing belt cover, on the timing chain cover, or to the side of the block on a bracket.

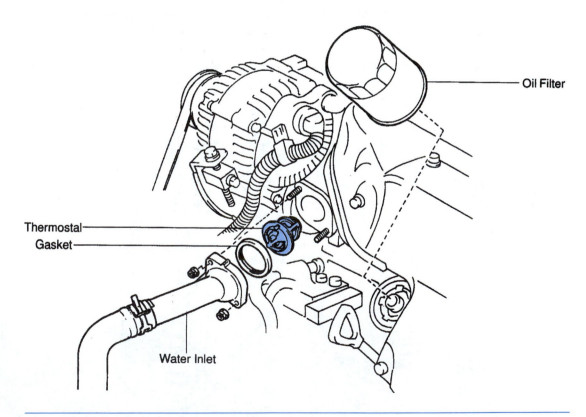

Figure 8-61 This thermostat is mounted slightly below the top of the engine block. (Reprinted with permission)

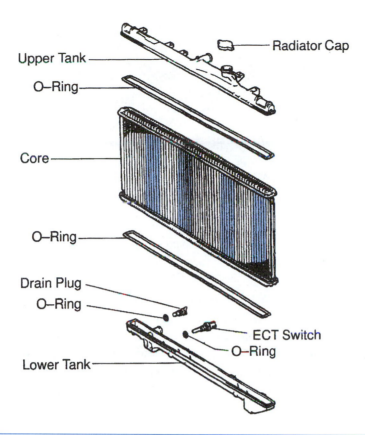

Figure 8-62 This is a vertical flow radiator. Most radiators are mounted at the front of the vehicle. (Reprinted with permission)

The radiator is either a cross flow or a vertical flow. The two types are very similar except for the direction of coolant flow (Figure 8-62). The coolant moves through small tubes down or across to the radiator's exit. The tubes are finned to expose as much surface area as possible to the passing airflow. A fan draws air from the front of the vehicle over the radiator fins and exhausts the air over the engine block. As the coolant exits the radiator through the lower hose, the water pump picks it up and pushes it into the engine block.

Most cooling fans are driven by electric motors controlled by a cooling fan switch mounted in a coolant passage or the PCM (Figure 8-63). When overheating is sensed, the switch or PCM closes the fan relay. This control of the fan's operation based directly on engine temperature reduces the load placed on the engine and saves fuel by using the fans only when necessary. Some vehicles use two fans, which may be referred to as cooling fans 1 and 2 or as a cooling fan and an air-conditioning fan. Since operating the air conditioning creates more heat in the engine, switching on the air conditioner also switches on a cooling fan. If only one fan is present, it will operate while a two-fan system will have one of the two working. It should be noted that under extreme, overheating conditions, the PCM could switch on both fans while switching off the air conditioning system.

Exhaust Systems

The original purpose of the exhaust was to clear the combustion chamber and quiet the noise of the engine. That same purpose remains but some components of the exhaust system are now used to clean the exhaust gases of harmful emissions.

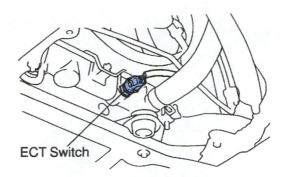

ECT Switch

Figure 8-63 This engine coolant temperature (ECT) switch sends temperature information to the PCM. The PCM controls the cooling fan, temperature gauge, and warning light. (Reprinted with permission)

Mufflers and resonators direct the exhaust through internal tubes and baffles to reduce noise.

The exhaust manifold is attached to the cylinder head with a separate opening or port for each cylinder (Figure 8-64). The manifold runners run together at some point shortly after leaving the cylinder head. The *exhaust pipe* extends from the manifold to a muffler (pre-1974) or to a *catalytic converter* (post-1973). **Mufflers** and **resonators** are still used to quiet the exhaust noises but are positioned after the converter. Between the converter, muffler, and resonators is a *tail pipe* made of stainless steel. A section of the tail pipe extends from the last muffler to the rear of the vehicle.

CAUTION: Unless specifically designed otherwise, the exhaust system must extend past the rear axle. Failure to do so may allow exhaust gases into the passenger compartment and cause injury or death.

For the best power and mileage, the exhaust must be tuned for the engine. Failure to exhaust gases quickly causes poor performance, particularly at higher engine speeds. Restricted or too-small exhaust pipes will not clear the combustion chamber for fresh air and fuel. A

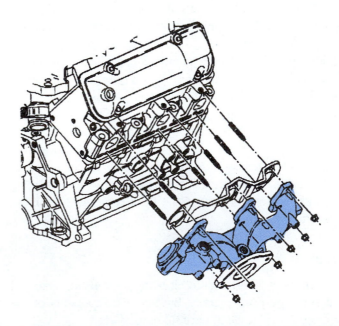

Figure 8-64 Note the short runners from the cylinder head to the exhaust manifold. Each runner is designed to exhaust the same amount of exhaust as every other runner. (Courtesy of Chevrolet Motor Division, General Motors Corporation)

fully-opened system with no noise or pollution devices is not necessarily the best means to achieve the most performance *unless* the system is designed or tuned to the vehicle and engine. In most areas, removal of noise-reducing devices is illegal. It is illegal anywhere in the United States to remove or disable the catalytic converter or other factory-installed emission devices.

Emission Control Systems

Shop Manual
pages 203–205

The burning of fossil fuel produces harmful emissions. Hidden within an engine's exhaust gases are three major chemicals or particles that harm the environment and living things. **Hydrocarbons (HC)** are particles of carbon left after ignition. HC is measured in parts per million of air. **Carbon monoxide (CO)** is a colorless, odorless, tasteless gas that can kill oxygen-breathing animals. **Nitrogen oxides (NO$_x$)** are emitted in the exhaust when combustion chamber temperatures exceed 2,500 degrees Fahrenheit.

At one time, there were six basic emission control systems used on cars and light trucks. Electronics and engine designs have eliminated some of them even though allowable emission limits have been lowered. The first four are still in use today while the last two have been replaced with better technology. In addition to the operating systems, a great deal of emission control is accomplished during the vehicle's engineering and design stage.

HC is the abbreviation for *hydrocarbons*, harmful particles found in vehicle exhaust.

CO is the abbreviation for *carbon monoxide*, a colorless, odorless, tasteless, poisonous gas found in vehicle exhaust.

NO$_x$ is the abbreviation for *nitrogen oxide* that may be in vehicle exhaust.

Design and Construction

The design of the engine and its internal components and operation determine the amount of harmful emissions produced. Even though no design alone can fully eliminate pollutants, the shape of the combustion chamber, valve timing, ignition timing, fuel and air delivery, and a host of other engineering details can significantly reduce emissions. Those details involve items such as the size of the air filter, placement of the spark plug, and even the type of oil filter.

Catalytic Converters

The two-way converter uses two metals, *platinum* and *palladium*, to reduce carbon monoxide and hydrocarbons through **catalyst action.** During the catalyst stage, the exhaust gases are heated to about 1,600 degrees Fahrenheit (878 centigrade) and the unburned fuel is burned off. The hydrocarbon (HC) molecules break apart, forming hydrogen (H) and carbon (C) atoms. This burning or oxidation of the exhaust also causes chemical changes in the carbon monoxide molecule by breaking it into oxygen (O) and carbon dioxide atoms (CO$_2$). Some of the oxygen forms with the released hydrogen to make H$_2$O or water. The best exhaust gas mixture at the tail pipe is H$_2$O and CO. A three-way converter does the same as the two-way but uses **rhodium** to help reduce nitrogen oxides (NO$_x$) (Figure 8-65).

Catalyst action is a chemical reaction between two or more elements. One element (catalyst) causes a chemical change in the other element(s) without changing itself.

Exhaust Gas Recirculating (EGR) Systems

The sole function of the EGR system is to lower *nitrogen oxide* emissions. NO$_x$ is formed in the combustion chamber when temperatures exceed 2,500 degrees Fahrenheit, The EGR valve allows some inert exhaust gases to enter the intake manifold and mix with the air/fuel (Figure 8-66). Since there is very little, if any, fuel in the exhaust gases, they occupy space in the combustion chamber, thereby preventing a full air/fuel intake. The lower amount of air/fuel available reduces the heat of the combustion and holds the chamber temperature low enough to block the formation of NO$_x$.

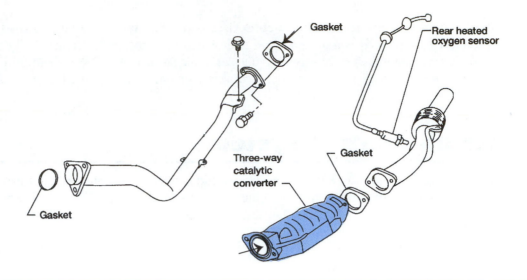

Figure 8-65 This type of converter helps reduces HC, CO, and NO$_x$. (Courtesy of Nissan North America)

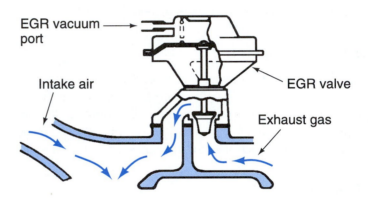

Figure 8-66 The EGR valve regulates the amount of exhaust gases allowed into the intake manifold. (Courtesy of Cadillac Motor Car Division, General Motors Corporation)

The EGR valve is controlled by the PCM which controls the amount of vacuum allowed to the valve's diagram. Some vehicles have a sensor that signals the PCM on the movement of the valve. This provides a feedback so the valve can be used more effectively. Other vehicles use a valve operated by a PCM controlled electrical motor.

Positive Crankcase Ventilation (PCV) Systems

The PCV system captures gases in the oil sump and routes them back into the intake system (Figure 8-67). The PCV valve is connected to a port on the intake manifold. The controlled vacuum draws gases from the top of the oil sump and feeds them into the intake for burning. As the gases are drawn out, a low pressure develops in the sump. Clean, fresh air is drawn down from the air filter into the sump where the air and sump gases are mixed before entering the PCV valve. The most common harmful pollutants found in the sump are unburned fuel (HC) and acids. The oil sump must have some type of pressure relief or the various engine seals and gaskets would leak oil. The PCV relieves the pressure without releasing the pollutants to the atmosphere.

Road draft crankcase ventilation was used prior to 1974. A pipe outside the block extended from the valve cover down to a point near the oil pan. The passing airflow would create a low-pressure area in the pipe and draw fumes from the engine.

Idling or deceleration

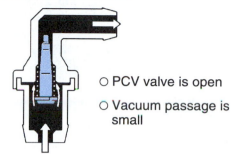

○ PCV valve is open

○ Vacuum passage is small

Normal operation

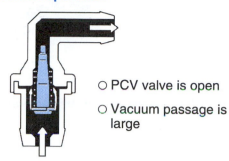

○ PCV valve is open

○ Vacuum passage is large

Figure 8-67 A restricted PCV valve will allow pressure to build within the crankcase and may cause seal and gasket failure. (Reprinted with permission)

Evaporative (EVAP) Systems

As the fuel lowers in the fuel tank, vapors are released that could escape into the atmosphere. To prevent this, the **EVAP system** contains and routes the fuel vapors to the intake manifold. The first component is the fuel tank filler cap. The cap is designed to allow air into the tank but prevents any vapor from escaping (Figure 8-68).

Vapor pickup tubes in the top of the tank are arranged so liquids cannot enter the system, possibly flooding the engine. The tubes route the vapor to a charcoal-filled canister located in or near the engine compartment. The canister acts as a reservoir to store the vapors until the PCM or other control device opens the purge valve. Intake vacuum is applied to the purge valve any time the engine is operating. If the PCM opens the purge valve, fuel vapors are drawn from the canister and mixed with the air/fuel in the intake manifold.

The newest EVAP systems are sealed tighter than the older ones. Sometimes, they cause a problem with engine operation. Loosening the fuel filler cap relieves the pressure and lets the engine operate. This is not a very common problem, but it is a point to remember if a vehicle comes in on a wrecker with a no-start problem.

The *EVAP systems* on 1998 and newer vehicles are designed to prevent engine operation if pressure or vacuum inside a fuel tank exceeds limits. Opening the fuel tank cap will allow the engine to start.

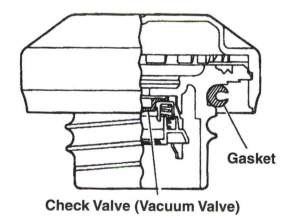

Gasket

Check Valve (Vacuum Valve)

Figure 8-68 This is the only type of fuel cap that should be used with vehicles equipped with the latest EVAP system. (Reprinted with permission)

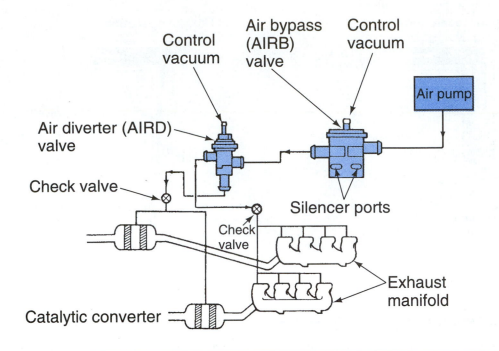

Figure 8-69 This is an older system and has been eliminated on many newer passenger cars and light trucks. (Courtesy of Ford Motor Company)

Secondary Air Injection Systems

The *secondary air injection* works with the catalytic converter. Since the converter needs heat to work correctly, secondary air injection supplies the necessary oxygen to promote burning of any fuel in the exhaust.

A crankshaft-driven, high volume, low-pressure, air pump supplies the air (Figure 8-69). The air is directed by a valve mechanism. The air may be pumped into the exhaust manifold or the exhaust pipe before the converter, into the front of the converter, or diverted to the atmosphere. The direction of flow depends on the type of secondary air-injection system, the type/make/model of the engine, and engine load. The only time air should be diverted to the atmosphere is during engine deceleration. During deceleration, the air/fuel mixture in the exhaust can become very rich. Adding oxygen to the rich exhaust could cause a backfire within the exhaust system and may damage exhaust components. The secondary air-injection system is one of the systems that has been replaced with better engine design on many vehicles.

Thermostatic Air Cleaners (TACs)

The TAC is used to quickly warm the incoming air during cold start and cold drive-away. It was used on carburetors and earlier fuel-injection systems. The air snorkel for the air filter has a vacuum-controlled valve (Figure 8-70). The vacuum is drawn from the intake manifold through a temperature-sensitive switch. The switch was generally mounted on or near the thermostat housing. When the engine is cold, the switch is opened to allow full vacuum to the TAC valve. The valve closes a small door to any outside air. Incoming air has to pass over an exhaust manifold stove, through a flexible tube, and into the air snorkel.

The exhaust manifold heats quickly and warms the air. When the engine reaches about 100 degrees Fahrenheit (38 centigrade), the switch closes off the vacuum and a return spring forces the door closed to warm the air. The TAC is another system that has been deleted because of new air/fuel mixture control technology.

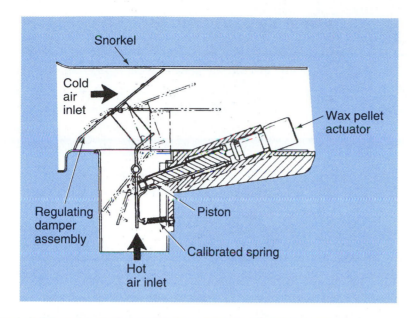

Figure 8-70 Most of the newest car engines no longer use this system, but many car and truck engines produced up to 1994–1995 still have some version of this system. (Courtesy of Chevrolet Motor Division, General Motors Corporation)

The Future of Internal Combustion Engines

Pollution, world population, and nonrenewable resources like gasoline have virtually doomed the vehicle engines used today. More people means more vehicles and the resulting pollution. As long as gasoline (a fossil fuel) is burned, there will be some harmful emissions. A possible replacement is an electrically-powered vehicle with rechargeable batteries. While the driveline and control devices are acceptable at this time, the life of the battery pack is not. A vehicle powered solely by batteries is not feasible with today's battery technology. However, research has provided data that can be used in other power applications. One is a hybrid-powered vehicle using an internal combustion engine in conjunction with an on-board battery pack and electric motor to operate the vehicle. Toyota has a hybrid-powered vehicle on the market in Japan and is expected to introduce it to the United States market in 2000. Other manufacturers have their own vehicles of this type under development, and they can be expected to make them available over the first decade of the twenty-first century. However, there will still be some pollution with the engines used.

Another possible system is the *fuel cell*. The fuel cell uses hydrogen and air to generate electricity, power the motor, recharge the battery, or power the motor only. Hydrogen atoms combine with oxygen atoms to create water and electricity. The water that is exhausted is claimed to be so pure that a person could drink it directly from the exhaust pipe. DaimlerChrysler has indicated that a fuel cell vehicle will be available for on-the-road testing in 2004. Other manufacturers and various government agencies in the United States and overseas are researching and developing alternative fuel engines. The entry-level technician working on today's engines will probably be working on fuel cell engines before his/her retirement.

Summary

❑ The major systems of a vehicle are: electrical, engine, driveline, steering, suspension, climate control, frame, and body.

❑ The internal combustion engine draws in large volumes of air, which is mixed with fuel.

❑ The two-stroke engine produces a power stroke for each revolution of the crankshaft.

❑ Air and fuel are drawn in during the compression stroke of the two-stroke engine.

❑ The rotary engine does not have the stop-go movement of a piston engine.

❑ The four-stroke engine has a distinct function for each stroke.

❑ The four-stroke engine completes intake, compression, power, and exhaust strokes in one cycle.

❑ Valves are used to open and close ports in the cylinder head.

❑ Valvetrains are either OHV or OHC.

❑ OHC engines may have two camshafts per cylinder head.

❑ The lubrication system cools, oils, and cleans the internal components of the engine.

❑ Synthetics and blends are made from chemicals rather than natural petroleum products.

❑ The starting system provides power to rotate the crankshaft which draws air and fuel into the cylinders.

❑ The charging system charges the battery and supplies the electrical current to operate the vehicle.

❑ Fuel is delivered to the intake manifold by a mechanical or electrical pump.

❑ EFI systems require an electrical fuel pump.

❑ Air is drawn into the intake manifold by engine vacuum and atmospheric pressure.

❑ The ignition system supplies the electrical spark to ignite the compressed air/fuel mixture.

❑ The ignition system must be timed to deliver the spark at the proper time in piston travel.

❑ The cooling system removes the heat from the engine block and cylinder head.

❑ In order to get fresh air/fuel into the combustion chamber, an exhaust system must quickly route the exhaust gases from the chamber.

❑ The catalytic converter reduces CO, HC, and NO$_x$ emissions.

❑ The EGR helps reduce NO$_x$ emissions.

❑ HC emissions are reduced using the PCV and EVAP systems.

❑ The secondary air injection system provides oxygen to assist the catalytic converter in reducing emissions.

❑ TAC systems warm the intake air during cold starts and cold drive-away.

❑ Internal combustion engines will probably be replaced with alternatively-fueled engines within the next 30 years or sooner.

Review Questions

Short Answer Essays

1. List the four strokes of an internal combustion engine, and explain the actions that happen on each stroke.

2. List the major systems of an automobile.

3. Explain how the lubrication system works.

4. List the major engine components that are pressure lubricated, and explain why pressure lubrication is used.

5. Discuss the differences between the carburetor and EFI fuel delivery systems.

6. Explain how ignition timing is controlled on a mechanical distributor system.

7. List and describe the actions of the cooling system and its components.

8. Explain the purpose of the exhaust system.

9. Discuss the TAC system.

10. Explain the elements and action of a catalytic converter.

Fill-in-the-Blanks

1. The EGR system is designed to help reduce _____ emissions.

2. The ideal exhaust emission would be _____ and _____.

3. On the compression stroke, the valves are _____.

4. The exhaust system quiets engine noise and helps _____ _____.

5. The ignition system may have one coil per spark plug or one coil for _____ _____ _____.

6. A 13:1 air/fuel ratio is a(n) _____ mixture.

7. EFI systems can change the amount of fuel injected by varying the _____ the injector is open.

8. The secondary air injection system supplies air to promote the _____ action of the catalytic converter.

9. The cooling fan is used to draw air over the _____ _____.

10. The _____ stroke increases the pressure on the air/fuel in the cylinder.

ASE Style Review Questions

1. The ignition system is being discussed. *Technician A* says a distributor is used on all engines. *Technician B* says the PCM calculates the timing of the spark. Who is correct?
   - **A.** A only
   - **B.** B only
   - **C.** Both A and B
   - **D.** Neither A nor B

2. *Technician A* says it is illegal to remove the muffler from the exhaust in all localities. *Technician B* says the catalytic converter was installed in 1974 cars. Who is correct?
   - **A.** A only
   - **B.** B only
   - **C.** Both A and B
   - **D.** Neither A nor B

3. Emission controls are being discussed. *Technician A* says the TAC captures fuel tank vapors. *Technician B* says the EVAP is used to route oil sump vapors to the intake manifold. Who is correct?
   - **A.** A only
   - **B.** B only
   - **C.** Both A and B
   - **D.** Neither A nor B

4. *Technician A* says the secondary air injection uses a pump similar to the one used in the cooling system. *Technician B* says the EGR system helps reduce NO_x pollutants. Who is correct?
   - **A.** A only
   - **B.** B only
   - **C.** Both A and B
   - **D.** Neither A nor B

5. The lubrication system is being discussed. *Technician A* says the camshaft may drive the oil pump.
   Technician B says the pressure relief valve is used to bypass oil around the oil filter. Who is correct?
   A. A only
   B. B only
   C. Both A and B
   D. Neither A nor B

6. *Technician A* says the battery can operate the vehicle's electrical systems for a short time.
   Technician B says the starter motor draws current from the AC generator. Who is correct?
   A. A only
   B. B only
   C. Both A and B
   D. Neither A nor B

7. *Technician A* says the carburetor requires an electrical fuel pump to maintain the fuel level in the carburetor bowl.
   Technician B says the camshaft drives the mechanical fuel pump. Who is correct?
   A. A only
   B. B only
   C. Both A and B
   D. Neither A nor B

8. The engine's theory of operation is being discussed. *Technician A* says valve overlap happens when the valves are not timed correctly.
   Technician B says an inference engine may have serious damage if the timing belt breaks. Who is correct?
   A. A only
   B. B only
   C. Both A and B
   D. Neither A nor B

9. *Technician A* says the air/fuel mixture enters the cylinder on the compression stroke on a two-stroke engine.
   Technician B says oil must be mixed with the fuel on a four-stroke engine. Who is correct?
   A. A only
   B. B only
   C. Both A and B
   D. Neither A nor B

10. Rotary engines are being discussed. *Technician A* says a rotary can produce more power because the piston does not have to start or stop.
    Technician B says the combustion chamber is round. Who is correct?
    A. A only
    B. B only
    C. Both A and B
    D. Neither A nor B

Drivelines

Upon completion and review of this chapter, you should be able to:

❑ Discuss the purpose of the driveline.

❑ Discuss using gears to change torque ratio.

❑ Describe the operation of a manual transmission/transaxle.

❑ Explain the purpose of the clutch.

❑ Explain the purpose of the torque converter.

❑ Describe the general operation of an automatic transmission.

❑ List and explain the components of drive shafts and axles.

❑ Describe the purpose and operation of the differential and final drive assemblies.

Introduction

The engine produces the power that drives the wheels. However, a direct connection between the engine and drive wheels is not a satisfactory arrangement. There has to be a method of delivering high torque to get the vehicle moving and then lowering the engine speed once under way. The transmission and other components of the driveline accomplish that task. For the purpose of this chapter, the term *transmission* refers to a RWD transmission and a FWD transaxle. The term *differential* refers to the RWD differential and the FWD final drive. Specific operational and design differences will be noted for each.

Purpose and Types of Drivelines

Shop Manual
page 217

The driveline delivers the power of the engine to the drive wheels. A transmission is used on most rear-wheel drives while a transaxle is used with front-wheel-drive vehicles. Some vehicles, like the older VW Beetles and Pontiac Fiero, used a transaxle mounted to drive the rear wheels. The 1999 Chevrolet Corvette has a transmission mounted near the rear drive wheels.

The primary difference between the transmission and the transaxle is the method of delivering transmission/transaxle output power to the drive wheels. The transmission uses a drive shaft, which drives the input gear of a differential. The differential turns the torque line 90 degrees and drives the axles. The axles are attached to the drive wheels.

A transaxle has an integral final-drive assembly. The final drive performs the same function as the differential except the transaxle output is delivered directly to the input gear of the final drive. There is no drive shaft in this arrangement.

The transmission and transaxle use gears to change the engine's output torque. A manual transmission requires the driver to select, through mechanical linkage, a gear that is best suited for the engine load. A typical gear setup is reverse, first, second, third, and fourth or direct drive. Some vehicles have a fifth gear.

An automatic transmission uses hydraulics, valves, sensors, and actuators to achieve gear changes. The driver selects a drive range and drives away. A typical drive range is reverse (R), which has one gear ration, and drive (D). Drive usually has three forward gear ratios and often has a fourth or overdrive gear. Each system has a neutral and the automatic has a position for park.

Gears must be used to control the amount of force being delivered to the drive wheels for control and traction.

Gears and Gear Ratios

Different **gears** and gear ratios are used to harness the torque of the engine as it is needed for different loads. The transmission and differential provide the means to do this.

Gears

Gears in a transmission are normally circular-toothed components. The teeth of one gear meshes or interacts with another gear. The diameters of the gears determine if the input torque is increased or deceased. If torque is increased, the speed is decreased and vice versa. A simple gear set consists of two gears working together (Figure 9-1). The gear delivering the torque into the gear set is the *drive gear*. The gear delivering the torque out of the gear set is the *driven*. Gear sets of three or more gears are called *compound gear sets* and may have more than one gear ratio within the set. Gear teeth may be on the outside of the gear, external, or on the inside of the gear, internal. Internal gears are usually found in planetary gear sets. Planetary gear sets are most commonly found in automatic transmissions but may be used for fifth or overdrive in a manual system.

Ratios—Torque and Speed

Ratios are computed by dividing the driven gear by the drive gear. A gear set with a 1-inch drive and a 2-inch driven has a gear ratio of two to one or 2:1. If the drive was delivering 10 foot-pounds of torque, the driven would be delivering 20 foot-pounds of torque to the output. The example here shows a doubling to the available torque (Figure 9-2).

Using the same example, we can show the speed ratio for this gear set. Speed ratios are computed by dividing the drive by the driven, in this case, one is divided by two or .5 to 1. The driven gear will turn half a revolution to each complete revolution of the drive. If the drive turned at 10 revolutions per minute, the driven would only turn 5 revolutions per minute. In some vehicles, this would be low gear: high torque with a low speed.

Direct drive or fourth gear in most vehicles has the transmission output shaft turning at the same speed as the input. The gear ratios between low and high are determined by the gears being used. An overdrive gear would have a larger drive gear than the output gear (Figure 9-3). Assuming that the overdrive drive gear is 2 inches in diameter and the output is 1

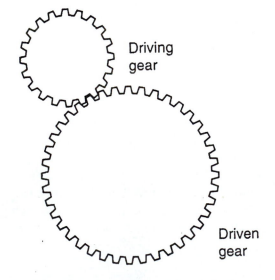

Figure 9-1 A small drive gear and a large driven gear increase the torque and decrease the speed.

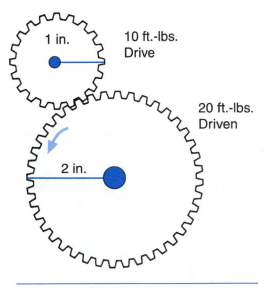

Figure 9-2 The 2:1 gear ratio in this gear set doubles the amount of torque available at the output.

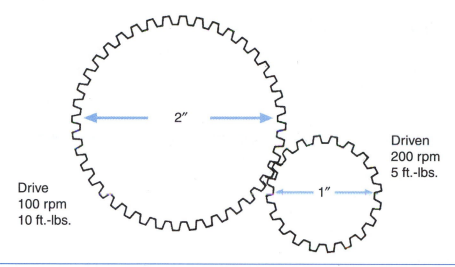

Figure 9-3 Using the small gear as the driven gear will result in higher output speed but less torque.

inch, divide the drive by the driven for a gear ratio of 2:1. In this case, each time the drive gear completed one revolution, the output gear would rotate two times.

Note that the high torque reduces the speed by the same proportion. This is true of any gear set. If the torque is high, the speed will be low. As the vehicle nears cruising speeds, different gear sets are selected to provide a high speed which, in turn, lowers the amount of torque being delivered to the drive wheels and lowers the engine speed.

Clutches

In order to apply the engine power smoothly to the transmission or transaxle, there has to be a device to couple and uncouple the two components. The most common clutch used in a typical vehicle is a manual clutch assembly for a manual transmission and hydraulic clutches within an automatic transmission. Hydraulic clutches will be discussed with automatic transmissions. The manual clutch assembly is made up of three sub-assemblies and the linkages (Figure 9-4). The sub-assemblies are the **pressure plate**, **clutch disc**, and **release bearing**.

Shop Manual
page 218

The *pressure plate* assembly is used to clamp the clutch disc to the flywheel.

The *clutch disc,* once clamped, provides the connection between the engine and the transmission.

The *release bearing* provides a point of contact between a stationary component (linkage) and a moving component (pressure plate assembly).

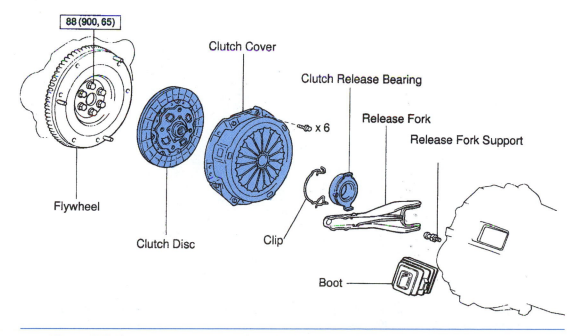

Figure 9-4 The major components of the clutch assembly. (Reprinted with permission)

Pressure Plates and Flywheels

The *flywheel* is driven by the starter motor during cranking to spin the crankshaft.

Bolted to the rear or output end of the crankshaft is a heavy, balanced **flywheel**, which is considered to be part of the engine assembly. According to one of Newton's Laws of Motion, anything in motion tends to stay in motion. The flywheel, once rotating, tries to keep rotating. This helps rotate the crankshaft and tends to reduce the vibrations generated by the start/stop movements of the pistons. The flywheel also provides a solid crankshaft-driven platform to mount the clutch's pressure plate (Figure 9-5). Some manufacturers refer to the pressure plate as the clutch cover. The flywheel is machined smoothly on its outward or rear face.

A steel frame supports the pressure plate, springs, and release levers or fingers (Figure 9-6). The frame is bolted to the flywheel. The pressure plate is finely machined on the flywheel side. High-tension springs apply force to hold the machined side firmly against the flywheel. This force clamps the clutch disc between the flywheel and pressure plate. The clutch is now engaged and engine power is transferred to the clutch disc.

Release levers extend from pivot points along the outer edge of the frame up and toward the center (Figure 9-7). If force is applied against the inner tips of the levers, the levers' other ends pull the pressure plate away from the flywheel and clutch disc. The clutch disc spins free and no power is transferred. The clutch is disengaged.

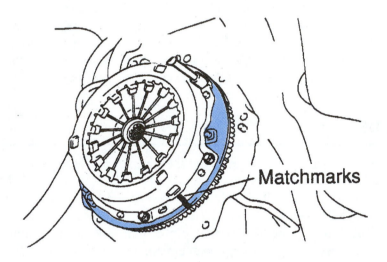

Figure 9-5 The pressure plate assembly is bolted to the flywheel. ((Reprinted with permission)

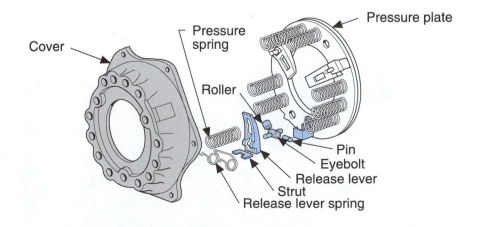

Figure 9-6 The pressure plate is moved rearward (disengaged) by the levers and clamped forward (engaged) against the flywheel by springs.

232

Figure 9-7 The release levers pull the pressure plate from the flywheel, thereby releasing the clutch disc. The machined flywheel side of the pressure plate is visible in this photo.

Clutch Discs

The **clutch disc** is an assembly of springs and friction materials (Figure 9-8). There are two friction areas, one on each side of the plate. Between the two opposing friction areas are cushion springs that help absorb the shock of engagement. Mounted around the center of the plate are torsion springs. The torsion springs are designed to reduce the sudden rotation impact of engagement. In the center of the plate is a splined hole that fits the splines of the transmission's input shaft. When the clutch is engaged and the clutch disc is rotating with the flywheel, power is transferred to the transmission at this point.

Material on the *clutch disc* is similar to the material used on brake pads and shoes.

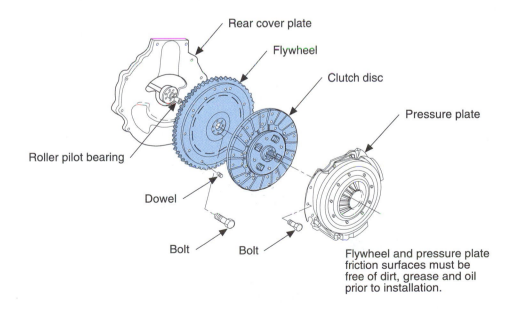

Rear cover plate

Flywheel

Clutch disc

Pressure plate

Roller pilot bearing

Dowel

Bolt   Bolt

Flywheel and pressure plate friction surfaces must be free of dirt, grease and oil prior to installation.

Figure 9-8 The friction material is laid at an angle to the radius of the clutch. (Courtesy of Ford Motor Company)

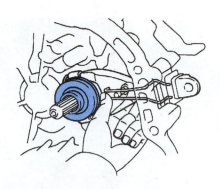

Figure 9-9 The release bearing applies the operator's force to the release levers to disengage the clutch. (Reprinted with permission)

Release and Pilot Bearings/Bushings

The release bearing, like all bearings, provides a point of contact between stationary and moving components (Figure 9-9). The outer shell of the bearing snaps into a clutch fork and moves forward and backward on a guide extending from the transmission. The bearing shell does not rotate. The rotating portion is mounted within the shell and rotates when it contacts with the release levers. When the clutch is engaged, most release bearings totally break contact with the release levers and spin to a stop. Other systems use a continuously-running bearing that maintains contact with the release levers at all times. This eliminates any free movement in the linkages. The bearing is sealed and permanently lubricated.

A **pilot bearing** or **pilot bushing** is installed in the rear of the crankshaft. The bearing or bushing is used to support the forward end of the transmission's input shaft. The pilot bearing is sealed and permanently lubricated. A pilot bushing is made of bronze and needs no lubrication.

A pilot bearing or bushing is not used with torque converters.

Clutch Linkages

Older cars and many present-day trucks use a manual linkage. The linkage starts at the clutch pedal and extends to the clutch fork at the **bell housing**. The linkage is a system of rods and ball joints that transfer the linear force of the driver's foot through the floorboard and around obstacles to apply a linear force to the fork (Figure 9-10). The fork extends through the side of

The bell housing may be part of the molded transmission case or it may be bolted to the front of the transmission case.

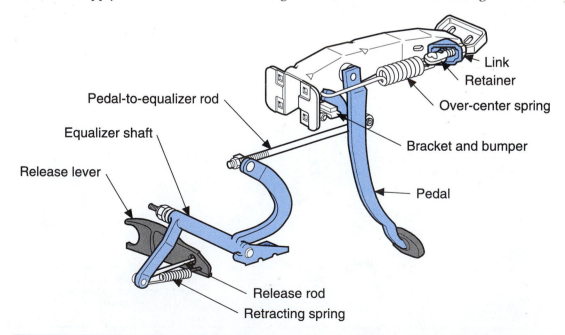

Pedal-to-equalizer rod

Equalizer shaft

Release lever

Link

Retainer

Over-center spring

Bracket and bumper

Pedal

Release rod

Retracting spring

Figure 9-10 A typical shaft and lever clutch control mechanism. (Courtesy of Ford Motor Company)

the bell housing and moves the release bearing. There are usually several springs to ensure that the clutch linkage returns to the disengaged position when the pedal is released. When the driver releases the clutch pedal, the clutch is engaged. Mechanical linkage usually requires periodic adjustment to compensate for clutch wear.

Hydraulically-Operated Clutches

New cars and many light trucks use a hydraulic system to apply the clutch (Figure 9-11). This requires less force from the driver, removes any free travel, and normally uses a continuously-running release bearing. The hydraulically-controlled clutch uses a small master cylinder to apply hydraulic pressure against a piston in the slave cylinder. The **slave cylinder** is mounted to the outside of the bell housing. The piston pushes a small rod that moves the clutch fork. A release spring pulls the slave piston to the release position when the clutch pedal is released. Hydraulic controls do not normally require adjustment for clutch wear. The hydraulic system keeps the release bearing in contact with the pressure plate fingers as the clutch disc wears.

The hydraulic clutch works similar to the hydraulic brakes but at much less pressure. The *slave cylinder* converts hydraulic pressure from the clutch master cylinder to mechanical action to operate the clutch.

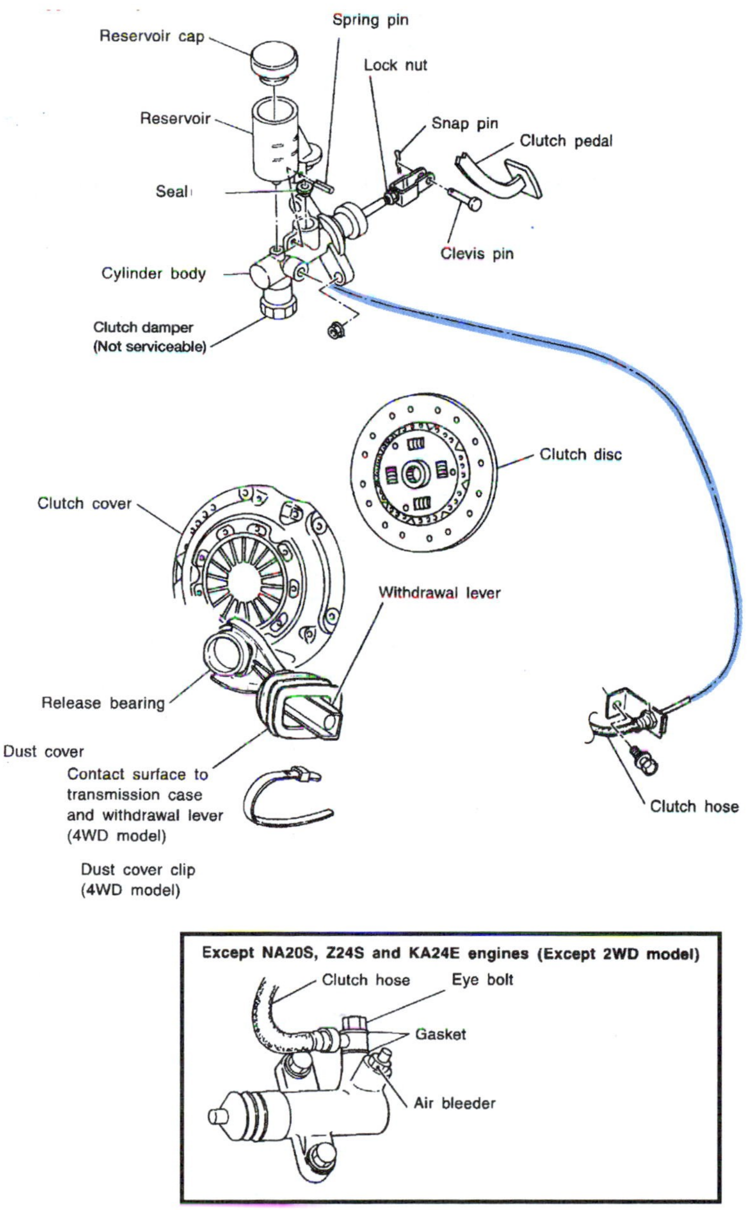

Figure 9-11 A hydraulic clutch with a manual transmission refers to the controlling linkage that operates the release bearing. (Courtesy of Nissan North America)

Some manufacturers use an internal slave cylinder mounted inside the bell housing. The release bearing guide is assembled as a two bore hydraulic cylinder. The inner bore fits around the transmission's input shaft. Inside the outer bore is the slave cylinder's piston shaped roughly like a doughnut. Fluid applied behind the piston directly applies the driver's force to the release bearing.

Clutch Operations

The clutch is released when the driver applies force to and pushes down on the clutch pedal. Through linkage or hydraulics, the release bearing is forced forward against the inner ends of the pressure plate release levers. The levers pull the pressure plate rearward, releasing clamping force on the clutch disc. The connection between the flywheel and the transmission input shaft is broken and no engine power is transferred to the driveline.

When the clutch is slowly released upward, the linkage allows the release bearing to move rearward. As the bearing moves, the release levers allow the pressure plate springs to clamp the clutch disc to the flywheel. Engine power is transferred to the transmission through the flywheel, clutch disc, and input shaft. The clutch should be released slowly to allow the connection to be made smoothly. A sudden release of the clutch pedal may cause damage to any part of the driveline, including the engine.

Shop Manual
pages 217–218, 223–225

The transmission *gear set* is matched to the type of vehicle. Passenger cars use lower ratios (less torque) while trucks use higher ratios (more torque).

Manual Transmissions and Transaxles

The manual transmission and transaxle provides a means for the driver to select a gear ratio that best suites the situation (Figure 9-12). The transmission houses a reverse gear set, neutral, and three, four, or five forward **gear sets**. This chapter will cover the basic five-speed, synchromesh, manual transmission and its components.

Housings

The transmission housing is made of cast iron or aluminum and will usually consist of at least two pieces (Figure 9-13). It is sealed at each end and may have a gasket between individual pieces. *Shifter rails* extend though the housing and are connected to the driver's shift lever (Figure 9-14). All of the gears are located within the housing. The major differences between the transmission and transaxle are the location and shape within the vehicle. Both use a similar gear layout and linkage and operate in about the same manner.

The two pieces of the housing are the *transmission housing* and the *extension* or *tail shaft housing* (Figure 9-15). All of the gears and some of the linkage are located in the transmission housing. The extension housing extends to the rear and covers the tail of the output shaft. Also enclosed in the extension housing is the speedometer drive mechanism. Some four- or five-speed transmissions may have the fourth or fifth gear in the extension housing.

Gear Arrangements

The gear layout is basically three shafts with gears placed along them (Figure 9-16). The **input shaft** and *input gear* form a one-piece unit. It extends from inside the transmission housing and is supported by the input shaft bearing. Behind the input shaft is the output shaft. The forward end of the output shaft is inserted into a pilot bearing mounted in a cavity in the rear of the input shaft.

The nose of the *input shaft* extends through the clutch disc and into the pilot bearing.

236

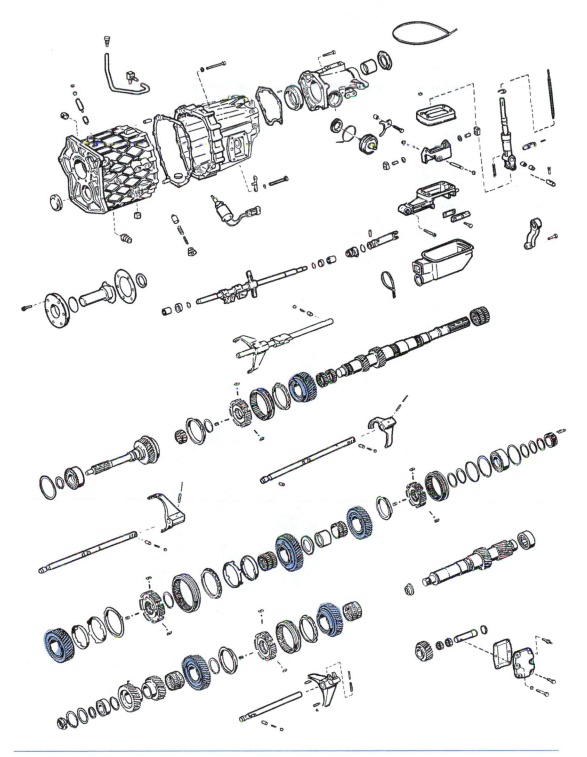

Figure 9-12 A layout of a typical five-speed manual transmission (Courtesy of Chevrolet Motor Division, General Motors Corporation)

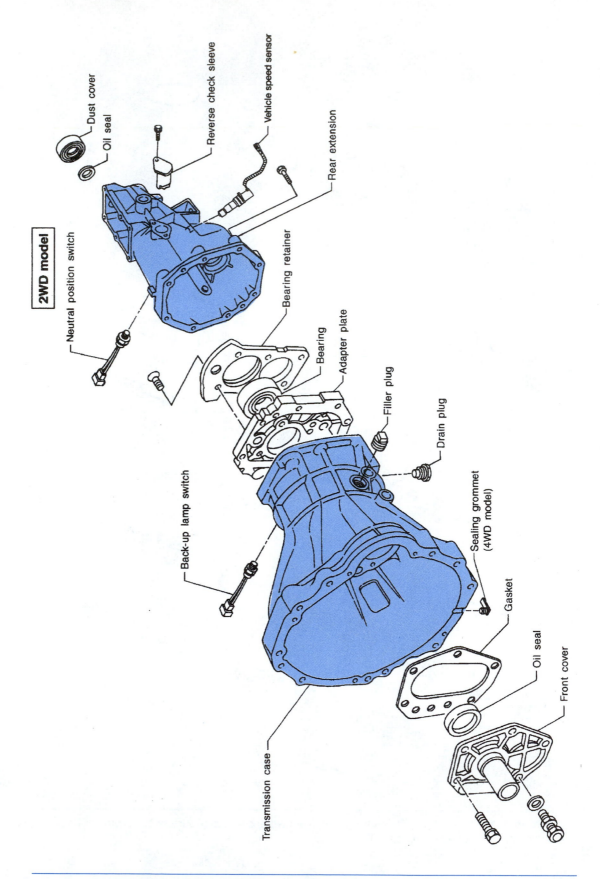

Figure 9-13 There are usually only two major housings in a manual transmission. This one has the transmission case and rear extension. (Courtesy of Nissan North America)

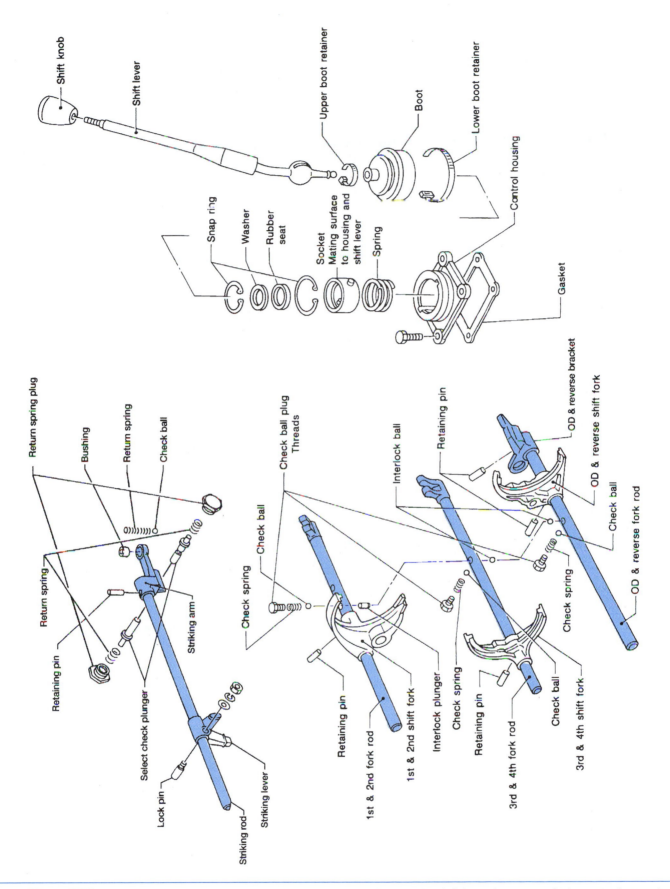

Figure 9-14 The shift rails or fork rods are moved by the operator through the shift lever. (Courtesy of Nissan North America)

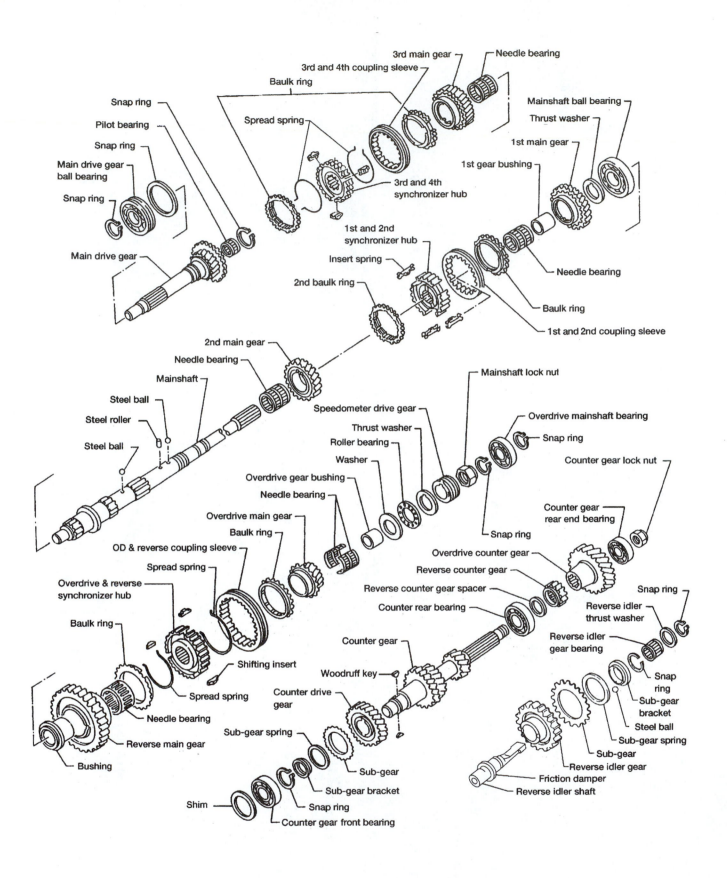

Figure 9-15 All of the gears are located within the transmission housing or the extension housing. (Courtesy of Nissan North America)

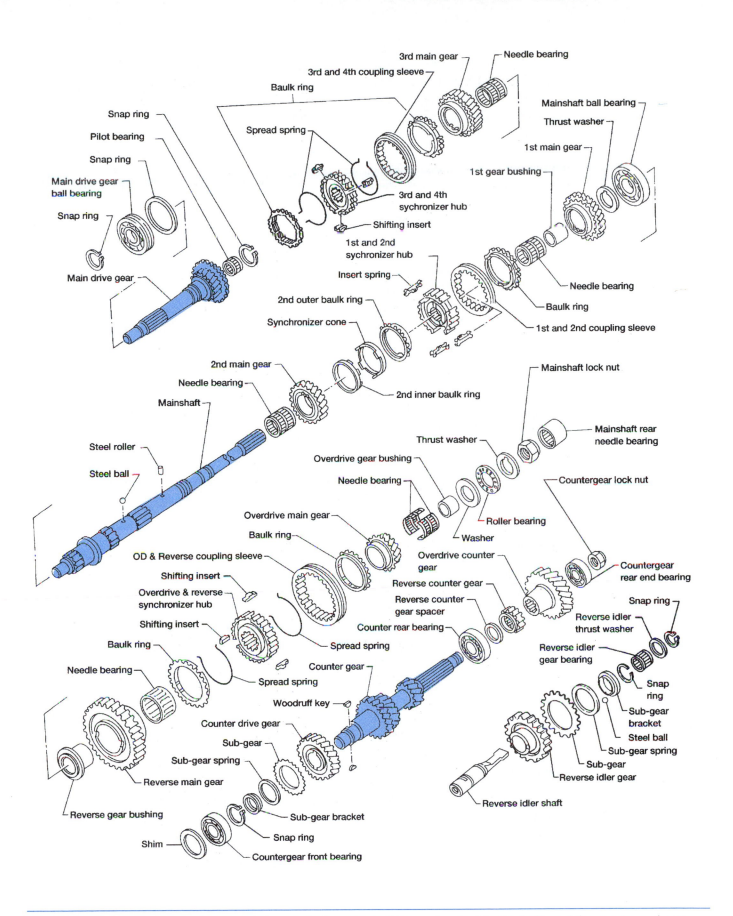

Figure 9-16 The shafts in this transmission are the input (main drive gear), output (main shaft), and the counter or cluster shaft. (Courtesy of Nissan North America)

First and second driven gears are bearing-mounted on the output shaft. Between the first and second and between second and third or direct are the **synchronizers**. Normally, the first driven gear is at the rear end of the shaft but within the transmission housing. (Figure 9-17). The reverse driven gear is directly in front of first or just behind first with second normally just forward of reverse. Some systems have a portion of the first/second synchronizer machined into a reverse driven gear. There is no actual third gear, but direct or third is obtained by locking the output shaft directly to the input shaft. Each forward gear is in **constant mesh** with a gear within the cluster gear. The first, second, and third gears turn freely on the shaft unless locked by the synchronizer. Fifth gear is usually outside the main transmission housing with fifth drive on the counter shaft and fifth driven on the output or main shaft.

Mounted to one side and aligned with the reverse drive and reverse driven is the *reverse idler gear*. The idler gear is a sliding gear in that it must be moved along a shaft to mesh with the two reverse gears. It is not a synchronized gear. The vehicle should be stopped and the clutch disengaged before selecting reverse. Once in place, the idler is driven by the reverse drive gear and drives the reverse driven gear (Figure 9-18). The idler gear rotates in the opposite direction of the drive gear. This causes the input rotation to change direction and the output shaft rotation is reversed.

Between first and second gears and between third and the fourth gears are synchronizer assemblies (Figure 9-19). They are splined to the output shaft and are used to make gear changing easier and quieter. They lock the driven gear to the output shaft. The synchronizer assembly consists of a hub, sleeve, two **blocker rings,** three centering locks, and lock springs. The hub is the portion splined to the output shaft. The sleeve moves along external splines on the hub. The sleeve also has teeth to engage the driven gear. The blocker rings engage the gear first and is used to bring the gear speed to synchronizer speed. The center locks and springs lock the sleeve in neutral and drive the blocker rings. A shifting fork fits over the synchronizer sleeve.

Cluster Gears

The **cluster gear** is a one-piece unit. There are usually five molded gears in the cluster, which is mounted on bearings or a counter shaft (Figure 9-20). The forward (from front of transmission) gear is meshed with the transmission's input gear. The third and fifth gears in the cluster mesh and drive the second and first driven gears on the output shaft. The fourth cluster gear is the reverse drive gear. Notice that the reverse drive and reverse driven are not meshed. Also notice that the entire cluster gear rotates anytime the transmission's input shaft is rotating. This means that all gears meshed with the cluster gears are also turning. The transmission in figure 9-20 shows the gear marked "6" is the fifth drive gear. It spins free on the cluster or counter shaft until it is locked to that shaft with the fifth gear synchronizer. The fifth driven gear is locked to the output shaft, but can only deliver torque after the fifth synchronizer is engaged.

Linkages

The shift lever is operated by the operator and may be mounted on the floor or the steering column. Most new vehicles have the lever on the floor. The movement of the lever is transmitted through mechanical linkages to the shifting fork in the transmission or transaxle (Figure 9-21).

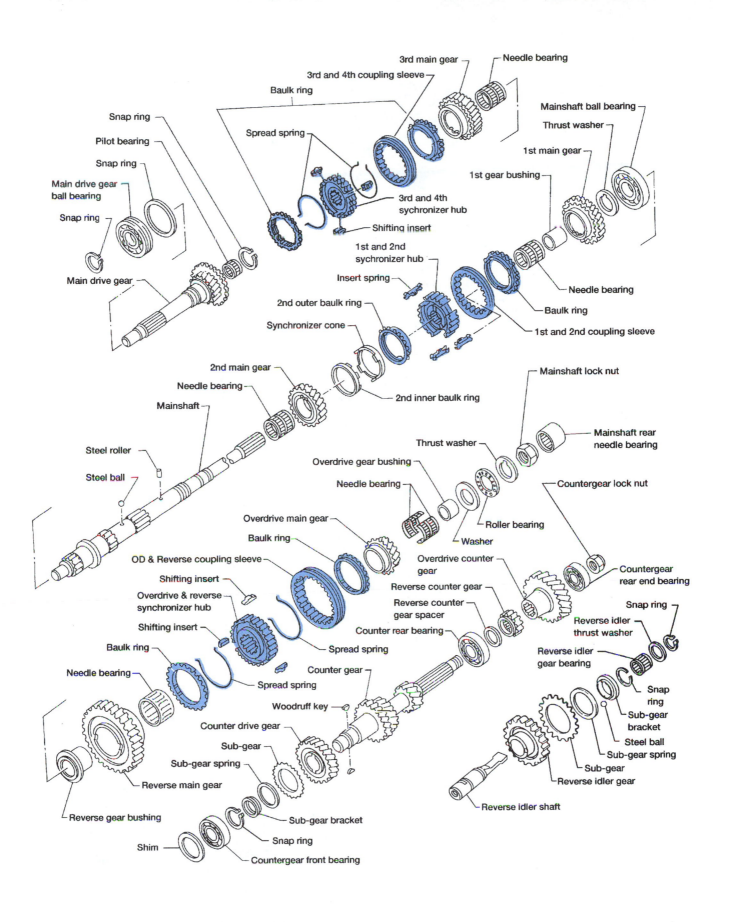

Figure 9-17 Synchronizers are provided for each forward gear. Most synchronizers serve two gears. (Courtesy of Nissan North America)

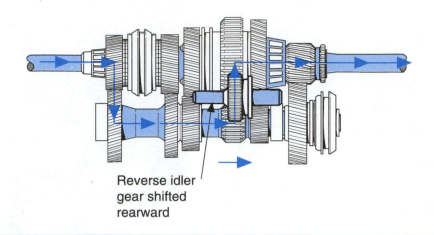

Reverse idler
gear shifted
rearward

Figure 9-18 The power flow in reverse goes from the input shaft/gear, to the cluster/counter gears, to reverse idler, to the reverse driven gear, and to the output/main shaft. (Courtesy of Ford Motor Company)

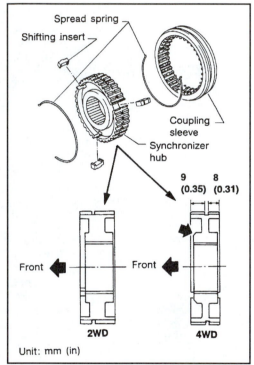

Figure 9-19 A basic synchronizer assembly. (Courtesy of Nissan North America)

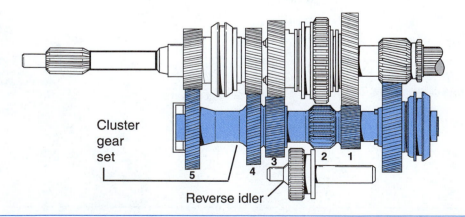

Cluster
gear
set

Reverse idler

Figure 9-20 This cluster gear is a basically a one-piece unit. The set shown has the fifth drive gear (highlight) mounted on the rear of the counter shaft.

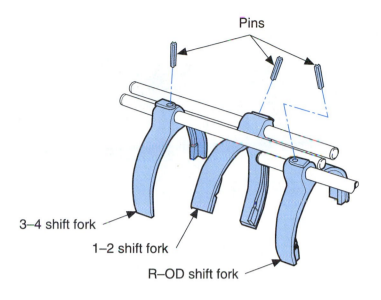

Pins

3–4 shift fork

1–2 shift fork

R–OD shift fork

Forks shown on shift rail

Figure 9-21 Note that each shift rail and fork can shift two gears. (Courtesy of DaimlerChrysler Corporation)

Operation

With the engine running, clutch engaged, and the transmission in neutral, the input shaft is turning and driving the cluster gear. The cluster gear is turning the first, second, and third driven gears (Figure 9-22). Since the synchronizers are not locked to the gears, the output shaft is not being driven. To select a gear, the clutch is disengaged and the shift lever is moved by the operator to the first gear position. The first gear synchronizer sleeve is pushed by the shifting fork into position with the first driven gear (Figure 9-23). As the clutch is slowly engaged, power from the engine is delivered to the transmission's input shaft and gear. The power flows through the input gear and cluster gear to the driven first gear. With the synchronizer in place, the

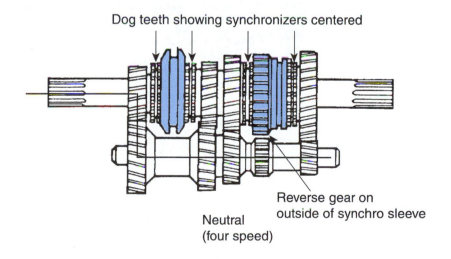

Dog teeth showing synchronizers centered

Reverse gear on outside of synchro sleeve

Neutral (four speed)

Figure 9-22 Even though the first, second, and third driven gears are being spun by the cluster gears, the synchronizers are not engaged. Thus, no power is delivered to the output shaft. (Courtesy of Volvo Car Corporation)

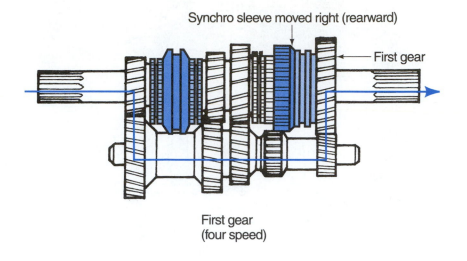

Synchro sleeve moved right (rearward)

First gear

First gear
(four speed)

Figure 9-23 The one-two synchronizer has locked the first driven gear to the output shaft. (Courtesy of Volvo Car Corporation)

power flow continues through the synchronizer assembly and rotates the output shaft. First gear has a gear ratio that provides high torque and low speed. The other forward gears are selected in the same manner. In direct drive or third gear, the 2-3 synchronizer locks the input and output shaft together giving a gear ratio of 1:1. In reverse, the two synchronizers are in neutral and the reverse shifting fork slides the reverse idler gear into position between the drive and driven reverse gears (Figure 9-24).

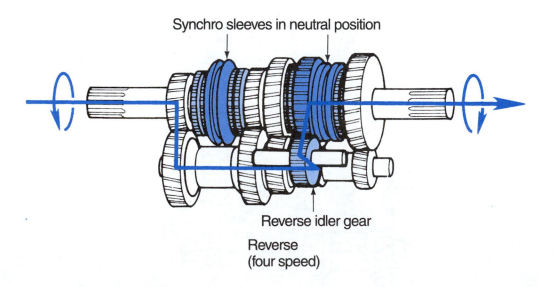

Synchro sleeves in neutral position

Reverse idler gear

Reverse
(four speed)

Figure 9-24 The reverse driven gear is part of the one-two synchronizer hub. Power in reverse is through the cluster to the reverse idler and through the synchronizer to the output shaft. (Courtesy of Volvo Car Corporation)

Torque Converters

The *torque converter* assembly replaces the clutch assembly when the vehicle is equipped with an automatic transmission (Figure 9-25). The converter is a fluid coupling. Automatic transmission fluid is used to transfer and multiply the torque from the engine. There are four subassemblies in the converter, three required for operations and another one to improve fuel mileage. Also, the flywheel may be replaced with a flexplate (Figure 9-26). The flexplate allows for the swelling of the torque converter during operation. Torque converters are almost the same for automatic transmissions and transaxles.

The torque converter is shaped like an inflated balloon that has been squeezed inward on two sides. It has enough weight to perform the same interia action as the flywheel. The front of the converter is bolted to the flexplate so the outer shell rotates with the crankshaft (Figure 9-27). The rear of the shell has an extended lug that slides into the transmission and locks to the transmission pump. The converter subassembly components are sealed within the shell.

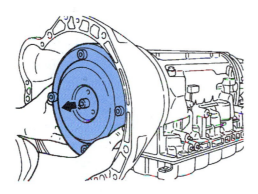

Figure 9-25 The torque converter is located inside the bell housing. (Courtesy of Nissan North America)

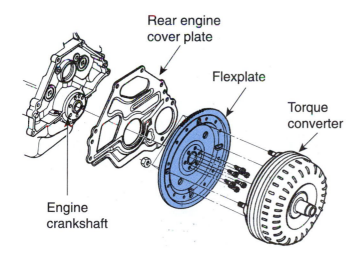

Figure 9-26 The flexplate is designed to allow the torque to expand and contract during operation. (Courtesy of Ford Motor Company)

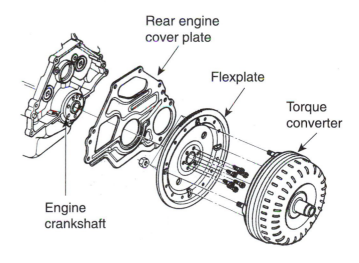

Figure 9-27 The torque converter is bolted to the flexplate which is bolted to the crankshaft. (Courtesy of Ford Motor Company)

Inside the shell are the **impeller**, **turbine**, and **stator** (Figure 9-28). The impeller is made of vanes that are welded to the inside of the shell. They are angled in the direction of rotation. When fluid is in the shell and the engine is running, the impeller directs the fluid against the turbine blades. Engine torque is delivered to the turbine, but since a fluid instead of a mechanical connection is used, some or all of the torque is lost and slipping occurs. If the vehicle is stopped with the engine idling, slipping is the desired condition. The fluid flow is classified as a *rotary flow* (Figure 9-29).

The turbine is bearing-mounted and free of a mechanical connection with the converter's outer shell. The turbine is splined-mated to the input shaft of the transmission. In some makes, this shaft is referred to as the *turbine shaft*. Blades on the turbine face against the rotation and opposite to the impeller vanes. As fluid from the impeller strikes the face of the blades, the turbine is rotated in the same direction as the outer shell but with much less speed and force. At this point, the engine torque is delivered to the input shaft of the transmission. However, the turbine could turn backward if a holding device was not in place.

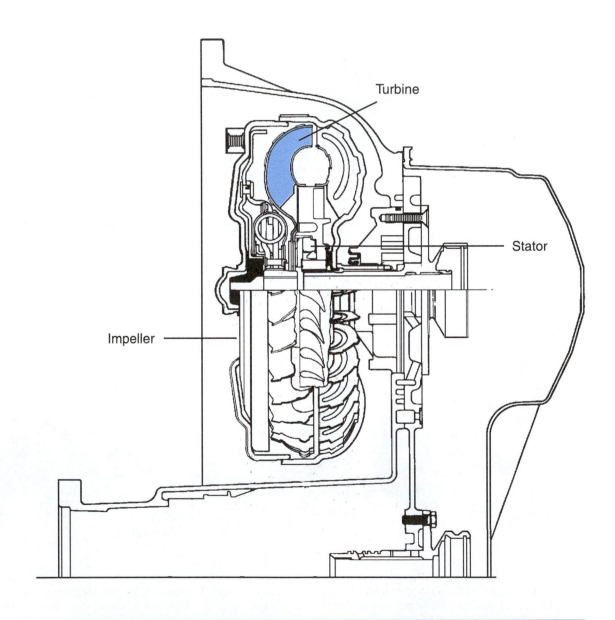

Figure 9-28 A torque converter must have an impeller, turbine, and stator for torque multiplication and creating a fluid coupling. (Courtesy of General Motors Corporation, Service Operations)

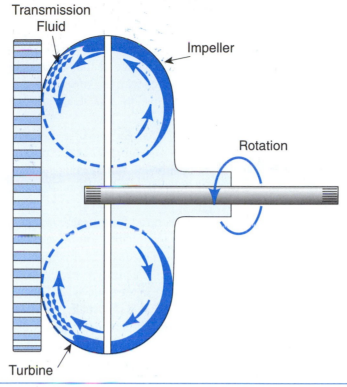

Figure 9-29 Rotary flow causes the turbine to spin.

The stator is the holding device. The stator is a *one-way clutch* that allows the **turbine** to rotate in one direction only (Figure 9-30). If the turbine attempts to rotate backward, the stator locks and the turbine stops moving or reverses its rotation. To accomplish this task, the stator is bearing-mounted and splined to a transmission extension that is similar to the release-bearing

The *turbine* can be forced in reverse if the impeller fluid strikes the backside of the turbine blades.

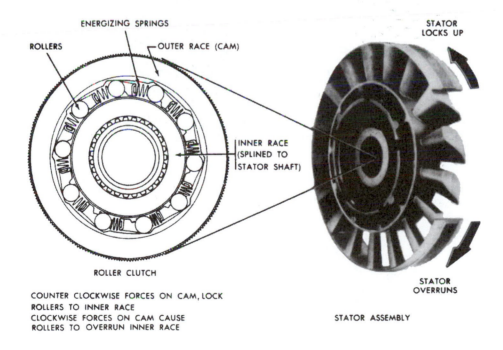

Figure 9-30 The stator will lock if the turbine tries to reverse rotation. (Courtesy of General Motors Corporation, Service Operations)

guide. A one-way clutch is a mechanical lock that is released or locked depending on the rotational direction of the turbine and stator. The outer edge of the stator locks into a cavity on the turbine. Under normal conditions, the stator rotates with the turbine. If the turbine moves backward, the stator locks drop into place and the turbine and stator are prevented from moving by the stationary stator mount.

In addition to its holding task, the stator has blades that redirect the fluid from the turbine. The stator blades are angled in the opposite direction of the turbine blades. As the fluid strikes the stator blades, it is forced back against the turbine blades. This creates higher pressures and increases the torque between the impeller and turbine. The fluid flow at this point is a **vortex flow** (Figure 9-31).

The impeller, turbine, and stator are all required for torque converter operations. The next component may or may not be included in a particular torque converter. It is the **lock-up clutch** (Figure 9-32). The lock-up clutch may be mechanical, hydraulic/mechanical, or electrical/hydraulic. The clutch, when operated, locks the turbine to the outer shell of the torque converter. When this occurs, the turbine and the input shaft are moving at the same speed as the crankshaft similar to direct in a three-speed manual. Usually, the lock-up clutch is not applied until the vehicle is in direct (high) gear and the engine load has decreased. Obviously, if the lock-up clutch was applied at a low speed, there would be no slipping within the torque converter and the engine would lug or stall. Almost every torque converter manufactured since 1988-1989 has a lock-up clutch of some type.

Without the stator and *vortex flow*, there would not be enough torque to power the gear train.

A *hydraulic/ mechanical* one-way clutch uses hydraulic pressure to activate the mechanical locks. An *electrical/ hydraulic* system uses electrical solenoids to control fluid to a hydraulic clutch.

A *lock-up clutch* locks the turbine to the outside of the torque converter and is not applied until the vehicle is in high gear and engine load has decreased.

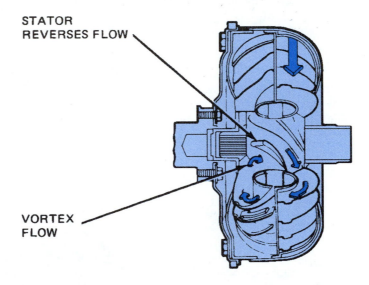

STATOR REVERSES FLOW

VORTEX FLOW

Figure 9-31 The vortex flow multiplies the amount of torque from the impeller by redirecting the fluid. (Courtesy of Ford Motor Company)

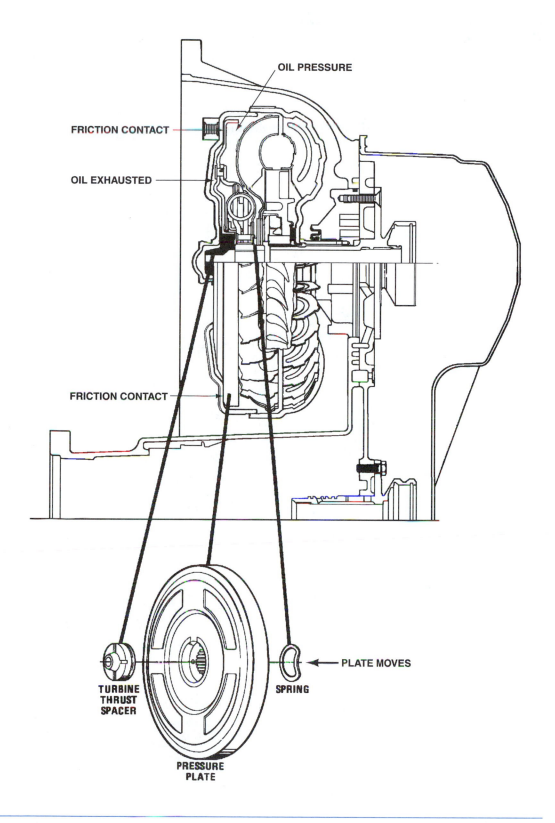

OIL PRESSURE

FRICTION CONTACT

OIL EXHAUSTED

FRICTION CONTACT

TURBINE
THRUST
SPACER

SPRING

PLATE MOVES

PRESSURE
PLATE

Figure 9-32 The torque converter lock-up clutch (TCC) makes a direct connection between the flexplate and transmission, providing a 1:1 gear ratio engine output and transmission input. (Courtesy of General Motors Corporation, Service Operations

Shop Manual
pages 218,
226–230

Some transmissions use three or four clutches in a system, which is based on the *Simpson gear train*.

Automatic Transmissions and Transaxles

Automatic transmissions and transaxles shift up and down through the forward gear ranges using hydraulic clutches, a valve body, and various controls. The driver selects park, reverse, neutral, and a forward range using a shift lever similar to the one use for manual transmissions. A hydraulic pump driven by the outer shell of the torque converter supplies the necessary fluid volume and pressure to operate the transmission.

There are many different transmission designs that are based on the **Simpson gear train**. A basic Simpson system uses two hydraulic clutches, two bands, two planetary gear sets, an overrunning clutch, and a valve body (Figure 9-33). This chapter will discuss the Simpson gear train design since it is basic and is the simplest to understand at this point.

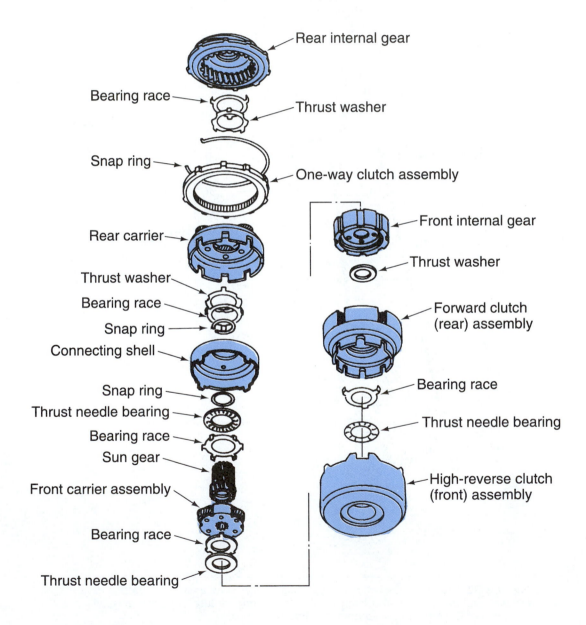

Figure 9-33 The Simpson gear train is the basic gear train for many automatic transmissions. The bands are not shown here. (Courtesy of Nissan North America)

Hydraulic Clutches

A hydraulic clutch or *clutch pack* is composed of a piston assembly, a drum, and a number of steel and friction clutch plates (Figure 9-34). The center hub of the drum fits around a gear or shaft and provides the mounting area for the piston. The piston is a thin, aluminum plate with a seal around the outer and inner edges. The piston is held to the bottom of the drum's bore with return springs and a retainer. Above the piston and within the bore are alternating steel and friction clutch plates. The steel plates have lugs at the outer edge to lock into grooves on the drum. The inner edge of the clutch discs is splined to fit either a gear or shaft. The plates are held in place loosely with a retaining snap ring.

As pressurized fluid is directed behind the piston, it moves forward and clamps the friction and steel plates together (Figure 9-35). This operation is similar to the manual pressure

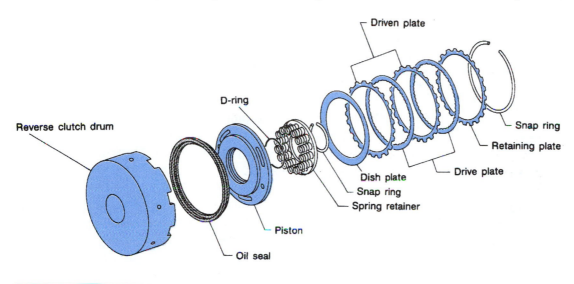

Figure 9-34 Components of a typical hydraulic clutch. (Courtesy of Nissan North America)

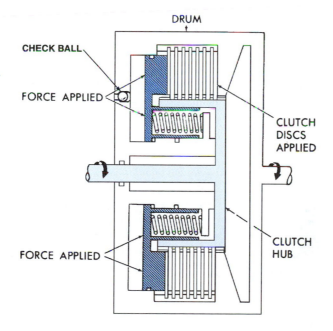

Figure 9-35 The pressurized fluid pumped in behind the clutch piston forces the piston forward (right in the drawing) and clamps the clutch discs. (Courtesy of General Motors Corporation, Service Operations)

plate clamping the clutch's clutch disc to the flywheel. In hydraulic clutches, input power may be driving the clutch discs or the clutch drum. As the plates are clamped together, they begin to turn together and the power flow is directed through the clutch pack from input to output.

Bands

The friction material on the *bands* is similar to the material used on the manual clutch's clutch disc.

Bands are wrapped around the outside of the clutch drum and act as a holding device. The band is anchored at one end while the other end is attached to a hydraulic *servo* (Figure 9-36). When fluid is applied to the servo, the band is applied and prevents the clutch drum from moving. If the clutch is not applied, the power flow passes through the clutch pack without any action. If the clutch is applied when the band is also applied, the gear or shaft splined to the clutch's clutch disc is held and prevented from rotating.

Planetary Gear Sets

The two *planetary gear sets* in a Simpson gear train are used together to obtain the correct gear ratio for a given situation.

A **planetary gear** has three gears. The outer gear, called a *ring gear,* is an internal gear. The gear teeth are on the inside of the ring. Fitted and meshed inside the ring gear is the *planetary carrier.* The planetary carrier is actually a set of three or four pinion gears mounted to a platform or carrier. The carrier is designed to allow the pinion gears' teeth to extend past both edges of the carrier. This allows the gears to mesh with the ring gear around the carrier and the *sun gear* mounted through the center of the carrier (Figure 9-37). The external sun gear extends completely or partially through the planetary carrier and meshes with the pinion gears (Figure 9-38).

Any of the three major gears in this set can be used as the drive or driven. However, in order for the set to work, one of the three must be held in some manner so the other two have a point to rotate around. For a quick example, use the following scenario. The ring gear is mechanically connected to the clutch pack drum. The planetary carrier is splined to an output shaft and an input gear drives the sun. The sun gear is also splined to the clutch discs of the clutch. The band is applied and holds the drum and ring gear in place. The clutch is released freeing the clutch discs. As power is applied to the sun, the sun rotates the pinion gears within the carrier. Since the ring gear is held in place, the pinions gears "walk" around the inside of the ring gear, thereby turning the planetary carrier. The carrier now turns the output shaft. The power flow moves from the input, to the sun, through the planetary gears, to the carrier, and to the output shaft (Figure 9-39).

Since the sun gear is smaller than the carrier, there is an increase in torque and a reduction in speed. The pinion gears are not counted in computing the ratio because they are used as idlers. If the power flow is reversed using the carrier as the input and the sun as output, the torque would decrease and speed would increase.

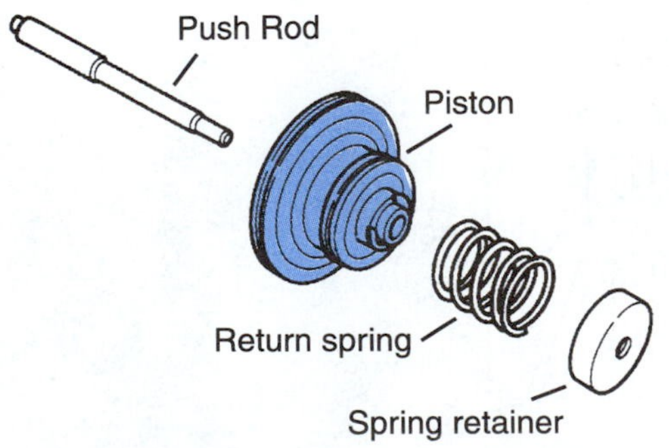

Push Rod

Piston

Return spring

Spring retainer

Figure 9-36 The hydraulic servo applies and releases the transmission's bands. (Courtesy of Nissan North America)

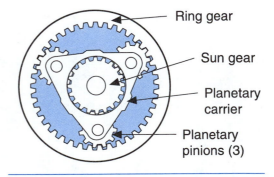

Ring gear

Sun gear

Planetary carrier

Planetary pinions (3)

Figure 9-37 A simple planetary gear set consists of a sun, planetary carrier and pinions, and ring gears.

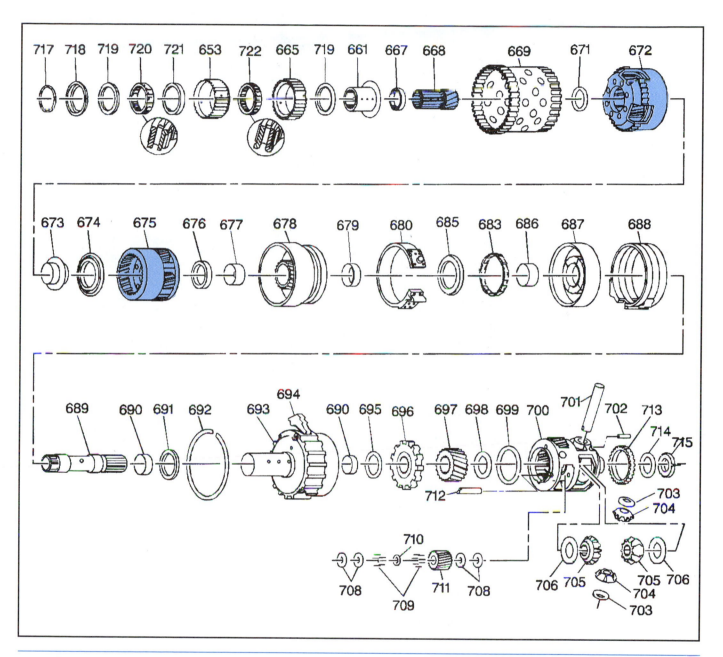

Figure 9-38 A sun gear is the input gear for this General Motors transmission connecting the input sprag clutch (665) to the input carrier assembly (672). (Courtesy of Chevrolet Motor Division, General Motors Corporation)

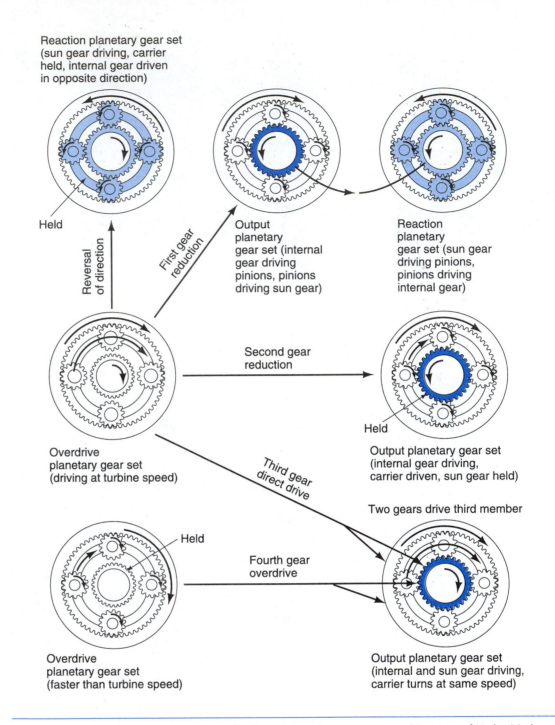

Figure 9-39 Possible gear and speed ratios using a planetary gear set. (Courtesy of Hydra-Matic Division, General Motors Corporation)

Overrunning Clutches

The overrunning clutch in most Simpson gear trains is a mechanical clutch. The inner ring or race is splined to a gear or shaft and the outer race is locked to the transmission housing or some other anchor. There are two basic types of overrunning clutches: *sprag* and *roller* (Figure 9-40). Between the two races there are diagonals that can tilt outward (long) to lock the races or tilt inward (short) to release the races. In a roller clutch, roller bearings are fitted between the races. One race has ramps. Springs hold the bearings in the release position. If the inner race tries to turn backward, the bearings roll up the ramps and lock the two races together, thereby locking the shaft or gear. The overrunning clutch is used in an automatic transmission to redirect the power flow.

256

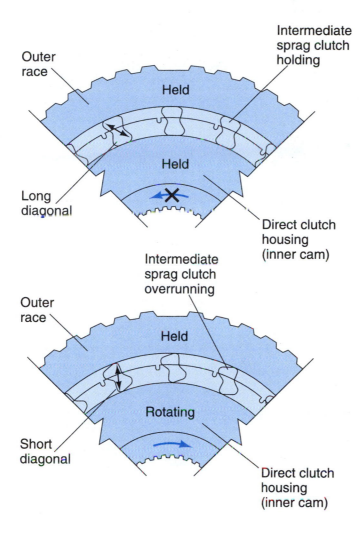

Figure 9-40 This sprag clutch uses diagonal pieces of machined metal to lock the races. (Courtesy of General Motors Corporation, Service Operations)

Valve Bodies

The valve body consists of many valves and fluid passageways (Figure 9-41). This chapter will not attempt to discuss the hundreds of designs and operations of the various valves and valve bodies. We will discuss the spool valve design, the major valves, and the three major controls.

Spool Valves

The *spool valve* is a highly-machined, steel rod that fits tightly and smoothly into a highly-machined bore in the valve body. The rod is grooved at different points along its length. The width of the grooves may be different. The remaining high points on the rod are called *lands* (Figure 9-42). There are fluid passageways entering the valve body bore at different points. With the spool valve inserted into the bore and a land moved directly over a fluid passageway, the fluid is blocked. Moving the spool until a groove lines up with the passageway, fluid flows through the groove and enters another passageway. In this manner, the fluid can be directed to or cut off from clutches, servos, other valves, or a control device. Springs or fluid move the spool valves. The major valves are the 1-2 and 2-3 shift, reverse, pressure regulator, pressure booster, and the manual valve. The operator controls the manual valve.

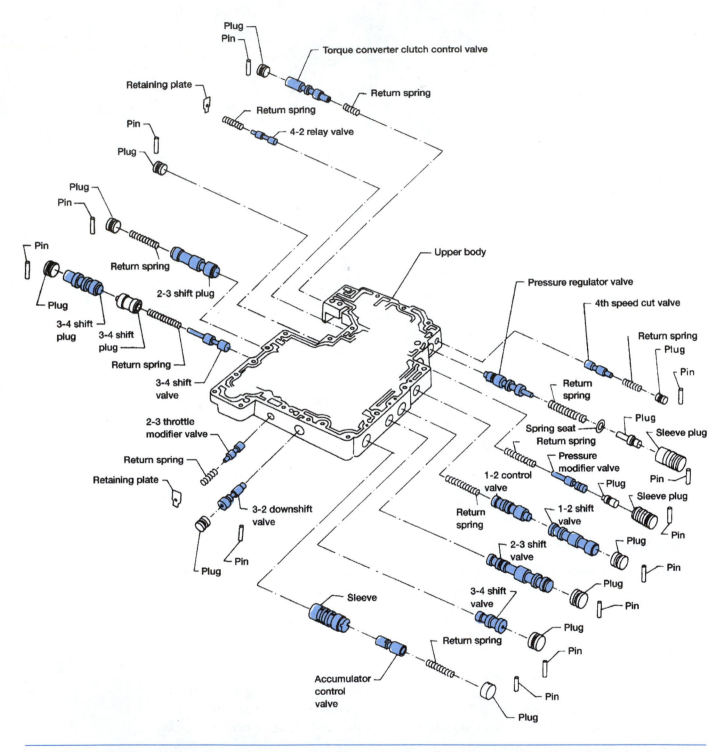

Figure 9-41 The valve body houses the valves and springs used to direct fluid. Note that all but three of the valves have return springs. (Courtesy of Nissan North America)

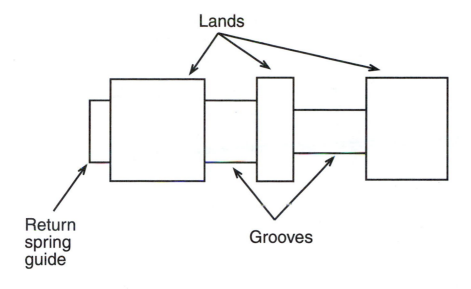

Lands

Return spring guide

Grooves

Figure 9-42 The spool valve is used to block or open fluid passageways in the valve body passages.

There are two primary control systems: throttle and governor. They work together to control up and down shifting through the forward gear ranges. The throttle system controls the position of the **throttle valve** (Figure 9-43). The **governor** controls the *governor valve* (Figure 9-44). Both valves are in the valve body. Shifting is the balancing of the two systems against each other. The throttle valve senses engine load while the governor senses the vehicle's forward speed. The vehicle starts in low gear with high throttle valve pressure and low governor valve pressure. This prevents an **upshift** to second gear. As the vehicle speed increases and engine load drops, the governor pressure increases and the throttle pressure decreases. As the governor pressure exceeds throttle pressure, the two valves move and fluid is directed to the 1-2 shift valve. The transmission shifts to second gear, engine load is increased, and throttle pressure rises to prevent upshifting to third until the next balance is reached.

Note that this is a simplistic manner of explaining the shifting actions of an automatic transmission. There may be several different valves actually involved in the shift process, but all are based on the relation between the vehicle speed and engine load.

The *throttle valve* senses engine load by mechanical linkage or by engine vacuum.

The governor pressure is generated by the *governor* mounted on the output shaft of the transaxle or by the electronic vehicle speed sensor.

As the *upshift* occurs, the pressure regulator and boost valve adjust the hydraulic pressure to ensure a firm clutch or band application.

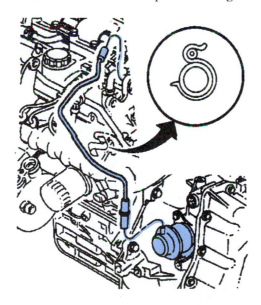

Figure 9-43 This throttle valve control uses engine vacuum (engine load) to move the throttle valve. (Courtesy of Chevrolet Motor Division, General Motors Corporation)

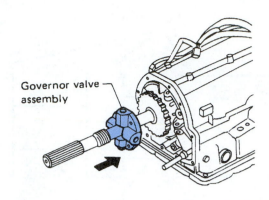

Figure 9-44 The governor uses the speed of the output shaft to move weights that control fluid flow. (Courtesy of Nissan North America)

There are many mechanical and operational designs for automatic transmissions. The newer transmissions/transaxles use electrical solenoids instead of spool valves to direct fluid and shift the gears. The solenoids can be used as shift solenoids or they can control the lock-up clutch in the torque converter. Most new vehicles use engine and transmission sensors to supply data to the PCM, which, in turn, operates the shift valves, torque converter lock-up clutch, and the many other devices needed to provide a smooth and comfortable ride. Almost all of the new systems have four forward gears and may drive all four wheels. Like the manual system, there is not much difference between an automatic transmission and transaxle. Both operate on the same theory and use very similar components. The space available for the transaxle did require some engineering to fit the components into a smaller housing.

Shop Manual
pages 219,
230–234

The typical *drive shaft* is a long, hollow tube with connection components at each end. It runs under the center of the vehicle from the transmission to the rear axle housing.

The *differential* is usually mounted within the rear axle housing.

Drive Shafts and Differentials

The **drive shaft** assembly provides the connection between the transmission's output shaft and the differential in the rear axle housing. The **differential** allows the drive wheels to rotate at different speeds when the vehicle changes direction.

Drive Shafts

A typical drive shaft assembly has two *U-joints,* a *slip joint,* and the drive shaft *tube* (Figure 9-45). The drive shaft is a hollow, balanced, steel pipe designed to withstand high rotational forces. Each end has a *yoke* to mount the U-joints. The rear U-joint connects the drive shaft to the companion flange on the differential. The front U-joint mates the drive shaft to the slip

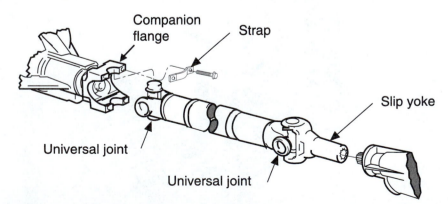

Figure 9-45 The drive shaft assembly transmits torque from the transmission to the differential. (Courtesy of Oldsmobile Division, General Motors Corporation)

joint. The slip joint is internally splined to fit over the output shaft of the transmission. The slip joint can slide in and out on the output shaft to allow for linear distance changes between transmission and differential. When the rear wheels go over a high spot, the differential moves upward, in effect, getting closer to the transmission. The opposite happens when the wheel drops into a hole. As the distance changes, so does the angle between the transmission and differential. The two U-joints rotate within their mounts to adjust for the different angles.

Differentials

The differential is mounted in the rear housing and uses a system of gears to let the two drive wheels turn at different speeds (Figure 9-46). As the vehicle rounds a curve, the outer wheel has to travel a longer distance than the inner wheel. The inner wheel has to reduce speed as the outer wheel increases speed. Both wheels use the center of the housing as the pivot point. A second purpose of the differential is to turn the power flow 90 degrees.

From the engine to the differential's **pinion gear**, the power is traveling basically in a straight line. At the rear axle housing, the power flow must be changed 90 degrees (Figure 9-47). The pinion drives a ring gear mounted to the *differential carrier*. The entire carrier rotates in the same direction as the pinion. Mounted within the carrier are two axle or *side gears* and two pinions. The pinions are attached to the carrier but can spin independently of each other. Meshed with and at right angles to the pinion are the side gears. A solid axle is splined through the center of each side gear. As the vehicle moves in a straight line, the differential carrier rotates and carries the pinion with it. Since the pinions are meshed with the side gears they rotate with the carrier and turn the axles. The pinions are not rotating or spinning at this time. During a vehicle turn, the inner side gear slows down and the two pinions begin to spin. The outer side gear speeds up for two reasons: the inner wheel's action and the actions of the

The input *pinion gear* is a long gear with spiral teeth. The pinion gears in the carrier are small, round gears with ten to fourteen teeth.

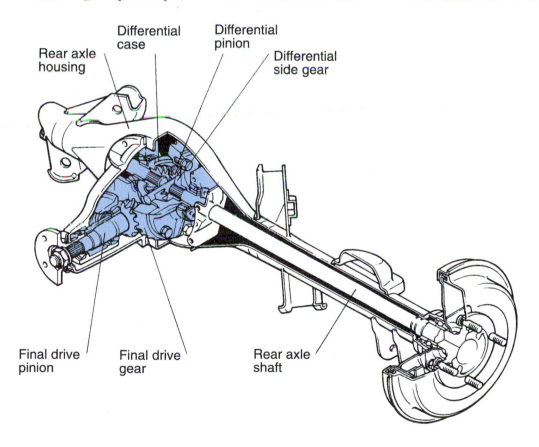

Figure 9-46 The pinions and side gears move with the case and transfer input torque 90 degrees to the axles. (Courtesy of DaimlerChrysler Corporation)

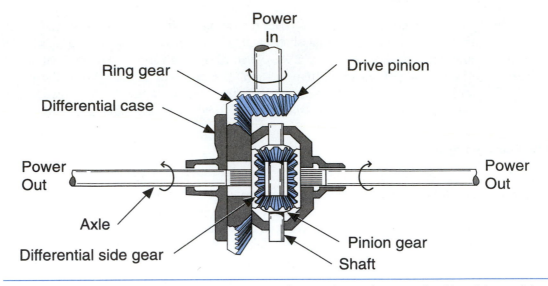

Figure 9-47 The differential changes the power flow 90 degrees between the drive pinion and the ring gear.

pinions. The two side gears change speed at the same rate. For example, if the inner one slows five RPMs, then the outer one must speed up five RPMs.

The differential carrier is mounted on bearings and must be adjusted for proper pinion gear and ring gear backlash. Some differentials may have a clutch that can lock or slow an axle (Figure 9-48). This is known by various names, but is actually a locking differential. Under

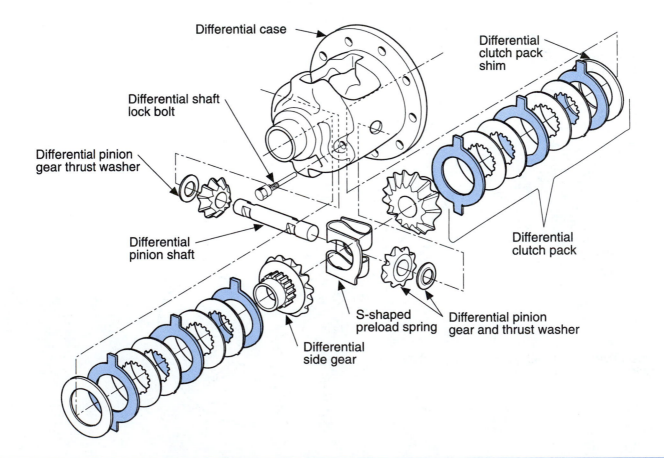

Figure 9-48 The clutches in a limited-slip differential are used to direct power from a spinning wheel to one with more traction. (Courtesy of Ford Motor Company)

normal circumstances the differential allows the wheels to speed and slow as needed. However, if a wheel is spinning, more power is sent to that wheel. The result is no vehicle movement. The locking differential uses a clutch to lock the spinning axle so the power is sent to the wheel that has the traction. Assuming that wheel has enough traction, the vehicle will move. As the traction improves on the spinning wheel, the power is more evenly divided between the two wheels.

Final Drives and Drive Axles

The *final drive* is the most important difference between the transmission and transaxle (Figure 9-49). Its purpose is the same as the differential and is mounted inside the transaxle housing. The ring gear is driven directly by the transaxle's output gear. There is no drive shaft used with transaxles.

Shop Manual pages 220, 234–239

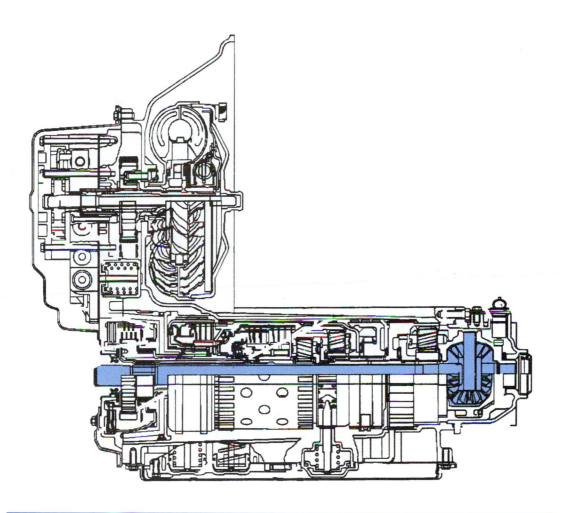

Figure 9-49 The final drive assembly is an integral part of the transaxle. (Courtesy of Chevrolet Motor Division, General Motors Corporation)

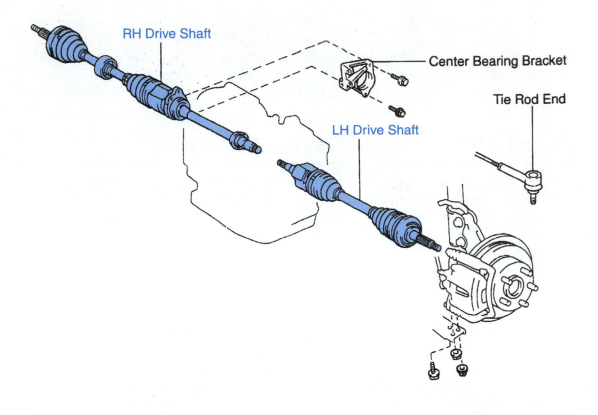

RH Drive Shaft

Center Bearing Bracket

Tie Rod End

LH Drive Shaft

Figure 9-50 The two drive axles in a FWD vehicle are usually different in length and some design features. (Reprinted with permission)

The FWD *drive axles* are also different from the rear wheel drive axles (Figure 9-50). RWD axles deliver power directly to the wheels that do not turn when negotiating a curve. With FWD, the front wheels are not only the drive wheels, but they are the steering wheels also. The drive axles must transfer the power to the wheels on sharp turns while moving over road irregularities. To do this, each axle has two constant velocity (CV) joints. The outer or outboard CV joint allows the wheel to be steered and lets the wheel and suspension travel over the road (Figure 9-51). The inner or inboard CV joint also allows for the changing vertical angles and adjusts for the linear distance between the final drive and the wheel (Figure 9-52). This joint performs the same function as the U-joint and slip joint on the drive shaft.

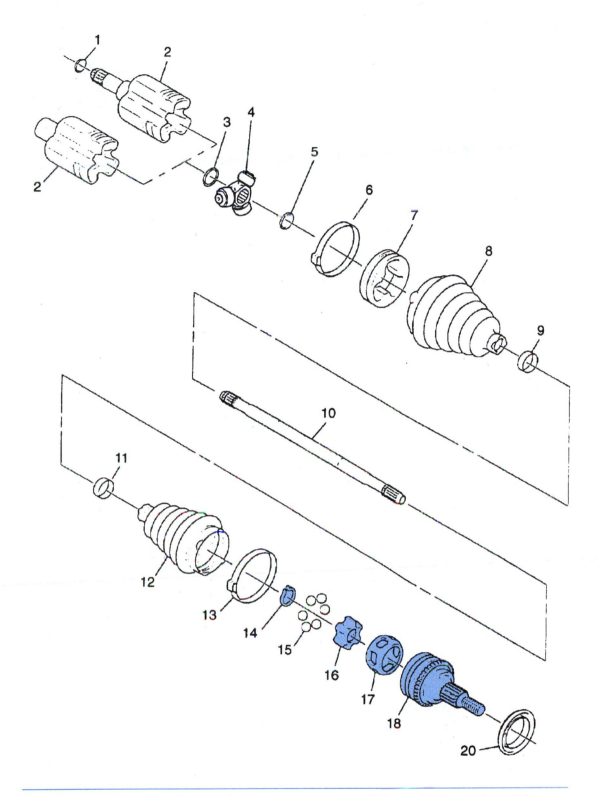

Figure 9-51 The outer CV joint allows the wheel to move with the road and steer the vehicle. (Courtesy of Chevrolet Motor Division, General Motors Corporation)

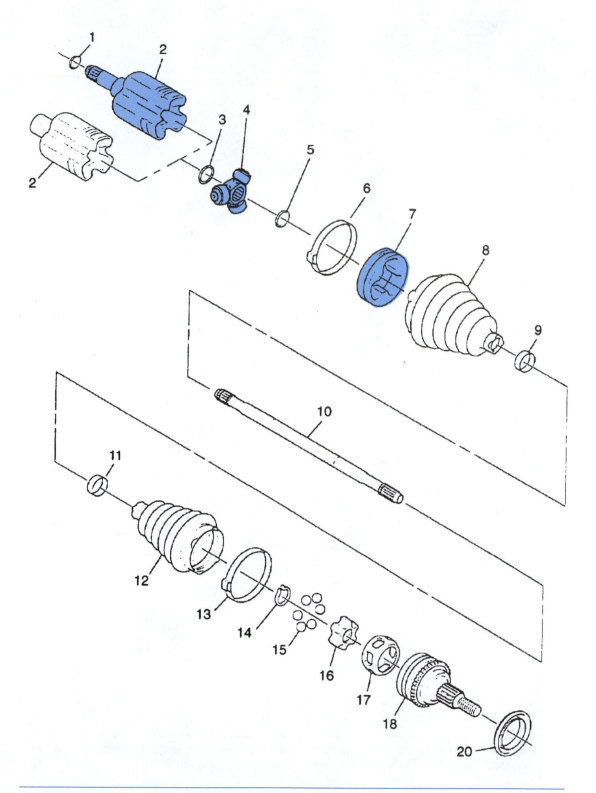

Figure 9-52 The slots in the inner tripod-type CV joint allow the axle to change in length and angle. (Courtesy of Chevrolet Motor Division, General Motors Corporation)

Summary

❏ The driveline is used to transfer engine torque to the drive wheels.

❏ Gear ratios are used to match the engine load with the vehicle speed.

❏ A clutch provides the driver a method to separate the engine from the driveline.

❏ The pressure plate clamps the clutch disc to the flywheel for transferring engine torque to the transmission.

❏ Manual transmissions/transaxles provide the necessary gear ratios to move the vehicle.

❏ The driver makes the gear selection in a manual transmission/transaxle.

❏ Torque converters are used to connect the engine to automatic transmission/transaxles.

❏ The impeller drives the turbine.

❏ The stator acts as a one-way clutch.

❏ A driver with an automatic transmission/transaxle selects the gear range.

❏ The automatic transmission/transaxle selects the forward drive gear automatically.

❏ The Simpson gear train uses two hydraulic clutches, two gear sets, two bands, and an over-running or one-way clutch.

❏ The bands are hydraulic/mechanical-holding devices.

❏ A typical drive shaft has two U-joints.

❏ The differential turns the power flow 90 degrees and allows the drive wheels to turn at different speeds during turns.

❏ Final drives are used with transaxles in place of differentials.

❏ FWD drive axles use an outer and an inner CV joint.

Review Questions

Short Answer Essays

1. Explain how a hydraulic clutch operates.

2. Describe the purpose and use of the torque converter stator.

3. Describe the two fluid flows in the torque converter during operation.

4. Explain how shifting is controlled in an automatic transmission or transaxle.

5. List and explain the purpose of each component of a drive shaft assembly.

6. Describe the purpose of the two CV joints commonly found in a drive axle.

7. Explain the purpose of transmission bands.

8. Explain how power is transferred from the flexplate to the transmission's input shaft.

9. Describe the actions of the 1-2 synchronizer during shifting from first to second gear.

10. Explain the purpose of the clutch release bearing.

Fill-in-the-Blanks

1. The _____ is located in the center of a planetary gear set.

2. The steel plates of a hydraulic clutch are fitted into grooves in the clutch _____.

3. The _____ of a spool valve closes the fluid passageway.

4. The transaxle powers _____ drive axles.

5. Movement of the _____ against the _____ of a clutch pulls the pressure plate away from the flywheel.

6. The rotary motion of the flywheel reduces the _____ of the engine.

7. The differential ring gear is directly driven by the _____.

8. If the drive gear is larger than the driven gear, then the speed of the driven is _____ than the speed of the drive.

9. The torque is increased if the _____ gear is smaller than the _____ gear.

10. If the sun gear is held and the ring gear is driven, then the torque must be _____.

ASE Style Review Questions

1. Automatic transmissions are being discussed. *Technician A* says the band drives the clutch drum. *Technician B* says a basic Simpson gear train uses three clutches. Who is correct?
   A. A only
   B. B only
   C. Both A and B
   D. Neither A nor B

2. A drive gear with 27 teeth and a driven gear with 9 teeth results in a gear ratio of:
   A. .33:1
   B. 3:1
   C. 3.3:
   D. 1:.33

3. FWD drive axles are being discussed. *Technician A* says the outboard CV joint provides vertical and horizon movement. *Technician B* says the inboard CV joint provides vertical and steering movement. Who is correct?
   A. A only
   B. B only
   C. Both A and B
   D. Neither A nor B

4. Manual transaxles are being discussed. *Technician A* says the reverse idler and reverse gears are connected with a synchronizer. *Technician B* says the reverse idler is moved with a shifting fork. Who is correct?
   A. A only
   B. B only
   C. Both A and B
   D. Neither A nor B

5. Differentials are being discussed. *Technician A* says the carrier lets the drive wheels travel at different speeds. *Technician B* says the pinions do not spin when the vehicle is traveling in a straight line. Who is correct?
   A. A only
   B. B only
   C. Both A and B
   D. Neither A nor B

6. Hydraulic clutches are being discussed. *Technician A* says the piston is applied by spring force. *Technician B* says the clutch discs are splined to a gear or shaft. Who is correct?
   A. A only
   B. B only
   C. Both A and B
   D. Neither A nor B

7. Final drives are being discussed. *Technician A* says the ring gear is driven by the transmission output shaft.
   Technician B says power is delivered to the final drive by a drive shaft. Who is correct?
   A. A only
   B. B only
   C. Both A and B
   D. Neither A nor B

8. Automatic transaxles are being discussed.
   Technician A says the transaxle pump is driven by the torque converter.
   Technician B says the shift valves control shifting. Who is correct?
   A. A only
   B. B only
   C. Both A and B
   D. Neither A nor B

9. Drive shafts are being discussed. *Technician A* says the slip joint corrects for distance changes between the transaxle and differential.
   Technician B says U-joints allows angular changes in the drive shaft's movement. Who is correct?
   A. A only
   B. B only
   C. Both A and B
   D. Neither A nor B

10. Planetary gear sets are being discussed. *Technician A* says that if the ring is driven and the carrier is the drive, the torque will decrease.
    Technician B says if the sun is driven and the carrier is held, the ring is the drive. Who is correct?
    A. A only
    B. B only
    C. Both A and B
    D. Neither A nor B

Suspension and Steering

Upon completion and review of this chapter, you should be able to:

❑ Discuss how the Laws of Motion affect the operation of the suspension and steering systems.

❑ Explain the relationship between vehicle frame, suspension, and steering

❑ Define the different suspension systems.

❑ Describe the purpose and operation of springs and torsion bars.

❑ Describe the purpose and operation of shock absorbers and struts.

❑ Describe the difference steering systems and their general operation.

❑ Explain tire ratings and markings.

❑ Define alignment angles and wheel alignment.

❑ Describe SAI, setback, and toe-out-on-turn angles.

❑ Describe caster, camber, toe, and wheel alignment.

❑ Explain static and dynamic tire balancing.

Introduction

With power delivered to the wheels, the vehicle must be suspended and steered. The steering system allows the operator to steer the vehicle within designed limits. The suspension system supports the vehicle, provides a better ride, and assists in vehicle control.

Laws of Motion

In Chapter 4, "Automobile Theories of Operation," we discussed some theories of physics that apply to vehicle design and operation. One theory was Newton's Laws of Motion. The laws are based on an object, in the absence of outside forces, performing in a given manner, such as moving at the same speed in a straight line. As in most systems on a vehicle, there are almost always some outside force at work.

Obviously, the portion of the Laws of Motion discussed do not consider the outside forces that move, stop, or turn an object. For example, friction tries to stop or slow an object while a force greater then the **mass** (weight) of the object is required to move or accelerate it. When designing a vehicle that must work within the Laws of Motion and other physics, related items like mass, force, torque, friction, and speed must be considered. The suspension and steering systems have components that are almost constantly moving, stopping, or changing direction during vehicle operation. The suspension and steering systems must be able to overcome the straight physics of motion so the vehicle can be controlled.

Frame, Suspension, and Steering Relationship

Proper operation of the suspension and steering systems depend on a straight and true frame. The frame, full or unibody, must be strong enough to support the suspension and steering as they operate over the road and under different load conditions. A bent or warped frame may affect the steering to the point that the driver may have problems keeping the vehicle under control. A weak frame can flex as it is stressed when changing steering and suspension angles. A weak frame may also allow the suspension system to transmit the shock to the body, break mounting points, and create steering problems. Frames that are misaligned may not be readily apparent until some other component fails or begins to show excessive wear.

Shop Manual
page 251

Mass refers to the total weight of an object.

Shop Manual
page 251

Suspension System and Component Purposes

The suspension supports most of the vehicle's weight and absorbs most of the various shocks from the bumps and holes in the road. But a second and just as important purpose is the control a properly-functioning suspension system offers a driver. A suspension system allows the wheel assembly to moves upward and backward or down and forward as it maneuvers over the road. This angle of movement absorbs some of the shock and reduces stress on the suspension and steering components.

The mass of the vehicle is divided into two weights. The **sprung weight** is the mass supported by the springs and usually includes everything mounted above the axles (Figure 10-1). It is important to remember that only the springs support the sprung weight. **Unsprung weight or mass** includes components such as the axles, brakes, tires, shock absorbers or struts, and the springs themselves. Vehicle engineers try to reduce unsprung mass by reducing the weight of the wheels and suspension components. Aluminum wheels are now being used on many vehicles instead of steel ones. Another way to reduce unsprung mass is to use an independent suspension system.

When the tire moves up *(jounce)* or down *(rebound)* over the road surface, a shock or jar is transmitted through the suspension system (Figure 10-2). For the most part, the shock is

Generally, the lighter, *unsprung mass* has to steer, support, and control itself and the heavier *sprung weight*.

Figure 10-1 With the exception of the wheel assemblies, everything shown in this figure is sprung weight. (Courtesy of DaimlerChrysler Corporation)

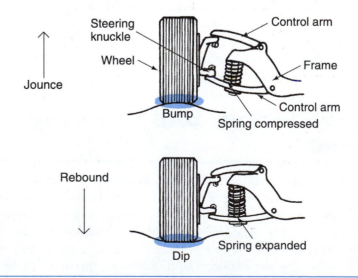

Figure 10-2 Most of the shock generated by the jounce and rebound of the wheel assembly is absorbed by the springs.

absorbed by the flexing of the tires and springs before it reaches the passenger compartment. However, a spring tends to rebound back and forth until its energy is expended. Since most roads do not have many long, flat surfaces, the vehicle would bounce up and down continuously unless the spring action was controlled.

A hydraulic cylinder known as a *shock absorber* or *strut* is used to control the motion of the springs. This cylinder allows the spring to give or extend enough to absorb most of the road shock. It controls further spring movement by **metering** fluid within the cylinder.

Ideally, the passengers would never feel the irregularities of the road, but the weight of a typical vehicle makes it practically impossible to eliminate all rough conditions. In addition to absorbing road shock, the suspension also helps control body lean during turns to help maintain vehicle control. *Sway bar* or *anti-roll bars* are used to reduce lean (Figure 10-3). As the vehicle is turned, its **center-of-gravity** and weight shifts toward the outside of the turn, making the vehicle unstable. The springs and sway bar must be stiff or strong enough for the driver to retain control. Some vehicles have a form of the sway bar mounted at the rear of the vehicle.

The suspension system is a compromise of all the factors involved so the driver experiences a smooth ride and control the vehicle at the same time. Newer and more expensive suspension systems use electronics to control *air springs* for a smoother, more controllable vehicle.

Metering is the controlled flow of a liquid or vapor. Turning the valve on a faucet is a means of metering water.

The center point of gravity is the balancing point of an object. When a vehicle is in a turn, its *center-of-gravity* moves toward the outer side of the vehicle, which tends to lift the tires on the inside of the curve.

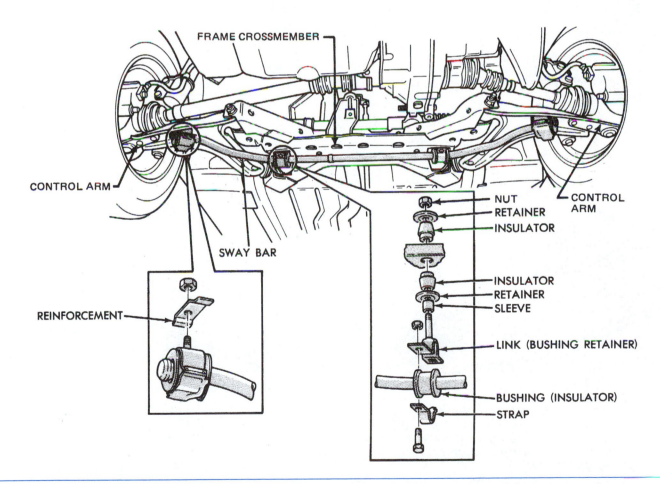

Figure 10-3 The sway or stabilizer bar is anchored to the body/frame with each end attached to the lower control arms. (Courtesy of DaimlerChrysler Corporation)

Shop Manual
pages 252–256

Payload is the amount of weight a vehicle is designed to carry including the operator. The term "gross payload" includes the total weight of the vehicle including all passengers and the payload.

A *coil spring* is a rod of steel that has been shaped in a spiral.

Torsion bars are splined at each end to fit the mounting brackets.

A *shock absorber* cannot support weight because of its valve design.

Types of Suspension Systems

Suspension systems can be generally classified as *non-independent* or *independent*. Today most non-independent systems are found on heavy trucks, primarily because of their rear-wheel drive and **payload** capacity. The most popular suspension for cars is a four-wheel, independent system while most light trucks use independent in the front and non-independent on the rear axle.

Independent Suspensions

An independent suspension is designed to allow one wheel to follow road contours without affecting the operation of the other wheels (Figure 10-4). Independent systems allow for a smoother ride while providing good support for the vehicle. Older independent systems used a spring and control arm.

A **coil spring** is placed between one control arm and the frame to absorb road shock and hold the tire to the road surface (Figure 10-5). Some vehicles use a **torsion bar** in lieu of the spring (Figure 10-6). A **shock absorber** is attached to the frame and to the control arm.

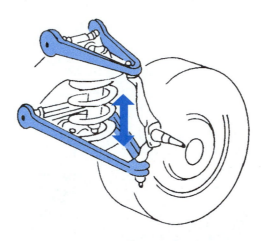

Figure 10-4 The control arms allow the wheel assembly to follow road contours without affecting the other wheels. (Courtesy of Federal-Mogul Corporation)

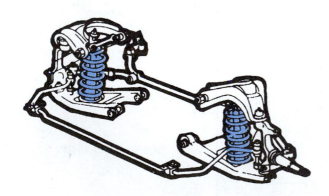

Figure 10-5 The coil spring is trapped between a control arm and the frame. In this illustration, the spring rests on the lower control and the bottom of the frame (not readily visible). (Courtesy of Hunter Engineering Company)

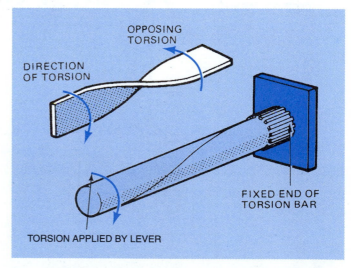

Figure 10-6 Twisting of the torsion bar acts like a coil spring.

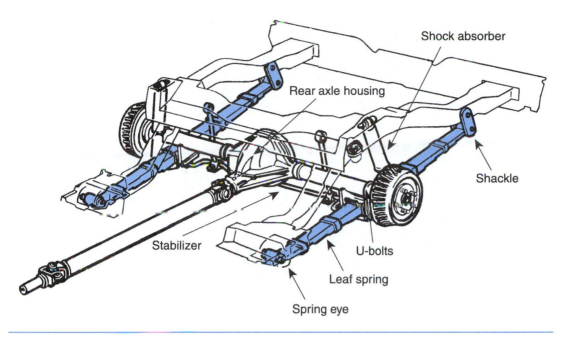

Shock absorber
Rear axle housing
Shackle
Stabilizer
U-bolts
Leaf spring
Spring eye

Figure 10-7 Only the top leaf is attached directly to the frame. (Courtesy of Chevrolet Motor Division, General Motors Corporation)

Non-Independent Suspensions

Non-independent suspension uses a solid or rigid connection between the two wheels on an axle. This is usually the rear-drive axle assembly. Many four-wheel-drive vehicles also have a rigid axle housing at the front. In a non-independent system, any movement of one wheel is transmitted to the opposite wheel. In most cases, **leaf springs** are used in this suspension system to support the sprung weight of the vehicle. Some non-independent systems use coil springs instead of leaf springs. Leaf springs are attached at one end to the frame and run over or under the axle (Figure 10-7). Shock absorbers are mounted between the axle and the frame.

McPherson Struts

A third, more recent suspension system uses **McPherson struts.** The system is an independent type that uses struts instead of control arms. The McPherson strut is basically a shock absorber with mounting brackets for a coil spring (Figure 10-8). The strut assembly also eliminates the upper control arm by using a rotating mount at the top of the strut. The McPherson strut assembly provides better support of the vehicle, controls the wheel assembly better, and reduces the weight of the vehicle. McPherson struts can be found on all wheels, but are most common on the front wheels.

Electronic Suspensions

In recent years, *automatic ride control* has become popular on more expensive cars. It is an independent suspension system. Using sensors, an air compressor, air valves, and a computer, the strength of the individual springs can be adjusted. Air can be added to or removed from the air spring (Figure 10-9). The onboard suspension computer typically is programmed to raise the vehicle during low speeds and stops. The vehicle is raised at the halt so the door can clear any road curbing. At cruising speeds, the vehicle can be lowered to reduce air resistance and

The strength of a *leaf spring* can be changed by adding or removing leaves. A common practice is to add an overload spring on a truck to support a heavier payload.

McPherson struts are lighter than a coil spring and control but are almost maintenance free and can be replaced much more easily.

An electronic air ride system requires some events to occur in sequence, such as the ignition key turned off and the door opened, before the computer commands an action. Do not fault this system until the required events are performed in order.

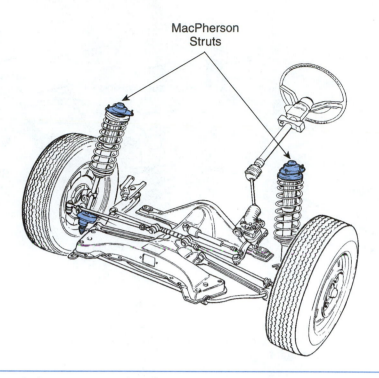

MacPherson
Struts

Figure 10-8 The strut mounts directly to the steering knuckle and the body. (Courtesy of McQuay Norris)

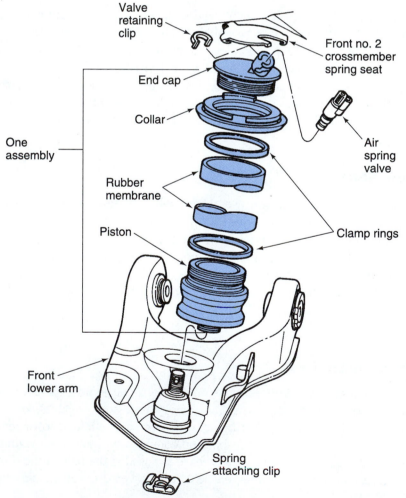

Valve retaining clip

Front no. 2 crossmember spring seat

End cap

Collar

One assembly

Rubber membrane

Air spring valve

Clamp rings

Piston

Front lower arm

Spring attaching clip

Figure 10-9 Components of a typical air spring. (Courtesy of Ford Motor Company)

provide better control through a lower center of gravity. On some automatic suspension systems, the driver can select a stiffer suspension for narrow, winding roads (for better vehicle feel) or a softer setting for wide, straight roads like the interstate road system.

The number of sensors and actuators in the automatic ride control depends on the type of system. Some ride controls systems only level the rear of a loaded vehicle (Figure 10-10). Full automatic systems have height sensors and air valves at each wheel. There is also an electrically-driven air compressor mounted in the trunk or engine compartment.

All of the automatic systems have a small, onboard computer (Figure 10-11). It collects and analyzes data from the height sensor, door switches, ignition switch, brake switch, speed sensor, and other sensors required by the computer's programming. The height valve signals the computer when the vehicle is at the desired height.

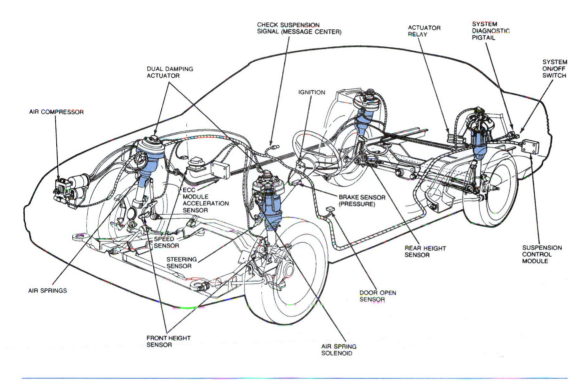

Figure 10-10 This fully automatic ride control will raise and lower the vehicle based on load, speed, and possibly road conditions. (Courtesy of Ford Motor Company)

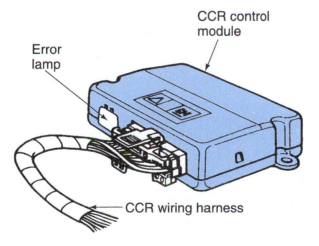

Figure 10-11 The computer command ride module performs like all vehicle computers by collecting data, comparing the data to a program, and commanding certain actions. (Courtesy of Oldsmobile Division, General Motors Corporation)

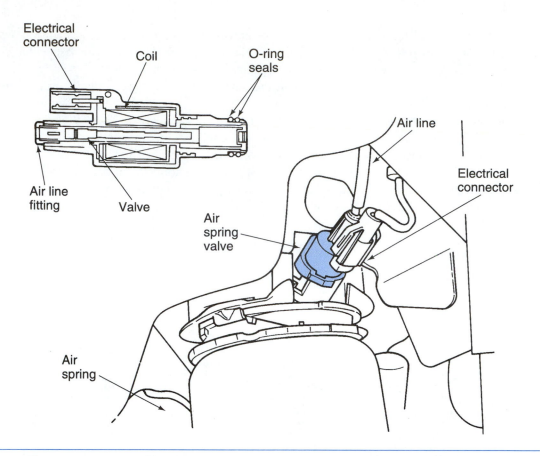

Figure 10-12 An air valve is located at the top of each air spring and is controlled by the ride control module. (Courtesy of Ford Motor Company)

The computer controls two primary actuators: *spring air valves* and the *compressor.* The air valves are solenoid-operated and can only be open or closed (Figure 10-12). An open valve allows compressed air into the spring or allows air out. The compressor can only be switched on or off by the computer. If the compressor is operating and the air valves are open, the spring will inflate, thereby raising the vehicle.

Suspension System Components

Shop Manual
page 257

⚠️ **WARNING:** Do not heat or bend steering or suspension components to make adjustments. The component may be weakened, which could result in failure. Damage or injury could also result.

The components of the suspension undergo a great deal of stress during operation. They must keep the tires on the road and maintain vehicle stability during all road and driving conditions. The suspension components are designed for durability with a minimum amount of maintenance.

Coil Springs

Coil springs are commonly used on the front of passenger cars and light trucks. They have also been used on the rear of many vehicles, particularly on the smaller types. They come in various strengths according to the type of vehicle and its maximum payload. The spring is made of a high-tempered, steel rod that is twisted into a coil. The diameter, overall length, and manufacturing treatment of the rod determine the strength or rate of the spring.

Variable rate coil springs are made so the spring can support more weight per inch of compression (Figure 10-13). Some variable springs use the same diameter wire and unequally-spaced coils. Other types have a wire machined in different diameters. The smaller diameter portion of the wire supports lighter loads. The vehicle tends to stay more level, even under moderate loads with variable-rate coil springs. This type of spring action provides a better ride at light payloads and better support as the vehicle is loaded.

Air springs have been used for many years on over-the-road trucks. The springs are actually large air bags that can be inflated or deflated to support different payloads (Figure 10-14). This ability to control the spring's capacity makes it possible to employ electronics to control a vehicle's ride. The bags are made of very tough, thick, flexible material similar to rubber. There is actually little wear on the bags unless they are under inflated. The main points of wear occur at the top and bottom where the bags are mated to the vehicle.

The wire in a *variable rate* spring may be tapered, truncated cone, double cone, or barrel-shaped.

Air springs are also used under the driver's seat on many over-the-road trucks to provide a comfortable ride.

Conventional   **Variable-Rate**

Figure 10-13 The variable-rate spring provides more vehicle control as the load becomes heavier. (Courtesy of Sealed Power Corporation)

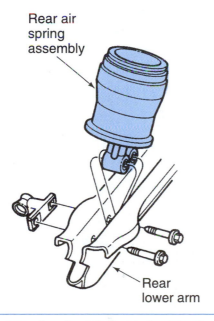

Rear air spring assembly

Rear lower arm

Figure 10-14 The bottom of the air spring is bolted to the control arm with the top end resting in a cavity in the frame. (Courtesy of Ford Motor Company)

Leaf Springs

Leaf springs are basically a set of long, curved, metal strips or leaves that are manufactured to bend at a certain rate. A leaf spring typically has three to five leaves (Figure 10-15). The leaves are placed on top of each other with the longest on top. Clamps are used to hold the stack of leaves in alignment with each other. One end of the top leaf is anchored to the lower side of a frame rail. The other end, usually the rear, is mounted in a mechanism that allows the top leaf to extend or shorten in length (Figure 10-16). This is necessary because as the leaf flattens under load, its ends move away from each other, thereby increasing the linear length of the leaf. The opposite action occurs as the leaf is bowed and becomes shorter. The flattening and bowing of the leaf spring absorbs the shock of road conditions much as coil springs do. Although leaf springs are usually stiffer than coil springs, they will still flex until all energy is expended.

A **monoleaf spring** has one steel leaf. It is thicker in the middle and tapers down to each end. The tapering provides a variable rate similar to that of a variable coil spring. The spring may mount to the vehicle in a style similar to a multi-leaf spring or it may be mounted transversely (Figure 10-17). Some vehicles use a monoleaf spring made of fiberglass. Shock absorbers are used with leaf springs the same as coil spring systems.

Leaf and coil springs are designed and manufactured to support the expected weight of the vehicle and its payload. However, standard coil springs tend to be slightly less supportive of the vehicle during turns. Coil springs tend to offer a more comfortable ride. Standard leaf springs are harsher in their movements, resulting in a rougher ride for the passengers and load but affording a little better support during a turn. Both types can be designed to provide a

Many General Motors Corvettes use a *monoleaf spring* that is mounted across the rear of the vehicle, giving the vehicle a fully independent suspension in the rear.

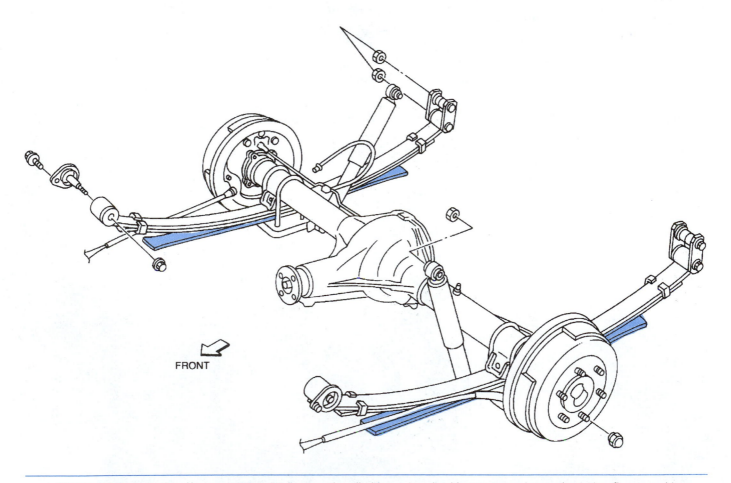

FRONT

Figure 10-15 Note the rear mounting point commonly called a "spring shackle." It can swing as the spring flexes, and its length changes. (Courtesy of Nissan North America)

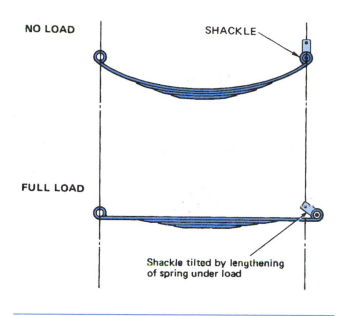

Figure 10-16 The rear shackle is pushed to the rear as the spring is flatten under load.

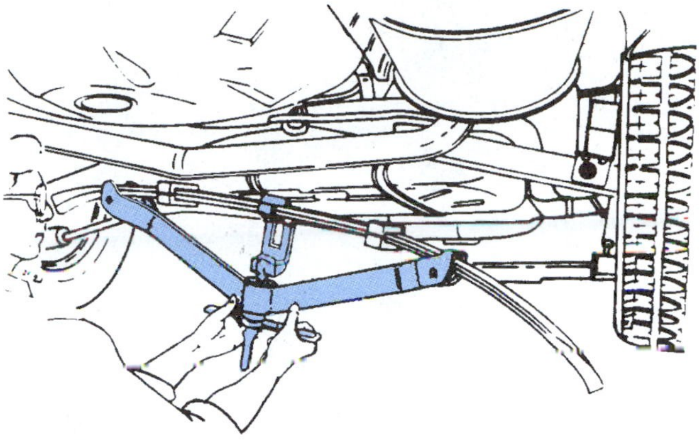

Figure 10-17 This mono-leaf spring is anchored in the center, allowing the two ends to act like an independent suspension. (Courtesy of Oldsmobile Division, General Motors Corporation)

smoother or firmer ride. However, the typical spring is usually a compromise between the two requirements.

NASCAR and other racing vehicles use specially-designed and manufactured coil springs on all four wheels. Each spring is designed for certain race conditions such as the type of track, condition of the track, and where the spring is mounted. Springs of this type are expensive and are impracticable for everyday use.

Torsion Bars

Torsion bars are long, round, steel bars that have been manufactured to twist and then return to their orginal condition. Torsion bars are anchored to the frame at one end and splined to the lower control arm at the other (Figure 10-18). As the control arm moves, the torsion bar is twisted and absorbs the shock of the movement. A typical torsion bar is usually stiffer than a coil spring, but less than that of a leaf spring when used in similar conditions. Many vehicles with torsion bar suspensions have an adjustment capability to manually raise or lower the vehicle's **ride height** (Figure 10-19).

The *ride or curb height* is the distance an "at rest" vehicle sits above the ground. It is measured at specific points on the vehicle.

Shock Absorbers

Shock absorbers do not support any weight of the vehicle. They are designed to control spring movement and are used in some manner on all suspension systems. The absorber is a long, extendable cylinder that houses several valves, a piston, and a piston rod (Figure 10-20). Hydraulic fluid in the cylinder is used to slow or stop the piston movement.

The lower end of the shock is normally attached to the lower control arm or axle. The upper end is attached to the vehicle's frame. As the control arm or axle moves with the wheel assembly, the upper or rod end of the absorber moves in and out of the cylinder. Inside the housing at the lower end of the rod is a valve assembly mounted within a piston. There is hydraulic fluid on both sides of the piston. The fluid must move from one side of the piston to the opposite side as the piston/rod moves up and down. The piston valve assembly meters the fluid's rate of movement (Figure 10-21). The resistance of the fluid against the piston assembly absorbs the energy of the spring. This restricts the oscillations of the spring and controls the up-and-down motion of the vehicle.

Special *shock absorbers* with air bags can be used to support a heavy load. The air bag cab inflates or deflates to keep the vehicle level.

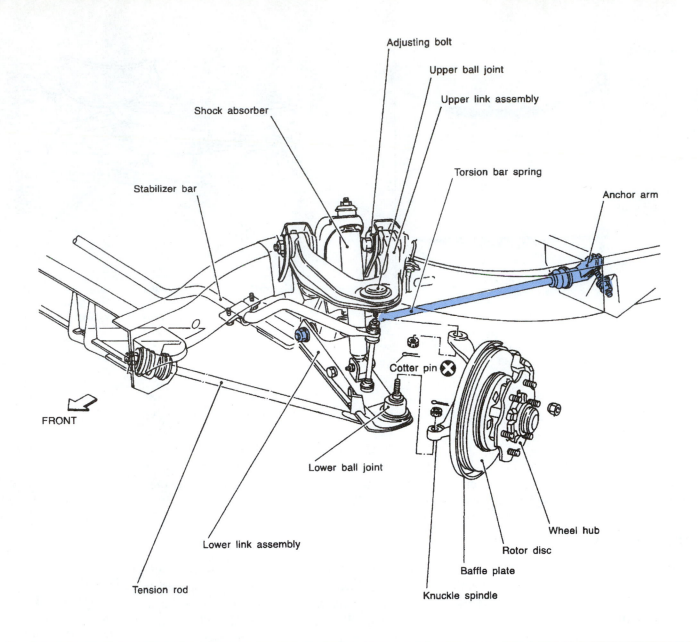

Figure 10-18 This is a typical torsion bar configuration. (Courtesy of Nissan North America)

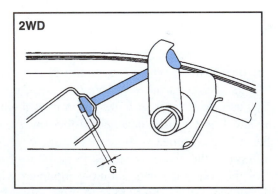

Figure 10-19 When adjusted to the measurement shown, the vehicle will be at the correct ride height. (Courtesy of Nissan North America)

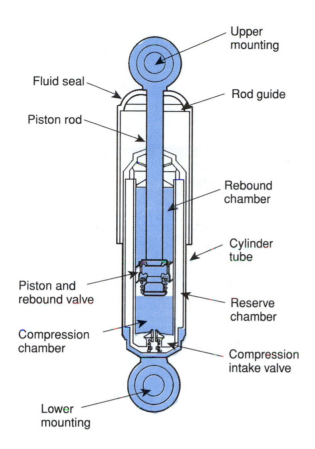

Figure 10-20 Fluid moves between the rebound and compression chamber to control spring movement.

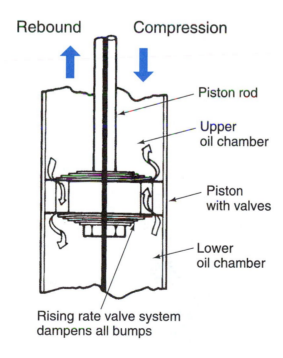

Figure 10-21 Metering valves in the shock absorber piston controls the speed of fluid movement. (Courtesy of Delco-Remy, Division of General Motors Corporation)

A problem with earlier shock absorbers was the delayed initial action of the absorber. The cylinder could not be completely filled with fluid. The fluid did not restrict piston travel until it was pressurized and forced to move through the valves. Newer shocks have pressurized nitrogen gas pumped into the cylinder after the fluid has been installed (Figure 10-22). This eliminates the initial lag, and the restricting action of the fluid is instantaneous upon any spring movement.

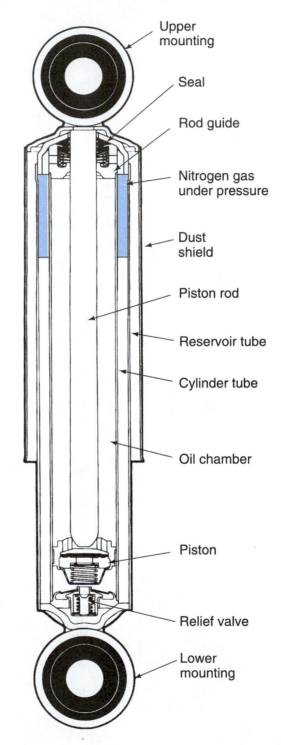

Upper mounting

Seal

Rod guide

Nitrogen gas under pressure

Dust shield

Piston rod

Reservoir tube

Cylinder tube

Oil chamber

Piston

Relief valve

Lower mounting

Figure 10-22 The pressurized nitrogen gas at the top is used to reduce shock absorber lag during initial spring action. (Courtesy of Delco-Remy, Division of General Motors Corporation)

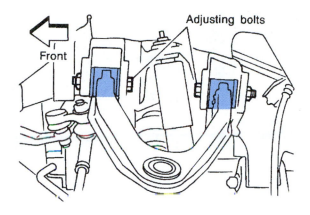

Figure 10-23 All control arms have flexible rubber–like bushings at the inner end to absorb and isolate road shocks. The bushings are usually made from neoprene with metal sleeves. (Courtesy of Nissan North America)

Control Arms

Control arms are the connection between the vehicle's frame and the steering knuckle. Control arms make it possible to use independent suspension systems. They may be located at each wheel or just on the two front wheels. On the rear of some vehicles, *trailing* or *radius arms* connect the wheel assembly to the frame. This type of arm will not keep the tire's tread flat on the ground. Other suspension components have to be used with trailing or radius arms to properly support and guide the wheels.

The control arms are mounted with rubber bushings at the frame and extend outward from the frame to the steering knuckle (Figure 10-23). The inner bushings allow the control arms to pivot up and down with the movement of the wheel. The SLA system has the shorter arm at the top. As the wheel moves upward over a hump, the shorter arm draws the top of the wheel inward. The same action happens when the wheel drops into a hole. This arc movement keeps the tire tread flat on the road. The two arms are designed to complement each other as the wheel moves over various road conditions.

If a coil spring is used, it is mounted between the frame and one of the control arms. This arm is the load-bearing arm and may be either upper or lower. At the ends of the control arms are ball joints that attach to the steering knuckle. The ball joints provide a pivot point for the steering knuckle to rotate during steering and while moving the wheel assembly over the road. Control arms used at the rear of the vehicle perform basically the same as front arms, although a steering operation is not usually present on the rear wheels.

Struts

The McPherson strut assembly houses a shock absorber internally and a coil spring mounted on the exterior (Figure 10-24). The coil spring mounted on the strut is not considered as part of the strut assembly. The shock absorber action of the strut is very similar to that of a regular shock absorber. The primary advantage of the strut is its use as the upper control arm and weight reduction.

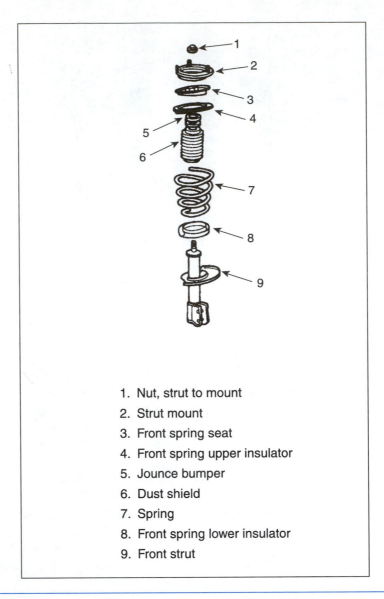

1. Nut, strut to mount
2. Strut mount
3. Front spring seat
4. Front spring upper insulator
5. Jounce bumper
6. Dust shield
7. Spring
8. Front spring lower insulator
9. Front strut

Figure 10-24 The strut provides an external mount for the coil spring. (Reprinted with permission)

Since it is mounted to the frame or body with a rotatable bearing at the top, the entire strut turns with the steering knuckle (Figure 10-25). The strut and spring are kept in alignment with the knuckle in this manner. The road shocks and steering stresses are more easily absorbed when all components are kept in line. A smaller and lighter-weight lower control arm can also be used. In addition to the suspension advantage, better steering control can be maintained for the same reasons. Pressurized nitrogen gas is also added to the strut to pressurize the strut's fluid.

The assembly and operation of the valves and piston in the strut is very similar to that of a shock absorber. The strut's rod is attached to the piston. The fluid within the cylinder must move from one side of the piston to the other. Metering valves control the rate of flow and restrict the oscillations of the coil spring.

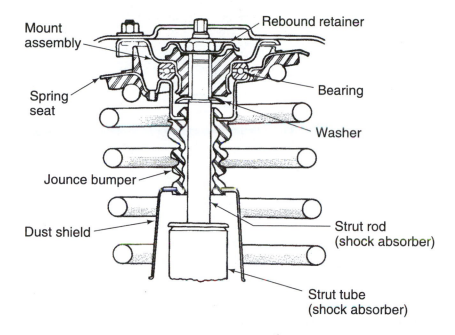

Mount assembly

Rebound retainer

Spring seat

Bearing

Washer

Jounce bumper

Dust shield

Strut rod (shock absorber)

Strut tube (shock absorber)

Figure 10-25 The strut's upper mounting makes it possible to eliminate the upper control arm. (Courtesy of DaimlerChrysler Corporation)

Sway Bars

Sway bars are designed to restrict or reduce vehicle lean during turns. The bars extend across the undercarriage of the vehicle from control arm to control arm. The sway bar is anchored to the frame or body and attached to each of the lower control arms with links (Figure 10-26).

As the control arm moves, it bends or deflects the sway bar. The bar resists and reduces the amount of lean. As the vehicle leans, the outer side of the vehicle drops, pushing the lower control arm and the end of the sway bar upward. The inner side of the vehicle lifts and the springs push the lower control arm and the end of the sway bar downward. The opposing actions at each end of the sway bar, along with the manufacturing process, help prevent excessive leaning of the vehicle. This gives better control to the driver. The springs, regardless of type, must also support the vehicle as it leans. The sway bar alone cannot effectively reduce vehicle lean.

Sway bars can be purchased in different strengths to provide better support during turns.

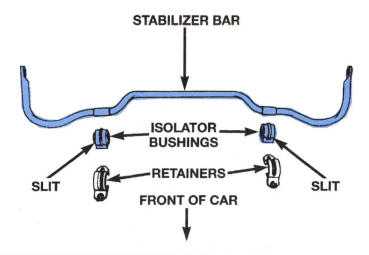

STABILIZER BAR

ISOLATOR BUSHINGS

RETAINERS

SLIT

SLIT

FRONT OF CAR

Figure 10-26 The isolation bushings prevent metal-to-metal contact between the sway bar and the body. (Courtesy of DaimlerChrysler Corporation)

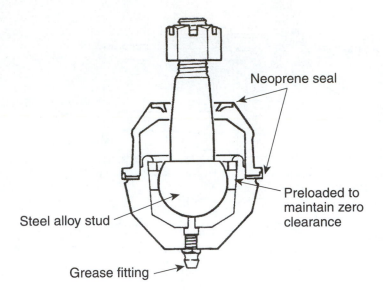

Figure 10-27 The ball joint makes it possible to steer the front wheels as they travel over the road. (Courtesy of Sealed Power Corporation)

Ball Joints

The ball joint is the connection between a control arm and the steering knuckle. It has a stud that is ball-shaped on one end and is treaded from near the ball to the end of the stud. The ball fits into a molded socket (Figure 10-27). The treaded end fits through a hole in the steering knuckle and is secured with a nut or through bolt. The ball can rotate within limits to allow vertical and rotational movement of the steering knuckle assembly. Each control has a ball joint. The ball joint is one of the few suspension items needing routine maintenance.

Types and Components of Steering Systems

Shop Manual
pages 259–266

⚠️ **WARNING:** Do not apply heat or bend steering or suspension components to make adjustments. The component may be weakened which could result in failure. Damage or injury could also result.

The steering system provides the driver with the means to steer the vehicle. There are two types of steering systems: *parallelogram* and *rack and pinion*. Power boost systems are available on both.

Parallelogram System and Components

The parallelogram steering system has been used for years on almost every vehicle made. The linkage in this system is designed to remain parallel to the control arms during all steering movements (Figure 10-28). There are several ball-joint-type devices connecting different parts of the linkage. The driver is connected to the steering linkage and gear through the steering wheel and steering shaft. The steering wheel and most of the shaft is mounted within the passenger compartment while the linkage and gearbox is located in the engine compartment or under the vehicle.

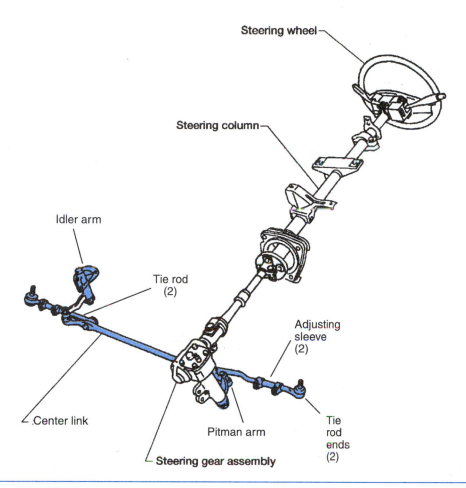

Steering wheel

Steering column

Idler arm

Tie rod (2)

Adjusting sleeve (2)

Center link

Pitman arm

Tie rod ends (2)

Steering gear assembly

Figure 10-28 The linkage shown could be mounted behind or forward of the front axle and is called rear or front steering. (Courtesy of Nissan North America)

The steering wheel allows the driver to move the linkage to steer the vehicle. Steering wheels on manual steering systems are larger in diameter than the ones used on a power steering system. This gives the driver some mechanical advantage during low-speed steering. The horn control button is located within the steering wheel. Some newer vehicles have switches for cruise control and other accessories on the steering wheel. The steering column supports and houses the steering wheel and shaft along with any switches and wiring mounted on or in the column (Figure 10-29). The steering shaft extends through the column to the steering gearbox.

The most common type of steering gearbox in the parallelogram system is the **recirculating ball and nut.** Large, steel balls are placed in grooves on the nut and *worm gear* (Figure 10-30). The steering shaft rotates the worm gear. As the worm gear turns, it uses the balls as a treading device to move the nut up and down the length of the worm gear. Teeth on the nut mesh with teeth on a *sector shaft*. The lower end of the sector shaft is splined to a *pitman arm*. The gear arrangement multiplies the force delivered by the steering shaft, thereby reducing driver effort.

The term *recirculating ball and nut* refers to the steel balls that move around in a continuous passage to move the nut that is meshed with the sector shaft gear.

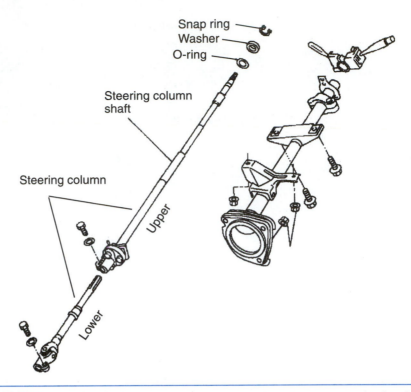

Figure 10-29 The steering shaft runs through the column from the steering wheel end to the coupler on the engine side of the floor pan. This shaft has upper and lower sections. (Courtesy of Nissan North America)

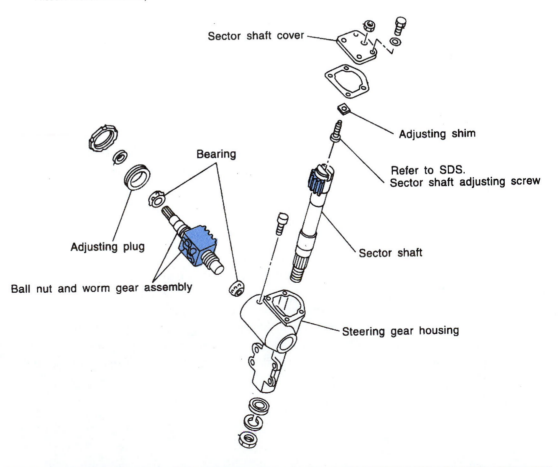

Figure 10-30 The balls located in the nut provide a means to reduce friction and increase the driver's turning effort. (Courtesy of Nissan North America)

The pitman arm is attached to a **center link.** An *idler arm* at the opposite end of the center link from the pitman acts as a rotating anchor for the link (Figure 10-31). Sockets similar to the ball joint attach the center link to the idler arm and two *tie-rods*. A threaded, split, pipe-like adjusting sleeve is used to connect a tie-rod end to each of the tie-rods. The tie-rod ends are attached to the steering knuckle with sockets. Any rotary movement of the worm gear results in a linear movement of the pitman arm, center link, tie-rods, and ends to turn the steering knuckle and the wheel assembly from side to side. The adjusting sleeve is used in adjusting the wheel during alignment. It is clamped to the tie-rod and tie-rod end (Figure 10-32).

In some vehicles, the *center link* is known as the "drag link."

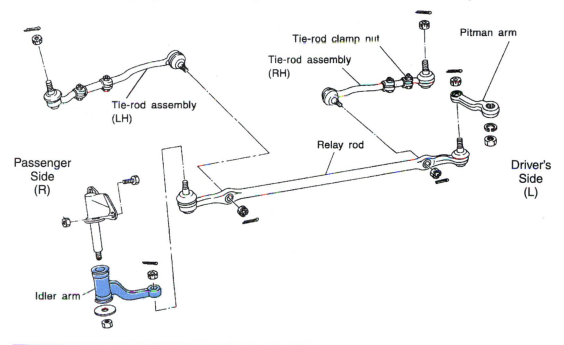

Figure 10-31 The idler arm supports one end of the center link and may not be used on some applications. (Courtesy of Nissan North America)

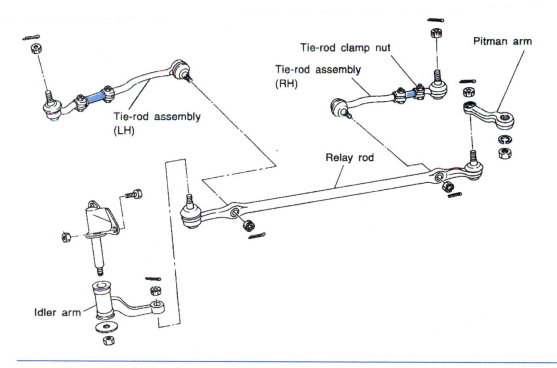

Figure 10-32 The threads on the tie-rod and tie-rod end are cut in opposite directions to allow the adjusting sleeve to act as a turnbuckle. (Courtesy of Nissan North America)

The steering knuckle is attached to the steering linkage and the suspension system. Attachment points at the top and bottom are used to attach the knuckle to the suspension. Another attachment point is used to connect the steering linkage. The steering knuckle may also have mounting holes or brackets for brake hardware.

Rack and Pinion Systems and Components

The steering knuckle on a rack and pinion system is very similar to the parallelogram system. The steering column and steering wheel are also the same for all practical purposes. The main differences are between the lower end of the steering column and the steering knuckle. The rack and pinion system is comprised of three major components: a rack, a pinion gear, and the linkage. The linkage also remains parallel to the controls arms during all phases of steering.

The rack is a long, straight bar of steel with teeth along a portion of one side (Figure 10-33). At each end of the rack are the inner tie-rod sockets. The tie-rod sockets perform the same function as the tie-rods and are attached to tie-rod ends. The sockets can pivot, allowing the outer end to follow the vertical movement of the knuckle and wheel assembly. The tie-rod end can be adjusted during wheel alignment procedures. The rack fits and is sealed in a housing where the rack teeth mesh with the pinion teeth (Figure 10-34).

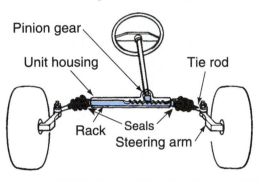

Figure 10-33 The rack extends through the housing which is sealed at each end. (Courtesy of Sealed Power Corporation)

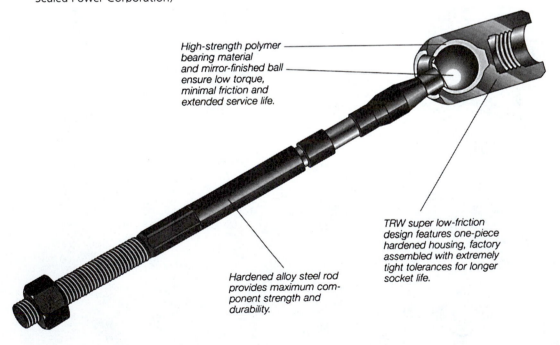

High-strength polymer bearing material and mirror-finished ball ensure low torque, minimal friction and extended service life.

Hardened alloy steel rod provides maximum component strength and durability.

TRW super low-friction design features one-piece hardened housing, factory assembled with extremely tight tolerances for longer socket life.

Figure 10-34 The inner tie-rod for a rack and pinion includes a ball/socket arrangement that is screwed onto the rack. (Courtesy of Federal-Mogul Corporation)

The steering shaft rotates the pinion. The pinion performs the same function as the worm gear. The pinion moves the rack left and right as the steering wheel is turned. The rack and pinion occupies less space, is lighter, has fewer parts, and generally gives the driver better road feel. The rack and pinion system has very little, if any, torque multiplication. Steering even a small vehicle in traffic with manual rack and pinion can be tiring.

Power Steering

Both types of steering systems can be equipped with power boost. The boost systems for each are very similar. A pump supplies pressurized fluid to increase the driver's steering force. The pump is equipped with a pressure relief valve to prevent over pressurizing the system. Automatic transmission fluid is used many times as a power steering fluid on older systems. Many newer models use specially-blended, power steering fluid. The service manual should be consulted before topping off power steering fluid.

A crankshaft belt-driven pump provides the pressurized fluid to the system. The fluid is delivered to the power steering valves through steel lines and flexible hoses (Figure 10-35). The power steering valve in a recirculating ball and nut gear is located at the top of the steering gear. The steering shaft rotates the valve to open a passageway to the correct side of the steering gear piston. Pressure applied to the piston assists the driver in turning the worm gear.

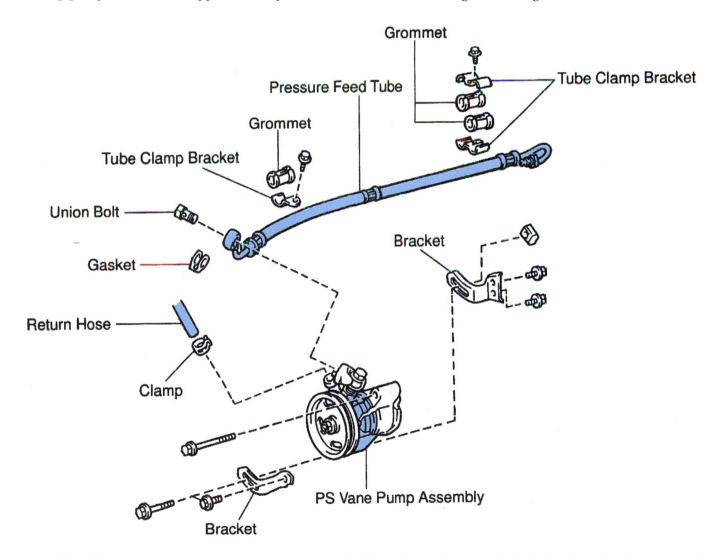

Figure 10-35 The liquid in the return hose is not under pressure so it can be fastened to the pump with a simple clamp. The feed tube requires pressure-type fittings. (Reprinted with permission)

Pressurized fluid on a rack and pinion is delivered to the valves located at the top of the pinion gear. As the shaft and pinion are rotated, a valve opens and directs fluid to the correct side of a piston mounted to or as part of the rack (Figure 10-36). In this system, the power boost is applied directly to the steering linkage. In the ball and nut system, the boost is delivered to the steering gears.

Most power steering systems have a power steering switch in the pressure line. The switch signals the PCM to raise the engine idle speed during low-speed conditions such as parking.

Wheels and Tires

Shop Manual
pages 266–272

Wheels

The tire is the contact between the vehicle and the road. Since the tire is *pneumatic* (air-filled), a wheel or rim is needed to support it. Most wheels are made of steel or an alloy of magnesium and aluminum. Steel wheels are made by welding the rim to the centerpiece. Magnesium and aluminum wheels are cast molded. Passenger car wheels are listed as *dropped center safety rims*. The rim flange is set above the center (Figure 10-37). Holes around the center fit over studs fitted into the hub, and lug nuts are used to mount the wheel to the hub. Wheels are measured in diameter (across the wheel) and width (between edges) (Figure 10-38). The tire is held to the wheel by air pressure that forces the *tire's beads* over the wheel's center humps

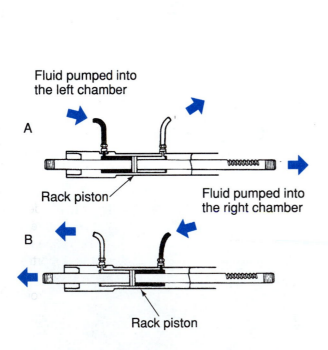

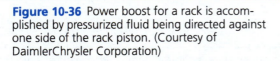

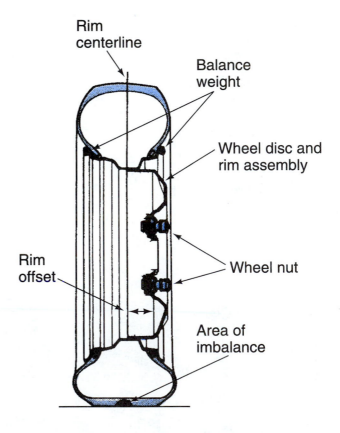

Figure 10-36 Power boost for a rack is accomplished by pressurized fluid being directed against one side of the rack piston. (Courtesy of DaimlerChrysler Corporation)

Figure 10-37 Note how the mounting portion of the wheel is set to the right of the rim centerline. (Courtesy of DaimlerChrysler Corporation)

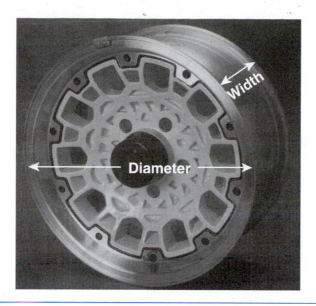

Figure 10-38 The diameter and width measurements of a wheel.

and seals them against the rim flanges. The two tire beads are tightly wrapped steel wires formed into loops slightly smaller in diameter than those of the rim flanges. Tires are matched to the wheel by diameter and width.

> ⚠️ **WARNING:** Radial- and bias-ply tires should *not* be mixed on a vehicle. Different-size tires could be mixed if the same-size tires are used on the same axle. However, it is not recommended. Mixing types and sizes can result in poor vehicle control and may cause damage or injury.

> ⚠️ **WARNING:** Do not reverse mount the rims (inside out). This changes the steering and suspension angles. Damage to vehicle components, particularly the steering and suspension, could result. The manufacturer's vehicle warranty may be voided when reverse mounting the rims.

> ⚠️ **WARNING:** Do not use tires that are different from the manufacturer's specifications in type and size. Damage to the suspension system and other vehicle components could occur. The vehicle's speedometer accuracy will be affected, as will engine and transmission controls. The antilock brake system may not function correctly with tires that are not within the manufacturer's specifications.

The present day tire is a result of research and development conducted at the racetrack and at tire manufacturing facilities. There are all types of tires, from the large racing slicks used for drag racing, to the small ones for lawnmowers and go-karts. Passenger car tires range from 13 inches to 18 inches in diameter. There are different types of **tread** and **sidewall** designs to meet almost all road and driver situations. Tires can be selected for different types of weather, payload, and driving conditions. Some vehicle manufacturers have normal, sport, or touring tires that can be placed on the same model car for different driving styles. Tires come in two basic configurations: *bias ply* and *radial ply*.

Tread is the molded portion of the tire that touches the road. It is supported by the *sidewall*, which extends from the bead to the tread area.

Bias Ply

Bias ply tires are no longer used on most vehicles. They can still be found on larger trucks and trailers. The fleet service technician may deal with bias ply tires regularly. The word "bias" refers to the direction the cords are laid under the tread (Figure 10-39). Cords run from one

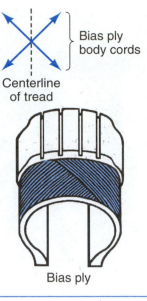

Figure 10-39 Bias cords make a stiff sidewall, but tend to lift the tread during cornering.

bead to the other forming the sidewall and supporting the tread. Bias ply tires have the cords installed at an angle to the radii of the tire and tread. The bias ply tire is stiff and can carry heavy loads. As the wheel tilts during a turn, the bias sidewall picks up the inner portion of the tread from the road, thereby reducing traction and increasing wear. Bias tires are not fuel efficient and can be rough riding because of the lack of flexibility in the sidewall.

Radial Ply

The radial tire is currently the most popular tire in use today. The cords are laid 90 degrees to the tread and parallel to the tire's radii (Figure 10-40). This provides a softer ride, better fuel mileage, and better traction. The radial sidewall stiffens the tread while maintaining a flat tire/road contact. As the wheel assembly tilts during a turn, the tire's sidewall flexes instead of pulling the tread from the road. The soft sidewall flexes more readily over various road condi-

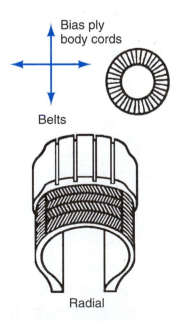

Figure 10-40 Radial cords cause a softer sidewall, allowing the tread to remain flat on the road.

tions. The tire's flexing sidewall results in a smoother ride and less rolling resistance. The lower resistance increases the vehicle's fuel mileage.

The greatest drawback to a radial tire design is the lack of driver education on its inflation characteristics. With a bias ply tire, it is fairly obvious that the tire is low on air pressure. However, with the soft sidewall of a radial tire, the tire tends to look a little underinflated even when pressurized to 35 psi. Many drivers do not recognize when the tire is low. As a result, many radial tires are worn out prematurely.

A tire that became available from several manufacturers in 1997 is commonly known as the run-flat or **extended mobility tire (EMT).** It has extra strong sidewalls that will support the vehicle for up to 125 miles. These tires are usually mounted on vehicles that are equipped with electronic tire pressure monitors to alert the driver of low tire pressure. Another type EMT has self-sealing capabilities, but it is not designed to run forever with a puncture. The tire was designed to seal the hole and hold air until a repair facility could be reached. With this tire, there is no loss of control that often happens during a sudden deflation of the tire.

Extended Mobility Tires (EMTs) are strengthened sidewalls to support a vehicle for up to 125 miles in the event of tire damage. Some have self-sealing capabilities that are temporary.

Tire Construction and Labeling

The tire is made up of several layers of different materials, each designed for a specific task. The cord layout was discussed in the last section. Other major components of the tire are the belts and tread.

Belts

Belts of reinforcing materials are placed under the tread for better support. The belts are laid under and run in the same direction as the tread (Figure 10-41). The belts help hold the tread in position on the **carcass** and gives better tread/road contact. The belts commonly used today are made from layers of steel wires. Belts are used in bias and radial tires.

The carcass is the completed tire minus the tread.

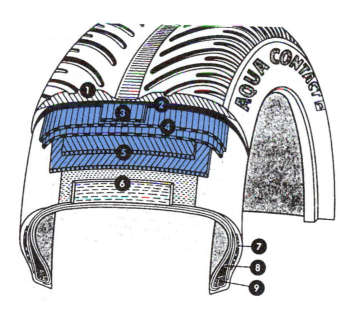

1. CAP tread
2. Tread base
3. Nylon wound aquachannel breaker
4. Two-ply nylon wound breaker
5. Two steel-cord plies
6. Two rayon-carcass plies
7. Double nylon bead reinforcement
8. Bead filler
9. Bead core

Figure 10-41 The belts support and hold the tread in place. (Reprinted with permission from SAE)

Treads

There are many different tread designs, each for a particular purpose. They can be molded to bias or radial carcasses, although some work better with one type or the other. *Straight treads* are most common on over-the-road trucks, which require traction for starting and stopping the vehicle with little concern for heavy cornering. *All-weather treads* are made for the typical light vehicle. They have good traction on dry or wet roads and have some mud and snow capabilities. The shortcoming of the all-weather tread is its ability to work well in most situations, but not perfectly in any. Like most equipment on the vehicle, it is a design compromise.

Another tread commonly found is the *mud and snow tire.* The tread consists of knobby lumps of rubber with large spaces between the knobs. This allows the mud and snow to be pushed out of the way so the knobs can reach a traction surface. The mud and snow tread is generally used only on four-wheel-drive vehicles and those areas of the world where mud or snow are common conditions. The northern parts of the United States have tire treads designed for snow only. While this tire works well in snow, it, like the mud/snow tread, is loud and wears quickly on dry pavement. A common solution in areas with regular snowfall is to use an all-weather tread and install tire chains when necessary.

A special tire is also available for ice. This tire has a mud/snow tread with small steel spikes protruding from the tread area. The spikes dig into the ice for traction. However, the spikes will not last long when driven on pavement without ice.

Each tire is equipped with tread wear indicators (Figure 10-42). Several solid strips of rubber run across the width of the tread area near the bottom of the grooves. When the tread wears to a depth where the indicators become readily visible, the tire should be replaced.

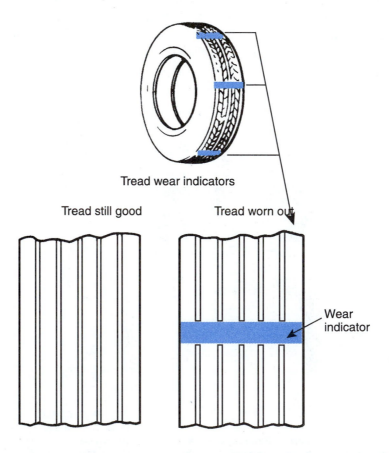

Tread wear indicators

Tread still good   Tread worn out

Wear indicator

Figure 10-42 The tire should be replaced if any wear indicator is even with the tread. (Courtesy of DaimlerChrysler Corporation)

Tire Labeling

Each tire is identified by markings etched or molded into the sidewall. The markings include the manufacturer, tire brand, tread size, and test certifications. The most recognized marking is the tire size.

The tire is sized in three measurements: *tread width, tire diameter,* and *aspect ratio* measured with the tire properly inflated. The size of the tire is one of the determining factors in its load capability. The tread width is the distance between the two edges of the tread measured in millimeters. The tire diameter is the distance between opposite sides of the same bead. The aspect ratio is the percentage of the height of the sidewall to the width of the tire (Figure 10-43). A typical tire size marking is:

P185/70R14

The first group, 185, means the tread is 185 millimeters wide. The letter "P" prefix designates a *passenger* car tire. The number 70 indicate the sidewall height is 70 percent of the tire's width or about 129 millimeters. The letter "R" indicates a *radial* sidewall. The last group, 14, is for a 14-inch diameter. This size tire fits a 14-inch rim with a width of about four to five inches. The tire could be fitted to a wider or narrower rim but the beads would not seal correctly and might result in a blowout.

A **Tire Performance Criteria (TPC)** number is molded into the sidewall near the tire size (Figure 10-44). This number indicates the tire's traction, noise, handling, rolling ability, and endurance characteristics. Replacement tires should have the same TPC number.

Other markings include the **Uniform Tire Quality Grading (UTQG)** which outline the tire's traction, tread wear, and temperature resistance. The grades are shown by letters and numbers stamped on the sidewall. Tread wear is based on a percentage of 100. Percentages range from 50 (poor) to 270 (long) relative wear. Resistance to skidding (traction) and temperature are listed by letters A, B, or C with A being the best and C the worst. High performance tires can be rated for speed using the codes F (50 mph) to Z (149 mph). The codes would show up as part of the tire size markings such as 225/70R 16 Q for a tire rated at 100 mph (Figure 10-45).

> The *Tire Performance Criteria (TPC)* is a number, molded in the sidewall, that lists important information about the tire.
>
> *Uniform Tire Quality Grading (UTQG)* is a system developed to provide the customer a way to compare, select, and trace tires.

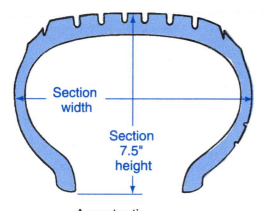

Aspect ratio

Vertical dimension is 75% of the horizontal dimension

75 - Series
75 - Profile ratio

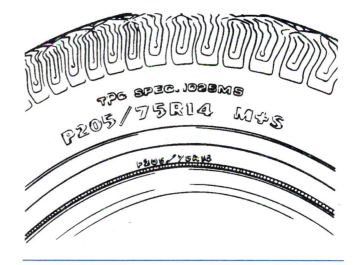

Figure 10-43 The aspect ratio has a direct effect on the vehicle ride height: the lower the ratio, the lower the ride height. (Courtesy of Oldsmobile Division, General Motors Corporation)

Figure 10-44 The TPC number identifies the type and performance of the tire. (Courtesy of Oldsmobile Division, General Motors Corporation)

TIRE SPEED RATINGS

Letter Symbol	Maximum Speed
F	50 mph
G	56 mph
J	62 mph
K	68 mph
L	75 mph
M	81 mph
N	87 mph
P	93 mph
Q	100 mph
R	106 mph
S	112 mph
T	118 mph
U	124 mph
H	130 mph
V	149 mph
Z	+149 mph

Figure 10-45 Tire speed ratings chart.

The Federal Department of Transportation requires certain information to be placed on the tire. The required information is:

- Tire size
- Maximum permissible inflation pressure
- Maximum load rating
- Generic name of each cord material in sidewalls and tread
- Actual number of plies in sidewall and tread if different from sidewall
- Tubeless or tube type as applicable
- Radial as applicable
- DOT manufacturing code listing the tire manufacturer, factory location, and date of manufacture

Wheel Alignment

Shop Manual
pages 272–275

 **WARNING:** Do not apply heat or bend steering or suspension components to make adjustments. The component may be weakened, which could result in failure. Damage or injury could also result.

A vehicle with perfect alignment of the four wheels will steer straight ahead and return to center after a turn. Wheels that are not in alignment with the vehicle and other tires affect the driver's control. Tires that are not in balance could also affect steering and ride. Alignment of the wheels places all wheels parallel to each other and the center or thrust line of the vehicle.

There are usually three angles that can be adjusted by the technician, but other measurements may affect tire wear and steering. **Steering axis inclination (SAI), setback angle,** and **toe-out-on-turns** can only be repaired or adjusted by replacing worn or damaged parts. SAI is the angle created by the steering knuckle and control arms or struts as they are installed in the vehicle (Figure 10-46). If the SAI is wrong, the damaged parts, including the frame, must be replaced or repaired.

Setback is the measurement of one wheel to the other wheel on the same axle (Figure 10-47). In other words, the two wheels on one axle should be straight across from each with relation to the frame/body. Extreme setback could cause steering and possibly suspension problems. Some vehicles may have enough adjustment capability to offset some error in SAI or setback. While this may prevent tire wear and retain good steering the vehicle will look as though it moves sideways. This "dog tracking" is a common condition on older, half-ton pickup trucks that carried loads across rough terrain and had a bent or twisted frame.

A bent or twisted frame can change the *setback, SAI,* and *toe-out-on-turn* settings without damaging suspension or steering components.

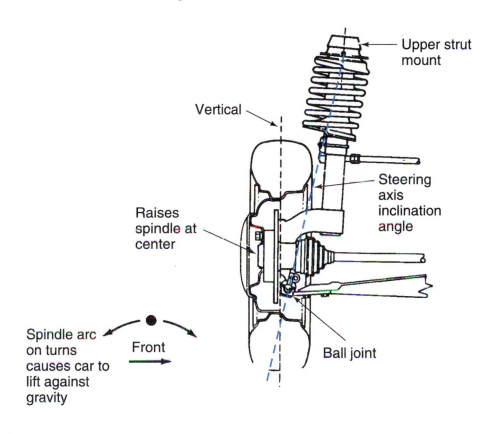

Figure 10-46 The steering axis inclination is permanently set by the installation of the suspension. (Courtesy of DaimlerChrysler Corporation)

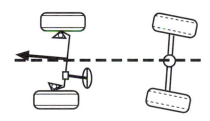

Figure 10-47 Setback occurs when the wheels or axles are not aligned to each other.

Toe-out-on-turn is required because the turning angles between the inside wheel and the outside wheel are different. When a vehicle travels around a curve, the inside wheel travels a shorter distance at a sharper angle than the outside wheel (Figure 10-48). The shaping and positioning of the steering linkage and knuckle sets the outside wheel's turning angle. If the toe-out-on-turns is wrong, some or all of the components must be replaced.

Three alignment angles that may be adjusted are *caster, camber,* and *toe.* Many new FWD vehicles are not equipped for adjustment of caster or camber or both. The rear wheels on many FWD vehicles can be adjusted for caster, camber, toe, or all three. All vehicles should be checked for four-wheel alignment even though an adjustment of the wheels may not be possible. A four-wheel alignment check gives the technician an idea of the frame or unibody condition by comparing the caster, camber, and toe on all wheels.

Caster is the forward or rearward tilt of the steering axis from vertical (Figure 10-49). Caster is measured as positive or negative. Most light vehicles require a positive caster or a rearward tilt of the top of the steering axis centerline. A positive caster tends to make the wheels steer straight and assists in the return of the steering to center after making a turn. Positive caster does require more steering effort when negotiating a turn.

Negative caster is used on large, heavy trucks to assist in turning the front wheels, but the wheels have to be steered back to straight. Poor caster will not normally wear the tire but may affect steering quality. Positive caster is usually expressed as a positive (+) number on the alignment machine.

Camber is the inward or outward tilt of the steering axis centerline (Figure 10-50). A positive camber tilts the top of the wheel away from the center of the car. Camber is used to place

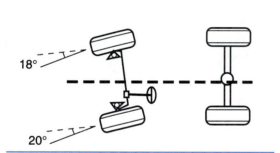

Figure 10-48 The outside wheel does not turn at the same angle as the inside wheel.

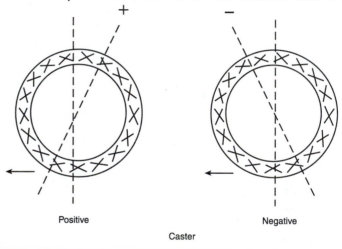

Positive   Negative

Caster

Figure 10-49 Positive caster is the rearward tilt of the steering knuckle or suspension when viewed from the side.

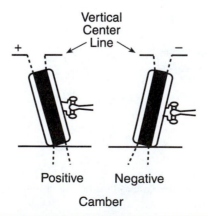

Figure 10-50 Camber is the inward or outward tilt of the wheel as viewed from the front.

302

the load of the vehicle at the center of the tire/road contact area. This prevents tire wear and assists in keeping the tires straight ahead. Poor camber will wear the tires and affect steering quality. Positive camber is usually expressed as a positive (+) reading on an alignment machine.

Toe is the difference in distance between the front edges and the rear edges of tires on the same axle (Figure 10-51). Older vehicles usually had some toe-in while most FWD cars have zero toe. All vehicles have some type of toe adjustment. *Toe-in* means the leading (front) edges of the tires are closer then the rear edges. *Toe-out* is the opposite. Toe sets the wheels straight ahead when the vehicle is loaded and moving. Note that the measurement is total toe, but toe adjustments are made at individual wheels so the wheels need to be set at equal toe. For instance, if the total toe is (+) 0.12 inch, each wheel would be set at (+) 0.06 inches. Poor toe wears the tires and affects the steering. Toe-in is usually expressed as a positive (+) number on an alignment machine.

A fourth measurement that may have to be adjusted before setting the caster, camber, and toe is the vehicle's **ride or curb height** (Figure 10-52). Many vehicles require the ride height to be measured and adjusted before beginning alignment procedures. The height of the vehicle affects the angles of the steering and suspension components. Earlier, in the discussion on torsion bars, it was noted that some torsion bars have an adjustment to change ride height. Other systems may require suspension components to be replaced to correct ride height. The amount of load a vehicle carries may change the ride height and suspension angles.

A vehicle that normally carries a heavy load, such as a large tool unit with repair equipment in the bed, should have a wheel alignment performed with the load onboard. The load tends, depending on its place in the bed, to add or remove weight from the steering wheels. Another example is a sales representative's vehicle, which usually carries numerous product samples and sales material in the trunk or rear seat. This removes weight from the front steering wheels. In each case, the load can force the alignment out of adjustment and result in excessive tire wear and poor steering.

When performing a wheel alignment, caster should be completed first, followed by camber. While caster and camber can be adjusted simultaneously on some vehicles, *toe is always adjusted last*. Toe may affect camber to some degree, but camber adjustments always affect the toe. Many vehicles require that a specified weight be placed in the passenger compartment to simulate the driver's weight. In theory, most vehicles are assumed to have a 150- to 175-pound driver without any other load.

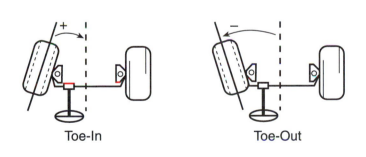

Toe-In   Toe-Out

Figure 10-51 Shown is the toe on a wheel that is half of the total toe.

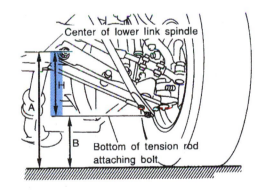

Figure 10-52 The ride height on this vehicle is calculated by subtracting the "B" measurement from the "A" measurement. Different vehicles have different measuring points. (Courtesy of Nissan North America)

Tire Balance and Rotation

Shop Manual
pages 269–272

A balanced tire provides a smooth ride, better steering, and smoother braking. An unbalanced tire may cause the vehicle to shake or vibrate at certain speeds and it may cause tire wear. Tires that are not rotated regularly will wear unevenly and shorten the tire's life. They will also become unbalanced quickly.

> **WARNING:** Do not move directional tires to the opposite side of the vehicle. The tire will not grip correctly. It will wear quickly and could cause an accident. Consult the service manual for directional tread tires.

> **WARNING:** Do not rotate specially designed wheels from their place on the vehicle. Some wheels are made to fit either a front or rear axle position. Moving this type of wheel to another axle may cause damage to the wheel, brake components, and suspension/steering components. Consult the service manual for this type of wheel.

Most vehicle and tire manufacturers recommend tire rotation about every 5,000 miles. All tires should be the same size and type. The direction of rotation depends on vehicle and tire manufacturer instructions, but generally, the rear tires go to the front axle and the fronts to the rear. Some manufacturers recommend that the tires be crossed on the vehicle as they are moved from axle to axle (Figure 10-53). Moving the tires to the opposite sides of the vehicle reverses the rolling movement of each tire. This helps even the wear across the tire. The tires should be balanced during the rotation.

Regular tire wear will cause some unbalancing of the tire. Rough road surface, striking curbs, or other road hazards can damage a tire or wheel, thereby causing an unbalanced condition. Balancing basically means adding weights to offset a heavy spot in the wheel assembly. Most wheels are balanced by the wheel manufacturer and will remain balanced unless damaged. If a wheel is bent or damaged in some way, it should be replaced. Even if a bent wheel assembly can be balanced, it will still wobble and cause a vibration. The wheel assembly is balanced using either static or dynamic procedures.

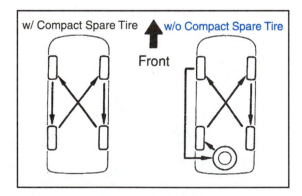

Figure 10-53 Note the difference in rotation when a full-size spare is used (right drawing). (Reprinted with permission)

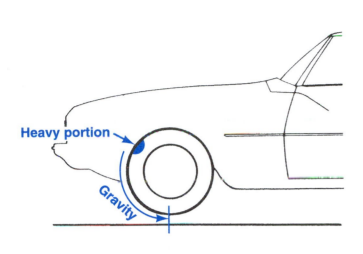

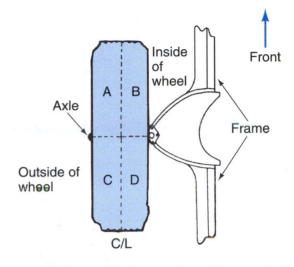

Figure 10-54 This wheel is statically unbalanced. It will rotate by itself until the heavy spot is at the lowest point.

Figure 10-55 The tire is theoretically divided into four quarters and each quarter is balanced to the others in dynamic balancing. A computerized balancer is required.

Static Balancing

A wheel assembly is considered to be **static balanced** if its weight is distributed evenly around the center of rotation as viewed from the side (Figure 10-54). The wheel will tend *not* to rotate if balanced. A *heavy spot* will cause the tire to rotate until the spot is resting at the bottom. The heavy spot would also cause a tramping or thumping noise as the vehicle was driven. Static balancing is performed by adding a weight equal to the heavy spot on the opposite rim of the wheel. This would stop the hopping or tramping of the wheel but it would not eliminate a side-to-side wobble if the wheel was not dynamically balanced.

Static balancing is usually done on heavy truck or trailer tires using a bubble balancer. *Dynamic balancing* is almost always used on cars and light trucks.

Dynamic Balancing

A wheel assembly that is evenly balanced in four quadrates (viewed from the front) is considered to be **dynamic balanced.** Not only is the tire evenly balanced around its center it is also balanced from side to side on the tire's vertical centerline. To accomplish this task, the front of the tire is divided into four, equal sections and each section is balanced to the other three (Figure 10-55). The entire wheel assembly is now balanced.

Alignment and Tire Balance Machines

There are two machines used by repair shops to align the suspension and steering and balance the wheel assemblies. Both are fairly expensive pieces of equipment and should be used only by trained technicians.

Shop Manual pages 269–275

Alignment Machines

Most shops use a special lift for wheel alignments. It is usually a drive-on lift with drop-down legs to hold the vehicle level during alignment (Figure 10-56). There are turntables for the front wheels so they can be moved for adjustment. Most lifts of this type have long rear portions on each side that can be unlocked to allow adjustment of the rear wheels. The lift may have jacks at the front and rear so the vehicle can be raised from the lift for compensating the **wheel sensors** and to make repairs or adjustments.

The *wheel sensors* must be compensated to allow for the angle of the wheel assemblies.

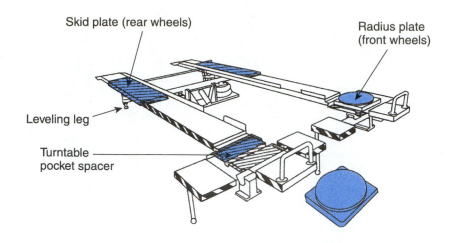

Skid plate (rear wheels)

Radius plate
(front wheels)

Leveling leg

Turntable
pocket spacer

Figure 10-56 The rear skid plate can move for adjustment of the rear wheels. The turntables are for the front wheels. (Courtesy of Snap-on Tools Company)

The *alignment machine* consists of the computer, receiver, and four wheel sensors (Figure 10-57). Older versions used electrical cables to carry the sensor data to the computer. The newest machines use sensors with radio transmitters to send the data to a receiver. The receiver translates the signal so it can be used by a personal computer. Most systems use Windows 95/98 to display the information in a readable form for the technician. The computer uses software that is designed and written for a particular brand of alignment machine. Hunter, Bear, and other machine manufacturers use the same basic data, but interpret and display it according to their software. Each

Figure 10-57 This alignment machine uses a 486 computer and Windows 95. (Courtesy of Hunter Engineering Company)

manufacturer usually includes some type of technician training package on its computer. Hunter Machine Company has video/audio segments that a technician can select to show exactly how an adjustment is done on a particular vehicle. Other manufacturers use similar video, audio, or graphs to show methods of adjustment. Most of the newer units also offer a vehicle safety checklist and a repair order that is completed on screen and then printed. The computer can store customer and vehicle data, before and after measurements, and other data the repair shop may desire. In this manner, the shop can recall data about a vehicle on the customer's next visit to the shop. Software can be added to connect the alignment machine to the cashier, service writer, other departments, and the Internet.

Tire Balancers

The tire balancer is another computerized machine used in most repair shops (Figure 10-58). The tire is mounted onto the machine's spindle and clamped down. Data on the wheel diameter, wheel width, and distance from the machine is entered into the computer using buttons or dials (Figure 10-59). The balancer is set up to spin the tire for a set time at a set speed. In this manner, it can locate the heavy spots and select the positions for the counter weights. The location of a **weight** is usually displayed using a stationary marker or arrow and a light that moves with the rotation of the wheel assembly. The light is aligned with the mark and the weight is installed at the top dead center of the wheel's rim. Weights may be added on one side or both sides of the wheel assembly.

The balancer can be used for static and dynamic balancing. However, most machines will only be able to balance tires up to a certain size and type. Some machines can balance a car tire up to 18 inches in diameter, but may not be able to balance a medium truck tire because of the tire weight and type.

Wheel *weights* are made of lead and should be treated and disposed of as hazardous waste when no longer usable.

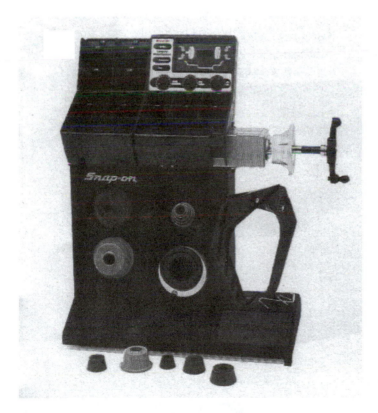

Figure 10-58 A typical wheel assembly balancer. (Courtesy of Snap-on Tools Company)

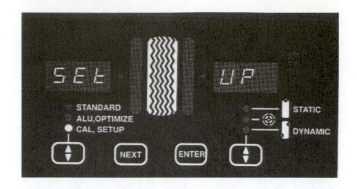

Figure 10-59 Tire information and type of balancing desired can be entered using push buttons. The data is displayed on the two screens ("SEE" and "UP"). (Courtesy of Hunter Engineering Company)

Summary

❑ The Laws of Motion set forth the principles of the movement of objects.

❑ The Laws of Motion directly affect the suspension and steering systems because of their normal operations.

❑ Independent suspension means one wheel can react to road conditions without affecting the opposite wheel.

❑ Non-independent suspension usually means the two opposing wheels are connected solidly to same axle.

❑ Control arms are used on independent suspension systems.

❑ Ball joints connect the control arms to the steering knuckle.

❑ Springs support the vehicle's weight while the shock absorber controls spring movement.

❑ Jounce is the upward movement of a wheel assembly; rebound is the downward movement.

❑ Parallelogram steering systems require a system of linkages that remain parallel to the control arms.

❑ Rack and pinion steering uses a direct linkage between the input steering gear (pinion) and steering knuckle to turn the wheels.

❑ Power assist units are used to reduce driver steering efforts.

❑ Wheels fasten to the hub and support the tire.

❑ The tire size includes tread width, diameter, and aspect ratio and determines the tire's load capacity.

❑ Tires are graded for traction, tread wear, speed, and temperature.

❑ SAI, setback, and toe-out-on-turns cannot be adjusted.

❑ The three possible alignment adjustments are caster, camber, and toe.

❑ Caster should be adjusted first, followed by camber, then toe.

❑ Camber adjustments affect the toe.

❑ Checking alignment should be done using four-wheel alignment procedures.

❑ The vehicle should be aligned while carrying its normal load.

❑ Tire balance can wear tires, cause the vehicle to shake or vibrate, and cause an uncomfortable ride.

❑ The wheel assembly can be balanced statically or dynamically.

❑ The alignment machine uses four wheel sensors and a computer to check the alignment of each tire individually and each tire to the center or thrust line of the vehicle.

❑ The tire balancer spins a tire to a set speed, uses a computer to calculate the heavy spots, and pinpoints the spot to add a weight.

Review Questions

Short Answer Essays

1. Describe the parallelogram steering system.

2. Explain how power assist is used on a rack and pinion steering system.

3. Describe the use of the short and long arm configuration in a suspension system.

4. Explain the purpose of caster and camber.

5. List the major components in a rack and pinion steering suspension.

6. Define the elements of the following data:

   P235/70R15

7. Define static and dynamic balance.

8. Describe the common purpose of coil springs, leaf springs, and torsion bars.

9. Explain the use of adjusting sleeves on a parallelogram steering suspension.

10. List the wheel data that is entered into the tire balancer.

Fill-in-the-Blanks

1. A(n) _____ is used to connect the pinion gear to the wheels.

2. Shock absorbers _____ the movement of _____ .

3. Tie-rod ends are attached to the steering _____ by a(n) _____ .

4. Control arms are mounted to the _____ and attached to the steering knuckle with _____ _____ .

5. Tire wear is not caused by misadjusted _____ .

6. A tire can be balanced using _____ or _____ .

7. The parallelogram steering system normally uses a(n) _____ gear system.

8. The control arm resting on the coil spring is the _____ _____ arm.

9. The _____ , _____ , and _____ can be adjusted using an alignment machine.

10. Pressurized _____ gas is used to cause immediate action in the shock absorber.

Terms to Know (con't.)

Rebound

Recirculating ball and nut

Sector shaft

Setback

Sidewall

Sprung weight

Static balance

Steering axis inclination (SAI)

Sway bar

Tie-rod

Tire bead

Toe

Toe-in

Toe-out

Toe-out-on-turns

Torsion bar

Tread

Uniform Tire Quality Grading (UTQG)

Unsprung weight

Variable rate springs

Worm gear

ASE Style Review Questions

1. The steering system is being discussed. *Technician A* says a parallelogram system uses a rack for connection to the steering knuckle.
   Technician B says the steering knuckle is used to mount the hub. Who is correct?
   A. A only
   B. B only
   C. Both A and B
   D. Neither A nor B

2. *Technician A* says a leaf spring gets longer when it is bowed.
   Technician B says leaf springs are normally used in a rigid suspension system. Who is correct?
   A. A only
   B. B only
   C. Both A and B
   D. Neither A nor B

3. Wheel alignment is being discussed. *Technician A* says ride height must be checked before beginning alignment procedures.
   Technician B says the total toe must be divided by two during the adjustment procedures. Who is correct?
   A. A only
   B. B only
   C. Both A and B
   D. Neither A nor B

4. Wheel alignment is being discussed. *Technician A* says caster adjustments may affect toe.
   Technician B says the toe adjustments should be made after the caster and before the camber. Who is correct?
   A. A only
   B. B only
   C. Both A and B
   D. Neither A nor B

5. Steering angles are being discussed. *Technician A* says setback results from the manufacturing and assembly of the vehicle.
   Technician B says SAI angles can be adjusted on some vehicles. Who is correct?
   A. A only
   B. B only
   C. Both A and B
   D. Neither A nor B

6. A vehicle with a bent frame is being discussed. *Technician A* says the SAI and setback angles are probably changed.
   Technician B says the caster, camber, or toe are probably affected. Who is correct?
   A. A only
   B. B only
   C. Both A and B
   D. Neither A nor B

7. The suspension system is being discussed. *Technician A* says a torsion bar suspension requires a shock absorber.
   Technician B says a leaf spring system does not use a shock absorber. Who is correct?
   A. A only
   B. B only
   C. Both A and B
   D. Neither A nor B

8. A non-independent suspension system is being discussed. *Technician A* says this type suspension is commonly used on rear-wheel-drive trucks.
   Technician B says coil springs can be used on this type suspension. Who is correct?
   A. A only
   B. B only
   C. Both A and B
   D. Neither A nor B

9. Tire wear is being discussed. *Technician A* says misadjusted toe will not cause tire wear.
   Technician B says misadjusted caster does not cause tire wear. Who is correct?
   A. A only
   B. B only
   C. Both A and B
   D. Neither A nor B

10. Wheel alignment is being discussed. *Technician A* says a good alignment will help the driver turn and steer the vehicle with little effort.
    Technician B says adjusting the toe before adjusting the camber could cause tire wear. Who is correct?
    A. A only
    B. B only
    C. Both A and B
    D. Neither A nor B

Brakes

Upon completion and review of this chapter, you should be able to:

- Discuss the application of Pascal's Law in brake systems.
- Explain the purpose and operation of the brake system.
- Discuss the use of power boosters.
- Explain the purpose and operation of master cylinders.
- Discuss and explain the use of control valves.
- Explain the operation of drum brakes.
- Explain the operation of disc brakes.
- Explain the purpose and operation of the parking brake.
- Discuss the purpose and general operation of antilock brake systems.
- Explain the basic operation of traction control systems.

Introduction

An object in motion tends to stay in motion. Obviously, our vehicles cannot always be in motion. They must be slowed and stopped at times. The brake system supplies the necessary force for the driver to slow or stop the vehicle. The system is built on the application of hydraulics theory.

A BIT OF HISTORY

The earliest automobiles used a mechanical linkage to force a block of wood, leather, or metal against the wheel for brakes. This is still used on some horse-drawn wagons.

Hydraulic Theory and Pascal's Law

Shop Manual
page 293

The automobiles produced in the early 1900s were stopped by using a mechanical linkage attached to what we now call *drum brakes*. The linkage required constant adjustment. The most that could be said for those brakes was that the top speed in those days averaged about five miles per hour. With heavier vehicles and higher speeds, mechanical linkage was not sufficient. Hydraulic brakes were first used in the 1920s.

Hydraulic Theory Review

As discussed in Chapter 4, "Automobile Theories and Operation," liquids cannot be compressed and make ideal transmitting agents when contained in sealed systems. In this manner, any force applied at one end would result in an equal force being applied at the other end. The force is transmitted by pressure buildup of the fluid within the system. The fluid's pressure can also be used to increase or decrease the forces at work.

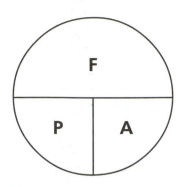

Figure 11-1 Force (ft.-lb.) is equal to pressure (psi) times area (sq. in.).

Pascal's Law states that pressure applied to fluid in a sealed system is transmitted equally in all directions and to all parts of the system. Pascal also developed a mathematical formula to calculate the use of pressure to transmit force. The formula is force (F) equals pressure (P) multiplied by area (A) (Figure 11-1). Force is measured in foot-pounds (ft.-lbs.) and pressure is measured in pounds per square inch (psi). Area is in square inches (sq. in.).

The use of Pascal's formula is similar to the one used when applying Ohm's Law in Chapter 4, "Automobile Theories of Operation." A major mistake during the first usage of Pascal's Law was mixing force and pressure during the calculations. An example of the correct formula is:

F = PA where P is 10 psi and A is 2 sq. in.
F = 20 ft.-lbs.

In this case, a 20 ft.-lb. force results when 10 psi are pressing a 2 sq. in. piston. Another method of the same formula is P = F/A.

P = F/A where F is 15 ft.-lbs. and A is 3 sq. in.
P = 5 psi

The formula can also be used to calculate **mechanical advantage.** This is similar to gear ratios where a small force can be increased to a higher force within the same system. In Figure 11-2, the .75 sq. in. input piston is being applied with a force of 100 ft.-lbs. Using the formula, the pressure within the system is 133 psi (P = 100/.75). Since the pressure is the same everywhere within the system, 133 psi are pressing on each square inch of the 3 sq. in. output piston. The force of the output piston is calculated by F = 133 by 3 or 399 ft.-lbs. This gives a mechanical advantage of 3.99 to 1 (399/100). Another method of calculating mechanical advantage is to divide the output by the input or 3/.75 = 4:1.

A *mechanical advantage* is the amount of force increased at the output compared to the input force.

Shop Manual page 295

The *master cylinder* converts the driver's mechanical input force to hydraulic pressure for the brake system.

Brake System Operation

The automotive brake system uses hydraulic theory to transfer the driver's input force throughout the system to an output force that applies the mechanical brakes. The components of a basic brake system include the pedal and linkage, the **master cylinder,** lines and hoses, controlling valves, and either drum or disc brakes at the wheels. Most systems also use a power boost to assist the driver's efforts in braking. The tires are not considered to be part of the brake system even though they provide the contact between the applied braking force and the road. If the tire is slick with little or no tread, or the road surface is slick, there may be no

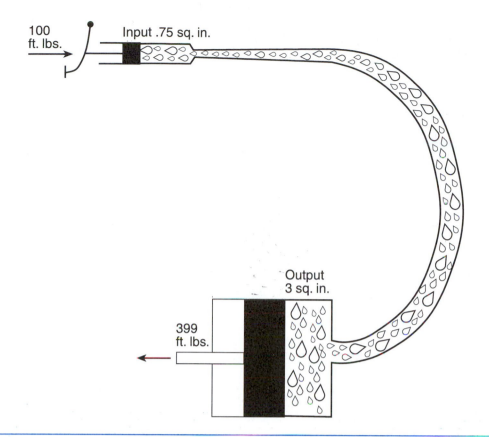

100 ft. lbs.

Input .75 sq. in.

Output 3 sq. in.

399 ft. lbs.

Figure 11-2 An output piston that is larger than the input piston can increase the input force.

slowing of the vehicle even when brakes are forcibly applied. An inspection of the tires' tread should be made when the vehicle is receiving brake repair or inspection.

Braking

As the driver presses downward on the brake pedal, the movement of the pedal is transmitted to the piston in the master cylinder (Figure 11-3). The piston pressurizes the brake fluid in the bore and attempts to force it through the brake lines. Since this is a sealed system the pressurized

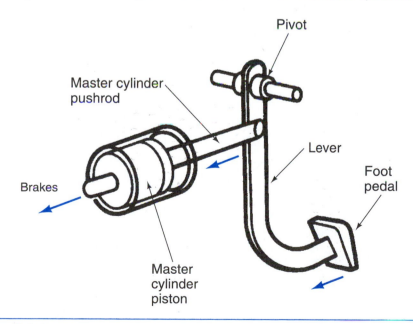

Pivot

Master cylinder pushrod

Lever

Foot pedal

Brakes

Master cylinder piston

Figure 11-3 The pressure generated in the master cylinder transmits the applied force to the wheel brakes.

Wheel cylinders work with drum brakes. Calipers are used with disc brakes.

fluid transmits the applied driver's force to the output pistons at the wheels. The pistons may be in a **wheel cylinder** or disc **caliper**. The pistons act against the drum brake shoes or disc brake pads, respectively.

Releasing

When the brake pedal is released, return springs at the pedal, inside the master cylinder, and at the drum *brake shoes* retract each of the brake components to the rest (release) position. Within the disc brake caliper, a *distorted square-cut O-ring* straightens when the brake pedal is released. As the ring straightens or returns to its normal shape, the caliper piston is pulled backward slightly. This reduces the contact of the pads with the rotor.

Power Boost

The diaphragm is a large, flexible disc stretched across the interior of the vacuum brake booster to divide it into two chambers.

Most vehicles use some type of booster on the brakes to reduce the driver's effort. The most common type used on passenger cars and light trucks is the *vacuum chamber* (Figure 11-4). The engine supplies the vacuum. A diesel engine requires the use of a vacuum pump to provide the vacuum. Movement of the brake pedal opens an air valve that allows atmospheric pressure to enter one side of the chamber. This extra force acting on the booster interior **diaphragm** increases the force exerted by the driver. Some vehicles use hydraulic systems to boost the driver's braking input.

A dual brake system is used on almost every vehicle now sold in the United States. The brake system has two hydraulic circuits, including a dual chamber master cylinder. In this manner, a leak or other failure on one system will not affect the other system. It will take additional space and effort to stop the vehicle, but the driver will still be able to maintain control.

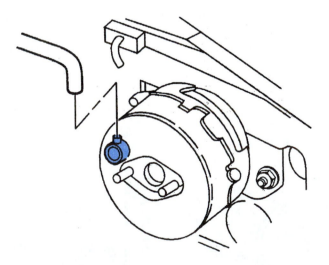

Figure 11-4 The check valve seals the brake booster vacuum chamber during engine shutdown or heavy engine load. (Courtesy of Chevrolet Motor Division, General Motors Corporation)

Brake System Apply Components and Operation

Shop Manual
pages 295–301

Each component of the brake is essential to the proper operation of the complete system. A failure of one component will affect the entire system in some manner, ranging from a substantially reduced braking action to complete system failure.

Pedals and Linkages

The pedal is connected to the master cylinder's piston with a *push rod* (Figure 11-5). If the system uses a power booster, the push rod is connected to the booster's diaphragm or valve to operate the assist. On most power boost systems, the pedal's push rod does not act directly on the master cylinder piston.

On some vehicles, the push rod must be adjusted for length when the master cylinder, booster, or pedal is replace. Better machining techniques have eliminated this adjustment requirement on many vehicles; however, some manufacturers still require an inspection of this mechanical connection when new components are installed.

The mounting for the pedal is generally a shaft from which the pedal hangs downward. A return spring is used to return the pedal to the fully up or off position after the brakes are released. Also mounted on or near the pedal shaft is the brake (stop) light switch (Figure 11-6). The switch is normally open. When the brake is applied, the pedal's shank or the push rod presses a plunger on the switch and closes the brake light circuit. At times, the position of the switch may have to be adjusted to work properly.

Modern vehicles have a multiple function brake light switch. The switch may be used to activate the *antilock brakes,* switch off the cruise control, and signal the PCM that the brakes are applied.

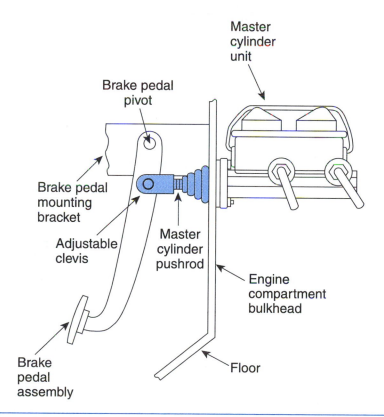

Figure 11-5 The force applied at the pedal is transmitted to the master cylinder's primary piston by the push rod.

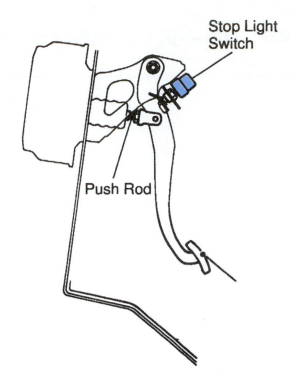

Stop Light Switch

Push Rod

Figure 11-6 This switch will only turn on the brake light. (Reprinted with permission)

Power Boosters and Master Cylinders

The master cylinder is the point at which mechanical force is changed to hydraulic pressure. The booster is placed between the brake pedal and master cylinder. It increases the driver's ability to apply force to the master cylinder piston.

The brake's power booster may be vacuum or hydraulic operated. The **vacuum brake booster** is connected to the engine's intake manifold or vacuum pump. The operating engine draws air from the vacuum chamber and maintains a vacuum in both sides of the chamber. A *check valve,* normally located at the connection of the hose and chamber, maintains the vacuum during engine shutdown or hard acceleration. A large diaphragm extends across the interior of the vacuum chamber. The brake's push rod is connected to the driver's side or rear of the diaphragm. A *power piston* is mounted to the front side of the diaphragm and extends forward to contact the master cylinder's primary piston (Figure 11-7). The power piston houses vacuum and air valves. During brake release, vacuum is supplied through the vacuum valve to the rear chamber of the diaphragm. This helps keep the diaphragm in the release position.

As the brake pedal is depressed (brake apply), the air valve is opened, thereby allowing atmospheric pressure to enter the rear chamber. At the same time, the vacuum valve closes the vacuum port to the rear chamber. The controlled opening of the air valve regulates the amount of pressure that can actually be applied to the diaphragm. This prevents full boost and brake lock-up during light braking. Since a vacuum is still applied to the front chamber, the diaphragm and power piston move forward, thereby increasing the driver's input force. The master cylinder piston is moved forward within its bore and a hydraulic pressure is created in the brake system. When the brakes are released, the rear vacuum valve is opened, the air valve is closed, and the applied vacuum and a return spring force the diaphragm to the center or off position.

Hydraulic boost is accomplished by directing a fluid under high pressure to the driver's side of a master cylinder. A *hydro-boost* system uses the power steering system to supply the pressurized fluid. Fluid from the power steering pump is routed through a valve system fitted between the brake pedal's push rod and master cylinder (Figure 11-8). The valve assembly is

A diesel engine requires a vacuum pump for the *vacuum brake booster* and other vacuum-operated devices.

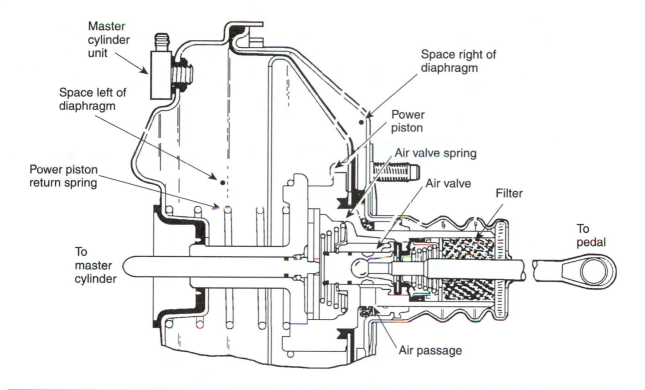

Master cylinder unit

Space left of diaphragm

Power piston return spring

To master cylinder

Space right of diaphragm

Power piston

Air valve spring

Air valve

Filter

To pedal

Air passage

Figure 11-7 The air volume through the air valve is dependent on the movement of the brake pedal. (Courtesy of General Motors Corporation, Service Operations)

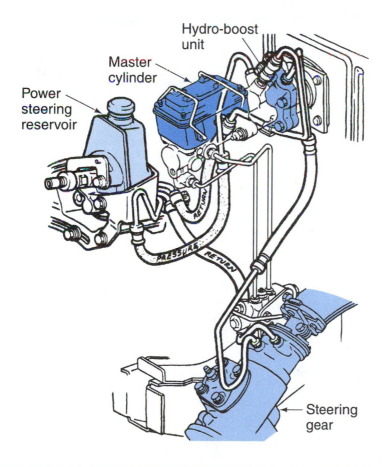

Hydro-boost unit

Master cylinder

Power steering reservoir

Steering gear

Figure 11-8 The power steering fluid passes through the hydro-boost valve when the brakes are not being used. (Courtesy of General Motors Corporation, Service Operations)

part of the master cylinder, but may be replaced individually on most units. As the pedal is depressed, the pushrod opens a valve that directs pressurized fluid to a power piston on the driver's side of the master cylinder. The valve's movement is proportional to the movement of the pedal. The further the pedal moves downward, the more the valve opens. The pressurized fluid forces a power piston and push rod to move the master cylinder piston. This, in turn, pressurizes the brake fluid and the brakes are applied.

When the brakes are released, a release valve is opened to allow the pressurized fluid to flow directly back into the power steering system. A small accumulator is filled with pressurized fluid in the event of a power steering failure. If there is not enough pressure within the boost system when the brakes are applied, the accumulator fluid is directed to the power piston and offers enough brake boost for two or three stops.

The General Motors *PowerMaster* system uses a special master cylinder. A small, electric-motor-driven pump supplies pressurized fluid to the power piston. The motor and pump are mounted under the master cylinder (Figure 11-9). Brake fluid is routed from the cylinder's reservoir to the pump. The fluid is pumped into an accumulator. Pressurized nitrogen gas on one

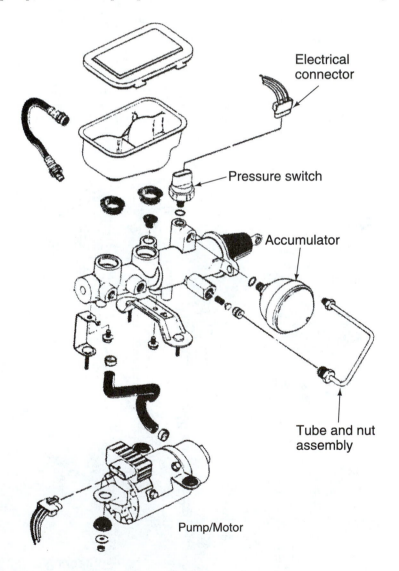

Figure 11-9 The PowerMaster motor and pump are attached to the lower side of the master cylinder. (Courtesy of General Motors Corporation, Service Operations)

side of the accumulator's diaphragm pressurizes the fluid. When the brakes are applied, a **valve** is opened between the accumulator and power piston bore at the driver's end of the master cylinder. The fluid acts against the power piston and assists the driver in applying the brake. When the brakes are released, a release valve is opened to allow the boost fluid to flow back to the reservoir. A pressure switch on the accumulator closes when the pressure drops below a set limit and turns the pump on. In this manner, pressurized fluid is always available for brake operation. As with the hydro-boost system, the accumulator stores enough fluid for two or three stops in case of pump or motor failure.

The movement of the valve *is proportional to the downward movement of the pedal.*

Master Cylinders

Single-piston master cylinders were installed on the orginal hydraulic brake systems (Figure 11-10). A single-piston master cylinder fed a single brake system. If there was a failure in any component, the complete system failed. The operation of the single-piston master cylinder works in the same manner as a dual-piston except for the extra components used in a dual system.

Dual or split brake systems have been required since 1967. This setup allows half of the brake system to be available to control the vehicle if there is a leak or some other failure. The most common system is a front/rear split where one master cylinder piston feeds the brakes on one axle. A **diagonally split system** has a front and a rear wheel fed by one master cylinder piston (Figure 11-11).

The diagonally split system *allows the driver to have braking power at each axle if a leak develops in the brake system.*

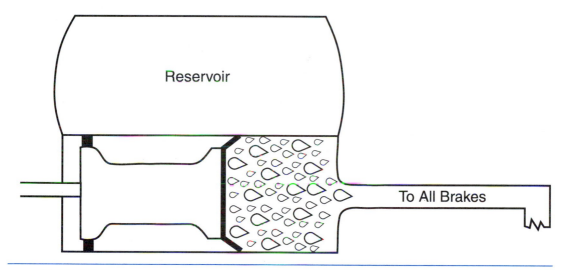

Figure 11-10 Single-piston master cylinders were outlawed from use on motor vehicles by federal laws and regulations.

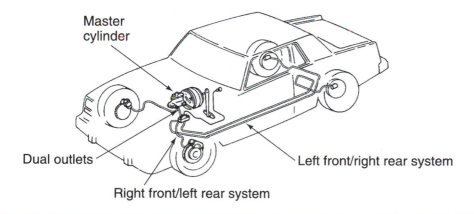

Figure 11-11 The diagonally split system requires a separate brake line from the master cylinder or combination valve to each wheel.

The split system requires a dual-piston master cylinder. The master cylinder is mounted on the forward side of the brake booster or directly on the firewall. The basic components of a dual master are the primary and secondary pistons, return springs, body, and reservoir.

Some master cylinders and reservoirs are molded into one unit, but most reservoirs are plastic and can be dismounted from the master cylinder. The reservoir is mounted to the top of the master cylinder. There are two chambers in the reservoir that separately feed the two bores of the master cylinder (Figure 11-12). The reservoir cover can be quickly removed to add fluid. Most reservoirs are translucent so the fluid level can be checked without opening the system. Some reservoirs have a fluid level sensor that activates a dash light to warn the driver of low fluid levels.

The master cylinder body houses the internal components and is drilled to accept the reservoir and brake lines. The threaded ports on the side of the body are connections for the brake lines and may have a metering orifice installed. There are also vent and replenishing ports drilled into the body that open into the bore and the reservoir. This allows fluid to flow to and from the bore as the pistons move.

The open end of the body's main bore is installed toward the firewall. Pistons and springs are slid into the bore and a locking ring is installed for retention (Figure 11-13). Some older master cylinders may have bleeder ports at the forward or closed end of the master cylinder, but most master cylinders are bled using the line ports.

The two pistons are the primary and secondary pistons (Figure 11-14). The primary piston applies the rear brakes and the secondary piston. As the brake pedal is depressed, the primary piston cup forces the fluid in the compression chamber to move into the brake line and against the secondary piston. As the pressure builds in the chamber and brake lines, the secondary piston moves, starts to pressurize the fluid in its compression chamber, and begins to apply the front brakes. If the vehicle is disc on front and drum on the rear, the initial movement of the primary piston moves the shoes out toward the drums. Return springs push the two pistons back to their release positions when the driver releases the pedal.

As the pistons move to their release positions, fluid in the spool area behind the piston head vents back into the reservoir. Fluid behind the head is needed to remove the pressure drop as the piston moves forward (Figure 11-15). This could cause the piston seals to leak and no pressure would be created in the brake lines. A replenishing port keeps the compression chamber filled so there is no air in the lines and fluid is available within the chamber for instant braking. When the brake is applied and the piston moves forward, the vent and replenishing ports are closed to the compression chamber.

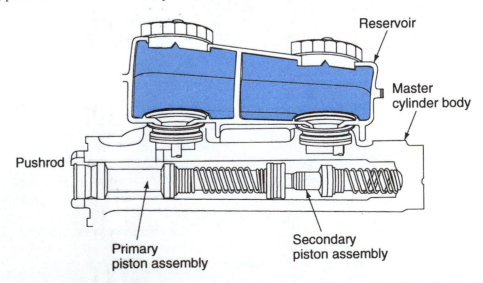

Figure 11-12 The dual-piston master cylinder requires a two-chamber reservoir. (Courtesy of DaimlerChrysler Corporation)

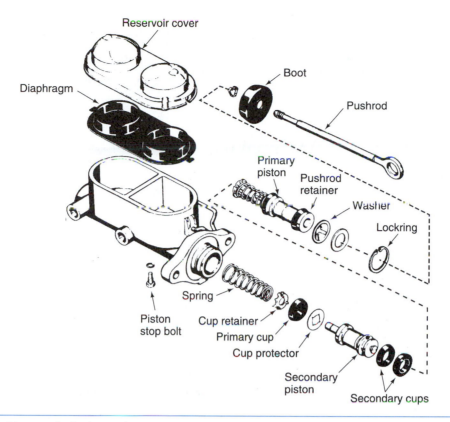

Figure 11-13 Master cylinder internal components are installed into the bore from the firewall end.

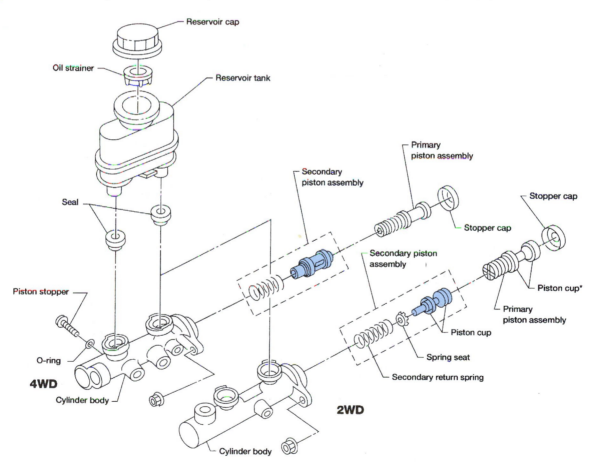

Figure 11-14 The secondary piston is positioned toward the front of the master cylinder bore. (Courtesy of Chevrolet Motor Division, General Motors Corporation)

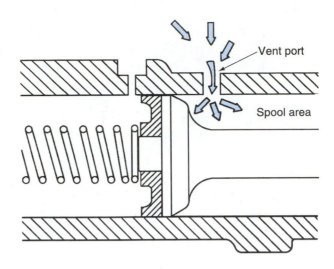

Figure 11-15 The vent port keeps the spool area full of fluid to prevent a pressure drop as the piston moves forward. (Courtesy of EIS Division of Parker Hannifin)

The dual master cylinder was required to allow the driver to retain at least some braking if there is a leak in the system. With the single-piston system, a leak anywhere in the system may cause a complete loss of braking. The dual system reduces that possibility. The *primary piston* has an extension screw that extends into the compression chamber that can apply the secondary piston. If a leak occurs in the rear brake system and the brake pedal is depressed, the primary piston moves in its normal manner. However, the fluid ahead of that piston leaks out and no pressure is created. The *secondary piston* will not move because of the lack of pressure. The screw on the primary piston will contact and push the secondary piston forward (Figure 11-16). In this manner, the secondary piston can charge the front brake system and the

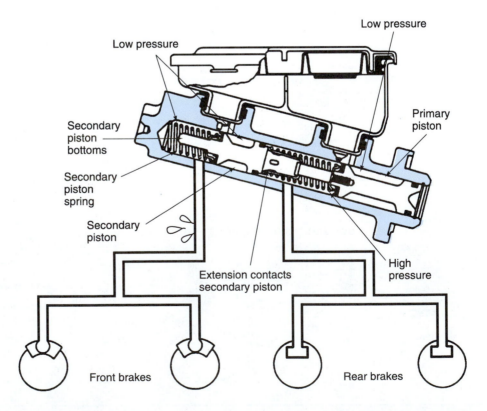

Figure 11-16 A leak in the front brakes will cause the secondary piston to bottom before the primary can build pressure. (Courtesy of General Motors Corporation, Service Operations)

driver will have some braking action available. If a leak occurs in the front system and the brakes are applied, the primary piston charges the rear brake system, but since the front circuit is leaking, the secondary piston must be pushed completely forward to its stop before pressure can build in the rear brakes. Both conditions will cause the brake pedal to drop drastically as the brakes are being applied.

Control Valves, Lines and Hoses, and Warning Lights

Shop Manual
pages 295–296

There are not too many valves in the automotive brake system, but a failure of one may cause a complete loss of brakes. Steel lines carry the brake fluid to the wheels and usually do not require any maintenance. Flexible hoses connect the steel lines to the individual wheels. This allows the wheels to move but does not bend or break solid lines. The hoses have to be inspected and replaced at intervals.

Metering Valves

Vehicles with front disc and rear drum brakes have to hold the front brakes off until the rear brakes move enough to contact the drums. The *metering valve* accomplishes this task. As the brakes are applied, the disc brake reacts instantly, but the rear brake must move far enough to contact the drum. In addition, enough pressure must be created to overrun the return spring on the shoes. If the front brakes were applied at the instance of brake pedal movement, they would probably lock down before the rear ones would begin to brake.

When the brakes are applied on a disc/drum system, the metering valve blocks fluid to the front until **pressure** builds in the rear system (Figure 11-17). This ensures the rear brake shoes are in contact with the drum before the front brakes begin to function. Once opened, the metering has no other function during the braking action. Operation of the metering valve could possibly be noticed during light braking or upon initial braking.

High *pressure* cannot be achieved until the shoes contact the drums, thereby creating a resistance to the shoes' movement.

Proportioning Valves

In order to control the vehicle during hard braking, it is necessary to apply more brake action at the leading axle than the trailing axle. Many brake systems may direct up to 90 percent of the braking action to the front wheels. During heavy braking, this could still cause the rear brakes to lock up. The proportioning valve divides the braking pressure between the front and rear axle to reduce or eliminate rear-wheel lock-up.

The **proportioning valve** is basically a block of metal drilled to accept line fittings, a valve, a piston, and a spring. The lines connected to the valve are from the master cylinder and rear brakes. The piston has a small and a large end with the small end at the valve. A valve stem connects the valve and piston (Figure 11-18). During normal braking, the spring holds the valve open and fluid is directed equally to each axle. As the vehicle's weight shifts forward during braking, the front wheel's road friction increases while the rear contact decreases. High pressure from the master cylinder during a hard stop enters the proportioning valve and presses on the large end of the piston, thereby forcing it to move. The valve stem moves with the piston and moves the valve to restrict the inlet port. This reduces the amount of fluid and pressure that is applied to the rear wheels.

As the pressure from the master cylinder changes, the *proportioning valve* operation also changes, thereby increasing or reducing the pressure to the rear wheels.

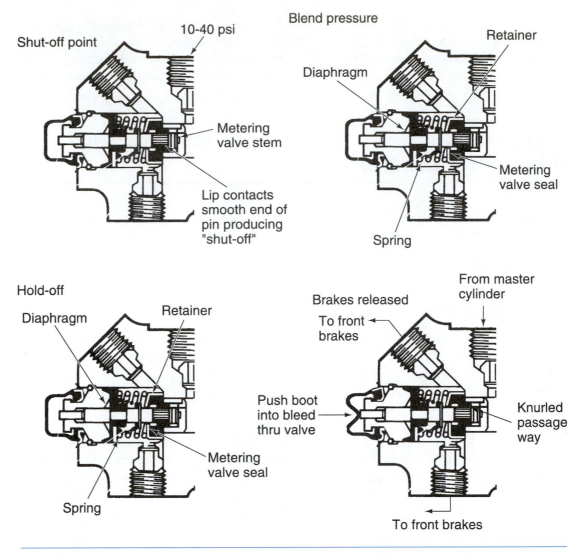

Figure 11-17 The metering valve holds off the front brakes until the rear shoes set to the drum. (Courtesy of General Motors Corporation, Service Operations)

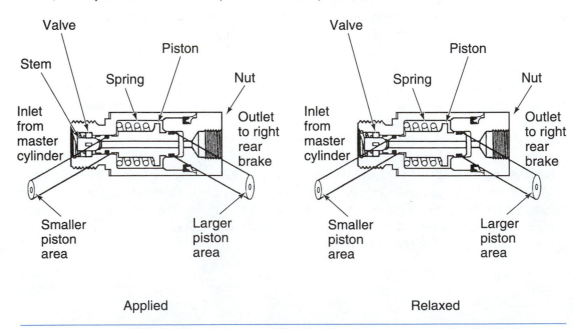

Figure 11-18 The proportioning valve reduces pressure to the rear brakes during hard braking. (Courtesy of General Motors Corporation, Service Operations)

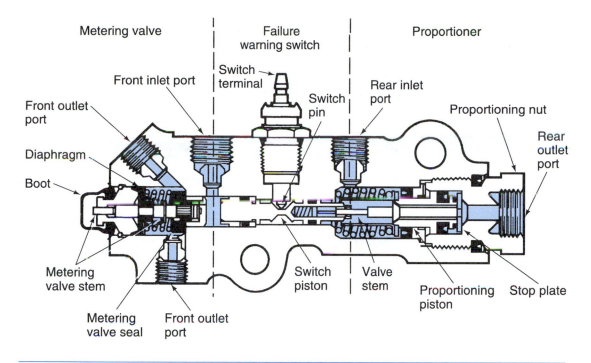

Figure 11-19 The combination valve can perform the functions of a metering valve and proportioning valve, and it turns on the brake warning light when necessary. (Courtesy of General Motors Corporation, Service Operations)

A *combination valve* is used on most vehicles. This combines the metering valve and proportioning valve into one unit (Figure 11-19). The operation of the two valves is the same as stand-alone units. The brake warning light switch may be mounted on the combination valve.

Load-Sensing Proportioning Valves

A truck is extremely light in the rear compared to the front. This is particularly important when the truck is being heavily braked. During braking, the truck's weight shifts to the front and the lightened rear wheels lose traction with the road. If the wheels lock up and skid, the truck can go out of control. A **load-sensing proportioning valve** helps prevent the lock-up by reducing the rear brake pressure during braking.

One type load-sensing valve uses the movement of the frame and rear axle to regulate pressure to the rear brakes. The valve bracket is mounted to the frame with a lever attached to the axle (Figure 11-20). As the rear of the frame rises during braking, the axle pulls down the lever. The valve attached to the lever restricts the amount of fluid and pressure being directed to the rear wheels.

Another type of load-sensing valve is mounted on the rear of the combination valve. It has a ball that moves with the tilt of the vehicle. As the rear of the vehicle rises during braking, the ball moves forward against a ball valve. The ball valve reduces the amount of fluid and pressure to the rear wheels.

Lines and Hoses

Brake lines are made of steel with flared fittings at each end. The steel lines connect the master cylinder to the valves and the valves to the wheels. But the steel lines do not extend completely to the wheels. Flexible hoses connect the steel lines at the frame to the brake system at

A heavily-loaded truck will force the lever in the opposite direction and allow more pressure to be delivered to the rear brakes under normal action. A *load-sensing proportioning valve* prevents lock-up by reducing rear brake pressure.

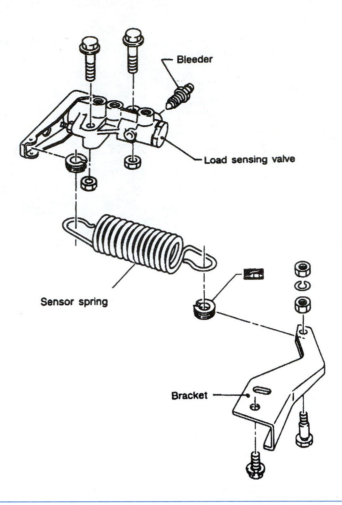

Figure 11-20 The frame-mounted, load-sensing, proportioning valve changes pressure to the rear brakes as the distance between the frame and rear axle changes. (Courtesy of Nissan North America)

the wheels (Figure 11-21). This is required because of the almost constant movement of the wheel assembly during driving. A steel line would quickly break if it were continuously flexing. While steel lines almost never require maintenance, the flex hoses should be inspected each time the wheels are serviced. They will deteriorate and bulge due to chemical actions and age. They can also have interval damage that is not visible.

Brake lines have two types of fittings. The older type is known as a *double flare* fitting. The newer type is an **ISO**-type flare (Figure 11-22). Each requires a matching fitting and cannot be interchanged. The ISO-type provides a better seal.

Warning Lights

The **brake warning light** is used to alert the driver to a problem in one of the systems. There is usually one red warning light located in the instrument panel. It is turned on by any one of three possible switches or sensors. The first switch is on the parking brake and reminds the driver that the parking brake is applied. The second, a brake fluid level sensor, is not present on all vehicles, but if used, alerts the driver to low brake fluid level in the master cylinder reservoir. The third is used on dual brake systems to warn the driver of a brake failure in one of the two brake circuits. It is usually mounted on the proportioning or combination valve. There are two types of switches that may be used in this warning circuit.

On one type, the switch plunger sets in a slot at the center of a long piston in the proportioning valve. The piston is held in the center by equal hydraulic pressures at each end of the piston. When a leak occurs and pressure drops on one end of the piston, the pressure at the

ISO are the initials for the International Standards Organization.

The red *brake warning light* may also be in parallel circuit with the low brake fluid sensor, combination valve, and the parking brake switch. A failure in either system will illuminate the warning light.

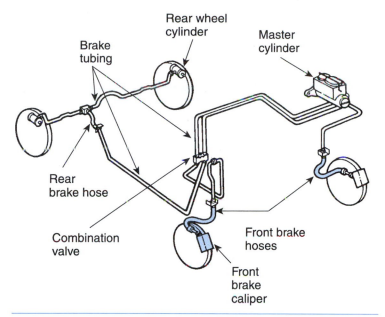

Figure 11-21 Flex lines allow the wheels and axles to move without bending or flexing the steel lines. (Courtesy of EIS Brake Parts)

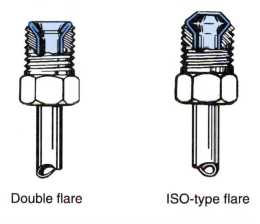

Double flare   ISO-type flare

Figure 11-22 Two types of fittings used on brake lines. (Courtesy of General Motors Corporation, Service Operations)

other end moves the piston out of the center. The shoulder of the piston slot pushes up on the switch plunger to close the light circuit (Figure 11-23). At the same time, the piston closes the outlet ports of the system that is leaking.

Another type of warning light switch is the hourglass-shaped type. The principle of operation is the same but the switch's ground contact extends into the slot. As the piston moves off center, the shoulder of the slot touches the contact and grounds the switch closing the circuit.

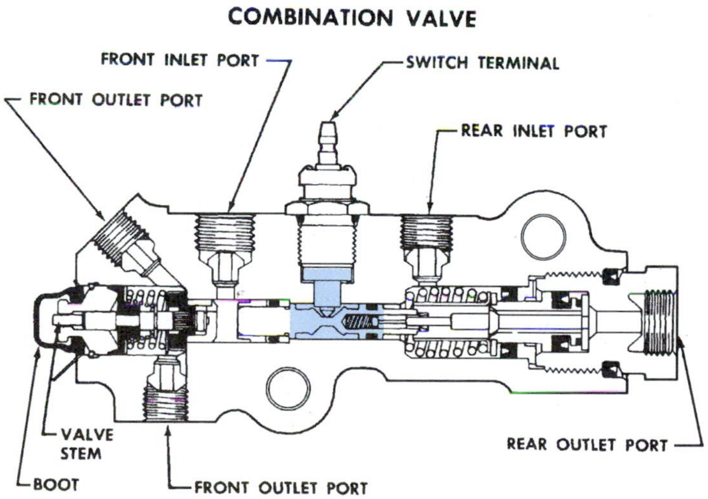

COMBINATION VALVE

FRONT INLET PORT —   — SWITCH TERMINAL

— FRONT OUTLET PORT

— REAR INLET PORT

— REAR OUTLET PORT —

VALVE STEM

BOOT

FRONT OUTLET PORT

Figure 11-23 The plunger is pushed upward to close the warning light switch if the piston moves from the center. (Courtesy of General Motors Corporation, Service Operations)

Drum Brake Components and Operation

Shop Manual
pages 297–300, 317

Asbestos was used for many years as brake *friction material*. Because of the health dangers of asbestos, most brake friction is made of organic materials or compounds.

The brakes on all vehicles consist of a **friction material** being clamped against a rotating device by hydraulic pressure. There are two basic braking systems based on how the friction material and rotating device interact with each other. The oldest system is drum brakes.

The wheel components in a drum brake system are the wheel cylinder, drum, shoes, adjuster, and mounting hardware (Figure 11-24). The hydraulic pressure is delivered from the master cylinder via steel and flexible lines. Once inside the wheel cylinder, the pressure is applied against two pistons.

Wheel Cylinders

The wheel cylinder has two pistons, a spring, sealing cups or seals, two links, and the cylinder body. The body has a machined bore for the pistons. A threaded hole in the bore provides a connection point for the brake line while a second threaded hole drilled into the bore is for bleeding air and fluid from the cylinder (Figure 11-25). The bleeder is always at the highest point of the bore when the cylinder is mounted on the vehicle. Within the bore are pistons, cups, and a spring. The spring is placed between the two pistons and cups to keep all internal components in place and in contact with each other during brake release. This spring does not return the pistons or cups to their starting positions when the brakes are released.

The two cups act as seals when hydraulic pressure is applied to the bore. The interiors of the two cups face each other and the center of the bore. The outside faces of the cups are flat against the inner face of its mating piston (Figure 11-26). As the pressure builds during brake application, the outer edges of the cup are forced outward against the bore to provide a more effective seal.

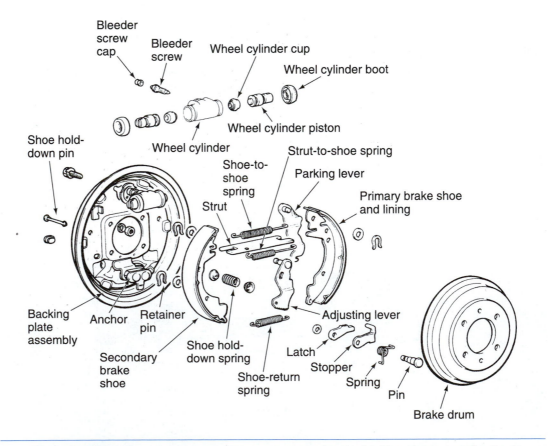

Figure 11-24 The drum brake components are mounted to the backing plate except for the drum. (Courtesy of Mitsubishi Motor Sales of America, Inc.)

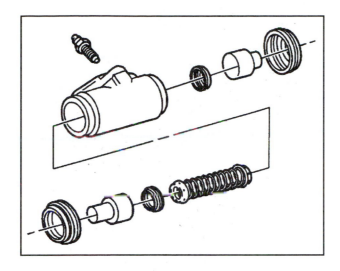

Figure 11-25 The wheel cylinder converts hydraulic pressure to mechanical action to apply the shoes. (Courtesy of Chevrolet Motor Division, General Motors Corporation)

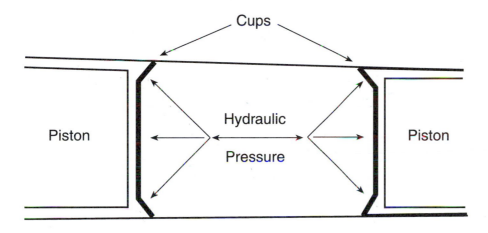

Figure 11-26 The hydraulic pressure applied against the cups seals the wheel cylinder's pressure chamber.

The pistons are placed at each end of the bore with the spring, cups, and hydraulic fluid trapped between them. The outward face of the piston is usually cupped to provide a mounting point for one end of a link. The links are the connecting point between the pistons and the brake shoes (Figure 11-27). Around each end of the cylinder is a boot, which serves as a dust cover to protect the bore. As hydraulic pressure is increased within the bore, the cups, pistons, and links move outward to apply force against the two brake shoes.

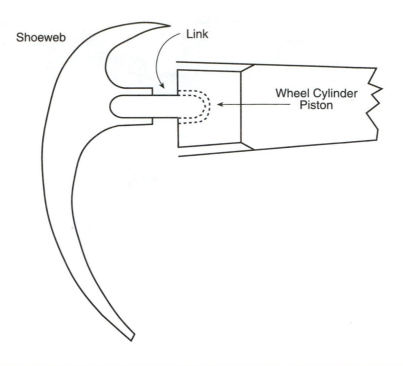

Figure 11-27 The link makes the connection between the wheel cylinder piston and the brake shoe.

Brake Shoes

The brake *shoe web* is the metal skeleton used to mount the friction material and attach the shoe to its mounting.

The brake shoes provide a mounting area for the friction material. The metal portion of the shoe is normally referred to as the **shoe web.** The friction material may be bonded (glued) or riveted to the web. Brake shoes are curved to match the inside wall of the drum (Figure 11-28). Return springs hold the upper ends of the shoes tight against the wheel cylinder links. Retainer springs attach the shoes to the backing plate, but allow the shoes to move out or in as the brakes are applied and released. The backing is bolted to the hub and acts as a rear cover and a mounting platform for the drum brake components. There are several methods to attach the shoes at the bottom, and the type of attachment at that point can help increase braking power.

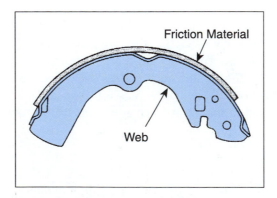

Figure 11-28 The friction is bonded or riveted to the web, which is curved to match the drum. (Courtesy of Nissan North America)

Brake Linings

The lining thickness, position on web, and shape may be different on the brake shoe set for one wheel. The secondary shoe may have a shorter and thinner liner than the primary shoe. The shoes must be properly placed during installation or a brake failure or excessive wear could occur.

> ■ **CAUTION:** Do not breathe brake dust. Some linings are made from asbestos which is classified as a hazardous material. Other linings contain metal or other materials which are also respiratory hazards. Breathing any brake dust could cause short- or long-term health problems.

> ■ **CAUTION:** Asbestos is a hazardous waste and most be disposed of in accordance with EPA and other agencies regulations. Improper disposal of asbestos can cause short- and long-term health or respiratory problems and can result in severe fines or other court-ordered actions.

Brake linings are made of heat-resistance material and are the part of the brake system that actually contacts the rotating drum. The linings may also have bits of copper, brass, zinc, and other materials to increase friction and prolong the life of the lining. This type of compound has replaced asbestos as a lining material.

Servo Action

Some systems use a stationary **anchor** to hold the lower end of the brake shoes in place (Figure 11-29). In this setup, the only force applied to the shoe for stopping is the hydraulic pressure within the wheel cylinder. This increases the stopping distance and requires more effort from the driver. Servo drum brakes are used on most vehicles.

A servo action uses the force from one operating member to increase the force applied to another operating member. In drum brake systems, each brake shoe receives equal force

The pressurized fluid trapped between the pistons and shoes acts as one *anchor* for the shoes.

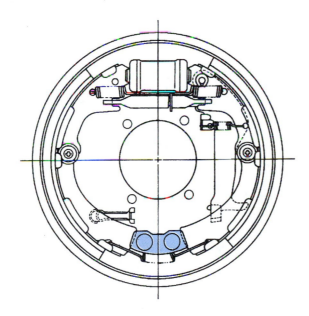

Figure 11-29 The brakes shoes are anchored at the bottom on this brake assembly. (Courtesy of Nissan North America)

from the wheel cylinder to push the shoe out against the drum. If the vehicle is moving forward, the front shoe contacts the drum and attempts to move with the rotating drum. An anchor would stop the shoe from moving, but with a servo system, this shoe is connected to the rear shoe. As the front shoe attempts to rotate, it transmits force through the linkage to the rear shoe. The rear shoe is now pressed against the drum by the wheel cylinder and the servo action. The upper end of the rear shoe cannot move inward because of the pressurized brake fluid trapped between the two wheel cylinder pistons. In this manner, the driver seats the shoes against the drum and the force of the rotating drum is also applied against the rear shoe. Stopping power is increased and the driver's effort is decreased. Note that the secondary or rear shoe performs most of the braking effort, however, the brakes may be set so the front shoe is actually the secondary shoe.

The vast majority of vehicles equipped with drum brakes usually have duo-servo action. The servo effect is possible in both forward and rearward braking with duo-servo systems. The technician who works on antique or classical vehicles may encounter a servo system where the servo action is only accomplished when the vehicle is braked during forward travel.

Brake Self-Adjusters

Drum brake systems require some type of mechanism to adjust the shoes outward as they wear. Older vehicles and some larger trucks require the technician or driver to adjust the brakes at regular internals, normally during routine, preventive maintenance. However, the majority of light vehicles use automatic self-adjusters. The adjuster is the connecting link between the upper or lower ends of the brake shoes (Figure 11-30). A spring holds the two shoes and adjuster in contact with each other. The adjuster has a starred wheel, which is threaded onto an adjusting screw and is normally turned by the movement of the shoes. An adjusting lever contacts the adjusting screw and is moved by the shoe as the brakes are applied. If the shoes are worn and have to travel more than a set distance for contact with the drum, the adjusting level moves further than usual. The additional motion of the lever causes a one or two tooth rotation of the adjuster. Additional adjustments are made each time the brakes are applied. If the shoe movement is short, the level cannot move far enough to actually turn the adjuster.

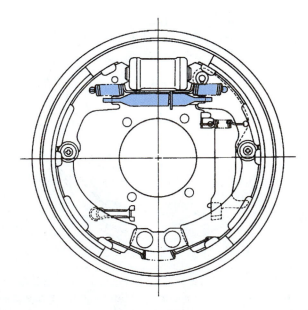

Figure 11-30 The automatic adjusting linkage connects the two shoes to provide servo braking. (Courtesy of Nissan North America)

Some newer systems use the application of the parking brake to adjust drum brakes. While this system works well, many drivers do not realize that the parking brake must be used regularly in order to adjust the brakes. The adjustment works the same way except for the operating member that performs the rotation of the adjusting screw. Most non-servo systems and many other vehicles have the adjuster at the top end of the shoes under the wheel cylinder. Most vehicles have a slot either in the drum or backing plate to allow a brake adjuster tool to be inserted for manual adjustment. This is primarily done to adjust the brakes after repairs have been completed.

Drums

The **drum** covers the brake mechanism and is attached to the axle or hub (Figure 11-31). The drum is drilled to fit over the lug studs. The interior friction area of the drum is machined to provide a smooth braking surface. The installation of the wheel assembly and lug nuts hold the drum in place. Some manufacturers use additional fasteners that hold the drum to the axle or hub. The friction area of the drum can be machined if grooved or otherwise damaged or worn. The inside diameter must be measured and compared to the manufacturer's specifications. A drum that is too thin may break during braking.

Disc Brakes

Disc brakes react quicker and operate differently than drum brakes. Disc brakes are pressurized and controlled the same way as drum brakes. The components of a disc brake are the pads, rotor, and caliper. Many vehicles use disc brakes at all wheels.

Pads

Disc **brake pads** are much smaller than drum brake shoes in area (Figure 11-32). The friction material is either bonded or riveted to a small metal pad. There are two pads per wheel, an inner and outer. The pads may be identical in shape and size but most have some differences between the inner and outer pads. Between the two pads is the brake disc or rotor.

The lining on a pad is composed of materials similar to the ones used on drum brake linings. However, a disc pad lining is much harder then the drum linings. The pad lining has to be harder because of its smaller size and the increased pressure or force applied to it during braking.

The *drum* and hub may be manufactured as a one-piece component.

Shop Manual
pages 295–300, 311–317

Since the *disc pad* friction area is smaller and there is no servo action, greater force is needed for braking. This is accomplished by using larger pistons than those used on drum brakes.

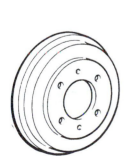

Figure 11-31 The drum fits over the lug studs and covers the drum brake assembly. (Courtesy of Mitsubishi Motor Sales of America, Inc.)

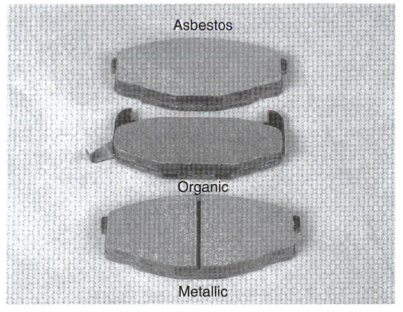

Asbestos

Organic

Metallic

Figure 11-32 Typical disc brake pads.

Rotors

Like the drum brakes, the *rotor* and hub may be manufactured as a one-piece unit.

The **rotor** is attached to the hub or axle and rotates with the wheel. Each side of the rotor is machined for smooth braking (Figure 11-33). As the brakes are applied, the two pads clamp against the rotor and slow it. When the brakes are released, the pads are moved slightly backward but they should never lose complete contact with the rotor. The rotor is subject to great heat during braking. Excessive and hard braking will cause the rotor to warp. The rotor can be machined if the warpage is not too severe and it is not too thin.

Calipers

The caliper is mounted at the wheel and most extend over the rotor in a U-shaped mounting. Most calipers are mounted to the steering knuckle using an adapter. The caliper may have one, two, or four pistons. The pistons in most calipers are larger than those found in a wheel cylinder. The caliper may be sliding or fixed. It is drilled to accept the line fitting and a bleeder valve.

The operation of the *sliding caliper* is similar to the actions of a wheel cylinder except the caliper can move as a unit. As pressurized fluid enters the piston bore, the piston moves out, pushing the inner pad against the rotor (Figure 11-34). The outer pad is on the opposite side of the rotor and held in place by the caliper portion that extends over the rotor. Since any action creates an equal and opposite reaction, the caliper moves away from the rotor. This action draws the outer pad against the outer edge of the rotor. More force applied against the piston and inner pad results in the same amount of force being applied to the outer pad. A sliding caliper may have one or two pistons mounted in the caliper.

A *fixed caliper* usually has four pistons, two on the inner side and two on the outer side of the rotor. It requires a much larger caliper and the additional pumping to fill, equalize, and bleed the additional cavities for the two outer pistons. As the fluid enters the cavities, the pistons are forced outward, thereby clamping the rotor. This is an older system and is not seen often on light vehicles.

There is no manual adjustment on a disc brake system and no return springs. Each bore is fitted with a square-cut O-ring near the outer end of each piston. The ring acts as a seal and adjuster. As the piston moves outward on each brake application, the O-ring distorts slightly

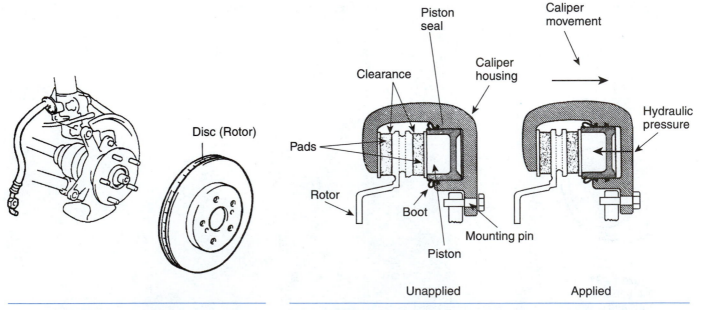

Figure 11-33 The rotor or disc provides the friction contact areas for the disc pads. (Reprinted with permission)

Figure 11-34 The sliding caliper moves on its mounting pins so equal force can be applied by both pads.

(Figure 11-35). When the brakes are released, the square ring returns to orginal condition while pulling the piston and pad back from the rotor (Figure 11-36). As the pads wear, the piston moves further outward and the distorted ring slides along the piston. When it recovers its shape after brake release, the piston is further down the bore (Figure 11-37). This action continually keeps the pads flat against the rotor and allows for the adjustment needed as the pads wear.

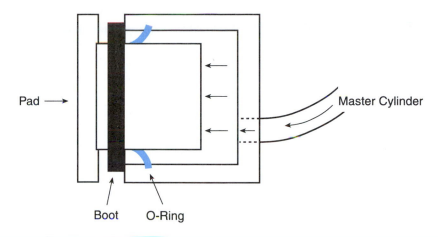

Figure 11-35 The square-cut O-ring distorts as the caliper piston moves outward.

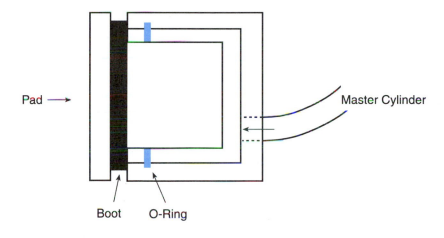

Figure 11-36 The O-ring's normal shape with brakes released.

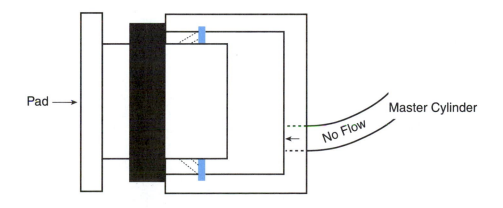

Figure 11-37 The square ring can only return the piston a short distance, which allows the piston to adjust for pad wear.

Shop Manual
pages 318–319

Parking Brakes

Commonly referred to as *emergency brakes,* parking brakes are designed to hold the vehicle still after it has been stopped. They function on the rear axle only in passenger cars and light trucks. A foot- or hand-operated lever near the driver is used to set the brake. A foot pedal, when used, is located at the left kick panel near the driver's left foot. If a lever is used, it is usually placed between the two front seats. A typical system uses cables to connect the driver's action to a lever or cam in the rear brakes to lock the shoes to the drum. Some rear discs have small drum/shoe mechanisms on the inside of the rotor for parking. Other rear discs have screws that force the pads to clamp the rotor when the parking brake is set.

Shop Manual
pages 311, 321

Antilock Brakes

Contrary to some beliefs, *antilock brake systems (ABS)* were not designed to reduce the stopping distance of a vehicle. They do prevent the wheels from locking and give the driver better control over the vehicle during hard or panic braking. Stopping distance is reduced to some extent by the fact that the wheels are not sliding and the tire maintains traction with the road. There are several different types of antilock systems. Some work only on rear wheels and are known as *rear-wheel antilock (RWAL).* RWAL is commonly used on light trucks and vans (Figure 11-38). The typical ABS system on a passenger vehicle is a four-wheel antilock. Four-wheel systems can be three or four channels. Three-channel systems control the braking at each front wheel and the rear axle set. Four-channel systems control the braking at each wheel (Figure 11-39).

Since this unit covers basic theories and operation, we will only discuss the basic components and operation common to all types of antilock systems. The components are the *speed sensors, pressure controllers,* and the *controller.*

Speed Sensors

In order to control brake pressure, the ABS controller must know the speed of the wheels. Speed sensors generate an AC signal each time a tooth on a **toothed ring** or **tone wheel** passes the sensor's magnet (Figure 11-40). The number of signal pulses sent to the module is used to calculate the speed of the wheel or axle.

A *toothed or tone ring* is a circular metal band or wheel with evenly spaced and sized protrusions.

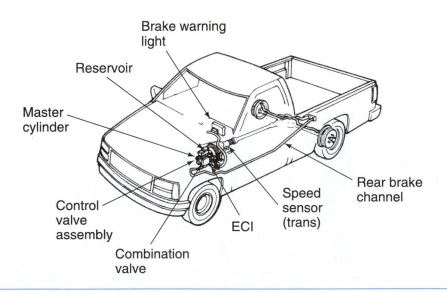

Figure 11-38 A RWAL brake system.

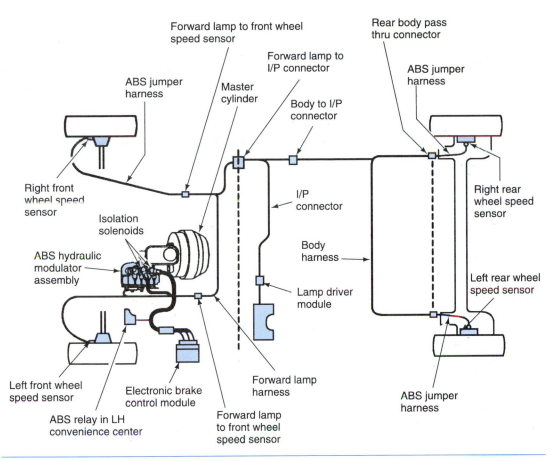

Figure 11-39 A typical four-wheel, four-channel, antilock brake system. (Courtesy of General Motors Corporation, Service Operations)

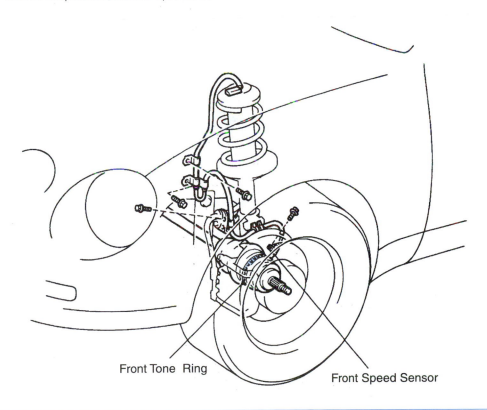

Figure 11-40 The tone ring on this FWD vehicle is located on the outer CV joint. (Reprinted with permission)

On RWAL and three-channel, four-wheel ABS, the rear speed sensor is mounted to measure the speed of the rear axle. This is done by placing the sensor at the output shaft of the transmission or at the ring gear in the differential (Figure 11-41). The toothed ring is placed on the output shaft or ring gear. The pressure to both rear brakes is controlled as one circuit. Each brake on the rear axle will receive the same amount of fluid and pressure. A three-channel system uses sensors at each front wheel in conjunction with the rear sensor. The front tone rings mounted on the brake rotor or the outer CV joint. The three-channel system can control pressure to each of the front wheels in addition to the rear brakes. It is typed as a four-wheel ABS.

Four-wheel, four-channel ABS controls the pressure to each wheel. This system will have speed sensors and tone wheels at each wheel (Figure 11-42). The pressure and fluid can be regulated independently of each wheel based on that wheel's speed.

Pressure Controllers

The devices actually performing brake pressure regulation are solenoids commanded by the antilock controller. There is a solenoid for each wheel or axle with a speed sensor. A RWAL system has one solenoid commonly known as an *isolation/dump solenoid* or brake actuator

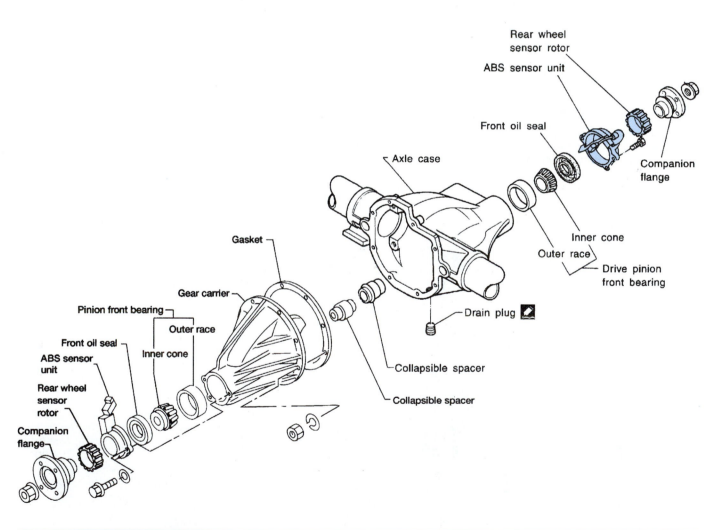

Figure 11-41 This rear antilock speed sensor is actually mounted inside the rear of the differential. (Courtesy of Nissan North America)

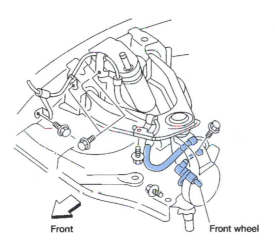

Front　　　　　　　Front wheel

Figure 11-42 A typical front wheel antilock speed sensor. (Courtesy of Nissan North America)

(Figure 11-43). While different systems use different means to actually control the brake pressures, they all use the same theory of operation. Each solenoid can isolate the wheel from the master cylinder and dump the existing pressure within that brake line. In this manner, no further pressure can go to the brake from the master cylinder. At the same time, fluid can be released from the brake line and reduce the braking effect. The released fluid is held in an accumulator for future use, if necessary. The accumulator is charged with compressed gas to keep the fluid pressurized. Some systems use an electric motor and screw to draw fluid from the brake line (Figure 11-44). The motor is reversed to force fluid back into the line. The solenoid can also pop the master cylinder isolation valve open to allow fluid into the brake line and solenoid as required.

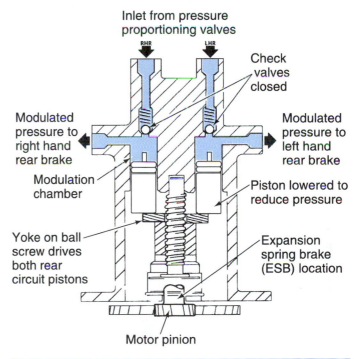

Inlet from pressure proportioning valves

RHR　　LHR

Check valves closed

Modulated pressure to right hand rear brake

Modulated pressure to left hand rear brake

Modulation chamber

Piston lowered to reduce pressure

Yoke on ball screw drives both rear circuit pistons

Expansion spring brake (ESB) location

Motor pinion

Second crossmember

Figure 11-43 The RWAL actuator may be referred to as the "isolator/dump valve." (Courtesy of Nissan North America)

Figure 11-44 The GM ABS actuator uses an electric motor and ball screw to relieve and charge brake lines. (Courtesy of General Motors Corporation, Service Operations)

ABS Operation

As the brake is applied, the ABS controller senses a signal from the brake light switch. At this point, the controller begins to monitor the speed of the wheels. If a wheel locks or stops rotating, the controller commands the solenoid for that wheel/axle to isolate the master cylinder and dump or relieve brake line pressure. With the brake partially released, the wheel rotates and the solenoid is turned off. The controller monitors all wheels equipped with a speed sensor and can control individual solenoids without affecting braking on the other wheels. The solenoid can be cycled thousands of times per minute. This can be an unnerving experience for the driver since the brake pedal and vehicle will pulsate and shudder as the solenoids are operated. The technician can help by informing the customer that this is a normal condition and the brake pedal must be kept depressed until the conditions requiring the panic braking no longer exists.

The antilock brake system has an **amber dash warning light** that can be lit by the controller if a defect is found. The only defect that the controller can detect is an electrical one. The controller cannot detect mechanical brake conditions such as a leaking wheel cylinder or excessively worn shoes or pads unless a mechanical condition directly affects a sensor.

The regular brake system will function normally if the ABS warning light is illuminated, but the ABS will not function. Certain brake failures may illuminate the red brake warning and the amber ABS lights at once. In this case, it is probable that the regular brakes have failed or have a fault and the vehicle must be serviced as soon as possible.

Some of the newer brake systems have a pressure sensor or pressure transducer to transmit brake system pressure to the control module. In this manner, the module can determine exactly how much force the driver is applying to the brake pedal. The control module can control lock-up better by comparing vehicle speed, wheel speed, and the pressure within the brake system.

The *amber ABS light* is not to be confused with the red brake light. The red light means the regular brakes may fail.

Shop Manual
pages 311, 321

Traction Control

Traction control is a method of redirecting or transferring power from a spinning drive wheel to a wheel with more traction. There are two basic types of traction control, each relying heavily on electronics. Dependable four-wheel ABS makes traction control a reality on many mid-price vehicles. However, for the driver who likes to accelerate quickly from dead stop, traction control can cause damage to the vehicle's driveline. Traction control is meant to assist on slick road surfaces not for mud-bogging or hot-rodding.

Antilock Brakes with Traction Control

In this type of traction control the brakes are actually applied to a wheel that has lost traction and the vehicle speed is below 30 mph. When the brake is applied to one drive wheel, the differential or final drive shifts power to the other wheel. Assuming that the second wheel has traction, the vehicle should move. However, extra components and software programming must be added to the ABS.

The basic additions include two pumps driven by an electric motor, a different component, pressure and pressure-reducing solenoids, relays, and the hydraulic and electric circuits. The added sensors are used to track brake fluid levels, brake pedal travel, rotational sensors (motors), pressure switches, and a thermal limiter to calculate brake pad temperature.

The controller antilock brake (CAB) and the ABS perform like all other ABS systems during normal and panic braking. The circuits are monitored for faults in the same manner also. However, when a drive wheel loses traction, the traction control program of the CAB operates the system as a braking device. The CAB monitors wheel spin and actuates the pump and solenoid to apply the brake at that wheel. Driveline power is diverted to the other drive wheel.

As the spinning wheel regains traction, the brake is released. The brake is only applied enough to divert the power and slow the spinning wheel. If the terminal limiter calculates the applied brake pad to be overheating, the traction control is turned off until the CAB resets the system.

Traction Control/Acceleration Slip

This system incorporates ABS with the engine and transmission control system to control wheel slippage. The system can perform in one or several of four modes at the same time. It should be noted that brake application is only done in conjunction with another mode. The modes are:

- Command the PCM to retard spark timing to reduce engine output.
- Force the throttle blade closed to decrease engine speed and output.
- Upshift the transmission to reduce delivered torque.
- Apply the brake on a spinning wheel.

On older systems without full electronically-controlled transmissions, the gear upshift is not possible. Like the previously discussed traction control and other electronic control systems, this system must be tailored, electrically and mechanically, to meet vehicle specifications. A traction control system for a Ford Crown Victoria will not work with a Lincoln Town Car even though the vehicles are very similar in weight and overall size. Each vehicle is set up to function with electronic differences that must be considered during the design and application of all systems.

Summary

❑ The use of hydraulics in the brake system allows force to be transferred with moving parts.

❑ The operator's force on the brake pedal is mechanically transmitted to the master cylinder's primary piston.

❑ The operator's force can be increased greatly by using a booster between the pedal and the master cylinder.

❑ A single-piston master cylinder may cause all of the brakes to fail if there is leak at any wheel.

❑ A dual-piston master cylinder allows the driver to have braking power on two wheels even if a leak is present on one wheel.

❑ The metering valve holds the front disc brake off until the rear drum brakes have moved to make contact with the drums.

❑ The proportioning valve prevents the rear brakes from locking during heavy braking by reducing the pressure.

❑ A combination valve houses the metering and proportioning valves.

❑ A brake warning light is used to alert the driver to a failure in the system.

❑ Wheel cylinders use two pistons and links to apply the brakes.

❑ Return springs on a drum brake force the shoes and pistons to their release positions.

❑ Most drum brakes use an automatic self-adjuster operated by the movement of the secondary shoe or parking brake.

❑ Disc brakes do not use a mechanical adjuster.

❑ Disc brakes are adjusted by the action of the square cut O-ring.

❑ Flexible hoses are used to connect the wheel cylinder or disc caliper to the steel line located on the frame.

❑ Parking brakes are used to hold the vehicle still but not for stopping the vehicle.

Terms to Know

Antilock brakes

Brake pad

Brake shoe

Brake warning light

Caliper

Check valve

Combination valve

Diaphragm

Drum

Fixed caliper

Hydro-boost

Isolation/dump solenoid

Load-sensing, proportioning valve

Master cylinder

Maxi-boost

Mechanical advantage

Metering valve

PowerMaster

Power piston

Primary piston

❑ ABS is used to prevent wheel lock-up during braking.

❑ The loss of ABS does not affect the operation of the regular brakes.

❑ Traction control uses the ABS or engine controls to control wheel spin during acceleration.

Review Questions

Short Answer Essays

1. Describe duo-servo action on a drum brake system.

2. Explain how disc brakes are adjusted for wear.

3. Discuss the purpose and operation of the master cylinder's primary piston.

4. Explain how brake pedal force is boosted with a hydro-boost system.

5. List and explain the purpose of the wheel-mounted components of a disc brake.

6. Briefly describe the actions taken by an ABS during a panic stop.

7. Explain the purpose of the proportioning valve.

8. Describe the operation of a frame-mounted, load-sensing, proportioning valve.

9. Discuss the operation of a metering valve.

10. Discuss the operation of combination valve-mounted, load-sensing, proportioning valve.

Fill-in-the-Blanks

1. The _____ brake shoe performs most of the braking.

2. The _____ system boosts the driver braking effort using an electric motor and pump.

3. The ABS warning light will not detect _____ conditions.

4. The adjustment of disc brakes is done by the _____ .

5. The master cylinder secondary piston provides fluid to the _____ brakes.

6. The metering valve controls fluid during _____ braking.

7. The proportioning valve controls fluid pressure during _____ braking.

8. A damaged steel line must be _____ .

9. A(n) _____ _____ connects the caliper to the steel brake lines.

10. If the ABS warning light is on, the regular brakes will _____ .

ASE Style Review Questions

1. The proportioning valve is being discussed.
   Technician A says a leak in the front brake should turn on the warning light.
   Technician B says the brake pedal will be much lower to the floor if a leak develops in the rear brakes. Who is correct?
   A. A only
   B. B only
   C. Both A and B
   D. Neither A nor B

2. The diagonally split brake system is being discussed.
   Technician A says one master cylinder piston will operate the brakes on a front and a rear wheel.
   Technician B says that if a leak occurs, the operator will still have two working brakes on an axle. Who is correct?
   A. A only
   B. B only
   C. Both A and B
   D. Neither A nor B

3. Disc brake calipers are being discussed. *Technician A* says adjustment for pad wear cannot be performed manually.
   Technician B says the square-cut O-ring automatically adjusts for pad wear. Who is correct?
   A. A only
   B. B only
   C. Both A and B
   D. Neither A nor B

4. The master cylinder is being discussed. *Technician A* says a leak at the front brake will stop the primary piston from building pressure.
   Technician B says the load-sensing proportioning valve is mounted at the rear of all master cylinders. Who is correct?
   A. A only
   B. B only
   C. Both A and B
   D. Neither A nor B

5. Split brake systems are being discussed. *Technician A* says the diagonally split system is set so a front brake and a rear brake are available for braking.
   Technician B says the front/rear split system has the rear brakes operating from the same master cylinder piston. Who is correct?
   A. A only
   B. B only
   C. Both A and B
   D. Neither A nor B

6. Disc brakes are being discussed. *Technician A* says the square-cut O-ring adjusts for pad wear.
   Technician B says a fixed caliper uses only one piston. Who is correct?
   A. A only
   B. B only
   C. Both A and B
   D. Neither A nor B

7. A two-square-inch input piston is being applied with 10 pounds of force. A four-square-inch output piston is being used. *Technician A* says mechanical advantage can be calculated using the formula $F = PA$.
   Technician B says the pressure in this system is 40 psi. Who is correct?
   A. A only
   B. B only
   C. Both A and B
   D. Neither A nor B

8. The metering valve is being discussed. *Technician A* says the movement of the valve determines rear brake pressure.
   Technician B says the valve adjusts rear brake pressure during initial braking. Who is correct?
   A. A only
   B. B only
   C. Both A and B
   D. Neither A nor B

9. Drums and rotors are being discussed. *Technician A* says the drum and rotor are mounted between the wheel and axle or hub.
   Technician B says drums are machined to match the brake pads. Who is correct?
   A. A only
   B. B only
   C. Both A and B
   D. Neither A nor B

10. Proportioning valves are being discussed.
    Technician A says the load-sensing proportioning valve may be mounted on the frame rail.
    Technician B says some proportioning valves use a ball to control the valve.
    A. A only
    B. B only
    C. Both A and B
    D. Neither A nor B

Auxiliary Systems and Climate Control

Upon completion and review of this chapter, you should be able to:

- ❏ Discuss the different and overlapping features of convenience and comfort systems.
- ❏ Discuss the warning systems found on most vehicles.
- ❏ Discuss some of the information systems available.
- ❏ Discuss the electrical circuits used with instrumentation.

- ❏ Discuss the interior and exterior lighting system.
- ❏ Explain the general operation of typical accessory systems.
- ❏ Describe the general operation of heating and air conditioning systems.
- ❏ Explain the purpose of the Clean Air Act and the control of refrigerant.
- ❏ Discuss restraint systems.

Introduction

The increasing use of electricity and electronics in the automobile has provided an almost unlimited range of accessories and informational devices. The car or truck can be equipped with systems that memorize seat positions for a certain driver, adjust individual temperature controls, provide anti-theft protection, and remote start the engine and heater on a cold morning. More and better protection devices for the passengers are possible with electronics. With the advances comes an increasing need for electronically-proficient technicians who can diagnose and repair sophisticated electronic circuits quickly and accurately. To the customer, the convenience of electronics brings the possibility of hugely increased repair costs.

This chapter, and its corresponding chapter in the Shop Manual, are not designed to make the entry-level technician an electronic or climate control repair person. Instead, the chapters will give the reader a working knowledge of the various auxiliary systems' operations and some basic diagnostic testing. Since some of the systems are married together, a failure of other systems may affect engine and transmission operation.

Passenger Comfort and Convenience

Shop Manual
pages 352–358

Back in the good old days of 1957 when this author received his driver's license, vehicle comfort consisted of a decent pad between the passenger and the seat spring! The only convenience item was blue oil smoke from the exhaust that killed half the mosquitoes in south central Virginia! Today, there are devices installed on the typical vehicle that were not dreamed possible in 1957.

Passenger comfort devices are items like heat, air, conditioning, padded and heated seats, and so on that make a passenger's ride in the vehicle enjoyable. Warning and information systems provide the driver with engine and vehicle operation and indications of problems before they get too serious. Systems like lighting, rearview mirrors, heated back glass, and automatic ride control give the passenger a comfortable ride while the driver has an almost unrestricted view of the actions around the vehicle. Of course, many of the so-called comfort items are required just to operate the vehicle. Comfort comes from the improved processes and equipment that accomplish the job.

Convenience features are those that make the comfort items easier or simpler to use. Power and heated outside mirrors eliminate the need to lower the windows on cold, rainy days. The driver can change seat positions during a long drive to relieve or at least postpone fatigue. There are driver seat systems offered today that will massage the driver's lower back. In the old days, if a driver got lost, the solution was to use a map or ask directions. Today, the punch of a button displays the vehicle's position on a small screen or digital display. In the old days, if a car broke down on a lone, dark road, the driver and passengers either slept in the vehicle or walked. Now, the just punch of a button summons help.

All of the features taken for granted today place a tremendous load on the battery and alternator, not to mention the training and operational costs to maintain nationwide repair facilities and technicians. The increased cost is passed to the customer in the vehicle's purchase price and maintenance. The only real drawback to the increased number of electronic devices offered is the cost of a mistake by the vehicle or part manufacturer, repair technician, or the owner.

Shop Manual
pages 341–343

A *light-emitting diode (LED)* is a diode that gives off light when current passes through it.

The term *heads-up-display (HUD)* was adopted because the driver does not have to look down to read operational data.

Instrumentation

Gauges and lights on the dash are the means used to convey information to the operator. They are usually in an area of the dash referred to as the *instrument panel*. The gauges may be *analog* or *digital*. Analog gauges use a needle moving over a fixed scale to display the data much like the hands on a clock (Figure 12-1). Other systems use numbers or digits that change based on sensor input (Figure 12-2). The numbers are displayed by **light-emitting diodes (LEDs).**

One system uses an illuminated band or bands against a fixed scale. The bands act like an analog gauge but are actually LEDs that are turned on and off by a computer module based on incoming data. The bands can be programmed as different colors to attract the driver's attention. For instance, the band representing vehicle speed could be programmed to turn red for any speeds over 70 miles per hour (mph). It should be noted that the red and amber warning lights are present in every dash regardless of the type of instruments used.

The **heads-up-display (HUD)** was adopted from military fighter aircraft. The information most commonly required is projected onto the windshield in the driver's line of sight (Figure 12-3). The information usually consists of vehicle and engine speed, oil and temperature conditions, and warning lights. In the future, the driver could possibly scroll through other data using steering-wheel-mounted controls. The object of HUD is to keep the driver's eyes and attention on the environment around the vehicle while providing necessary vehicle data.

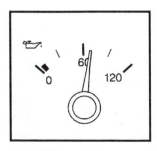

3800 V6 Engine

Figure 12-1 A typical analog oil pressure gauge. (Courtesy of Chevrolet Motor Division, General Motors Corporation)

Figure 12-2 The small dots of light seen in digital displays are the individual LEDs in the display.

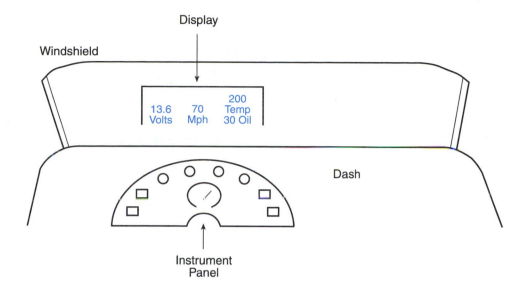

Figure 12-3 Some of the same data shown on the instrument panel can be projected onto the windshield with HUD.

Warning Systems

Warning systems and information systems (next section) would appear to be basically the same; however, they have distinct purposes. Warning systems are found on every vehicle and alert the driver to a fault or failure that could damage the vehicle or cause a loss of control. Information systems generally provide data that can assist the driver in some way.

Red and amber lights are used because of their visibility, relationship to danger, and eye-catching properties. Many times, gauges supplement the lights. Red means the danger is serious and failure to heed the **warning light** may cause serious damage or an accident. Amber lights indicate some repair is needed but the vehicle can be operated. Lights are used in place of or with gauges. Drivers tend *not* to check gauges like oil pressure or engine temperature until it is too late. The sudden glare of a red dash light will attract the driver's attention immediately.

Typical red warning **lights** are low oil pressure, high engine temperature, the alternator, door ajar, parking brake, or low brake fluid (Figure 12-4). On some vehicles, the oil pressure and engine temperature sensors or switches are wired in parallel to the same lamp. The parking brake and low fluid level sensors are also normally wired to the same lamp. A sensor or switch is mounted into the component/assembly being checked. Sensors will normally send data to the vehicle's PCM or another computer. The computer will turn on the warning light as

Shop Manual
pages 337–341

Warning lights replaced gauges during the late 1970s through the late 1980s. Many drivers wanted gauges instead of "idiot" lights. The industry has returned to the installation of gauges supplemented by warning lights.

Many times, the *lights* come on just before the engine or other components fail. Better use of electronics has improved the reliability of warning light systems.

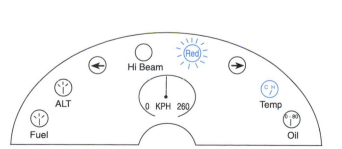

Figure 12-4 This lit warning light alerts the driver to a high engine temperature. Note the temperature gauge.

A grounding switch is a switch that, when closed, connects the circuit to ground.

needed. Older vehicles and some newer models use a simple switch that is opened or closed based on temperature, pressure, or other *trigger* conditions. The switch completes the circuit and the warning light is turned on.

The switch in some warning circuits is a one-wire **grounding switch** (Figure 12-5). In an oil pressure warning circuit, the switch may be held open or closed by oil pressure. When

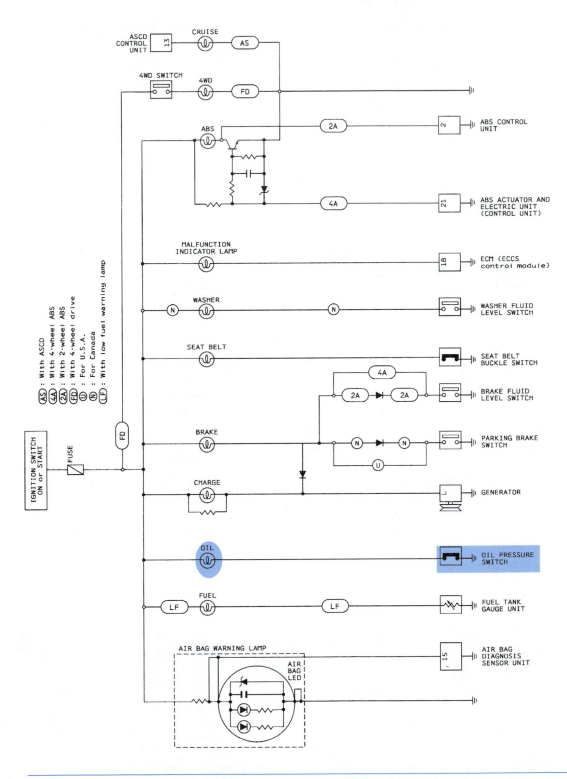

Figure 12-5 In this grounding switch, the oil pressure will hold the switch open. It is shown with the engine off (no oil pressure) and the switch closed. (Courtesy of Nissan North America)

pressure drops below a specific setting and allows the switch to function, the circuit activates. The alternator is switched on by low voltage in the sensing circuit. Mechanical switches may be used to control other warning lights.

A grounding switch at the parking brake pedal or lever usually switches on the red brake warning light (Figure 12-6). A fluid sensor in the brake's reservoir may also turn it on. A switch

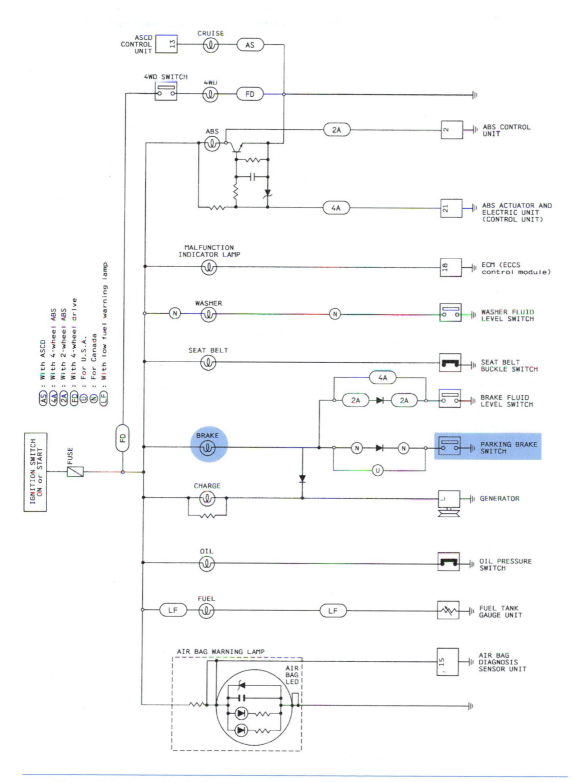

Figure 12-6 The parking brake warning light switch is closed by the movement of the parking brake pedal/lever. It is shown with the parking brake released. (Courtesy of Nissan North America)

Malfunction Indicator Lamp (Service Engine Soon Light)

SERVICE ENGINE SOON

Your vehicle is equipped with a computer which monitors operation of the fuel, ignition and emission control systems.

NOTICE:

If you keep driving your vehicle with this light on, after a while, your emission controls may not work as well, your fuel economy may not be as good and your engine may not run as smoothly. This could lead to costly repairs that may not be covered by your warranty.

Figure 12-7 A MIL light. Note the notice on the right. (Courtesy of Chevrolet Motor Division, General Motors Corporation)

The malfunctioning indicator light (MIL) was commonly known as the "check engine" light primarily because of the text "check engine soon" printed on the light's lens.

built into the doorpost or trunk latch controls door and trunk ajar lights. The door switch will activate interior lights.

The **malfunction indicator light (MIL)** is switched on by the vehicle's PCM (Figure 12-7). It indicates a problem with the PCM circuits for engine and transmission controls. Usually, the vehicle can be driven to a shop for the required repairs even though it may operate at a lower performance level. The most prominent amber light is for the antilock brake system. A lit ABS light means two things: there is a problem is the ABS electronic circuit and ABS operations have been terminated. The vehicle can be driven safety if the driver realizes that there is no antilock braking available. The regular brakes will work as designed. Other warning lights may include low fuel, electronic suspension failures, or problems with the seat belt, supplement restraint system (SRS), or the air bag. Most electronic systems that have some type of diagnostic capability will have some type of warning lights. The main point to remember is that a red light means immediate danger (except MIL) while other colors indicate a problem that does not present an immediate danger to the vehicle or passengers.

Many vehicles have audible alarms that activate when a warning light is lit. The alarms are usually sharp blips or buzzing designed to instantly alert the driver of a condition that may cause damage or is unsafe. Light and audible reminders are present on almost every vehicle for the seat belt, a key left in the ignition with the door open, or lights left on.

Shop Manual
pages 341–343

Information Systems

Information is conveyed to the operator on almost any vehicular operation desired. The data available ranges from a vehicle speed to its location on earth. The data originates from sensors on the vehicle or is received from high-orbit satellites.

The information is presented to the driver or passenger through digital or text display. Information such as vehicle operation is usually present whenever the ignition switch is on. The basics include vehicle and engine speed, oil pressure, engine temperature, A/C generator output, and fuel level. Analog gauges or digital readouts using LEDs can present the information in an easy-to-read format, normally a number.

A communication buss has all sensors, computers, and modules connected to shared circuits. In this manner, each computer and module can monitor data.

Newer vehicles may use a section of the PCM or a separate computer module to compute information. The 1999 Cadillac uses up to seventeen different computers. When a separate computer or module is used, it collects data from the PCM and other modules to calculate and present the information. A PCM may have a separate internal circuit to process and display the information, however, it is typical that separate modules share all data via a **communication buss.** A communication buss is similar to a computer network in a large office. Each computer is wired into the network and can communicate and share data with all other computers in the system. Additional information about the vehicle and its surroundings may include:

- Outside and inside temperatures
- Continuous or accumulated fuel mileage
- Time of departure and expected time of arrival
- Miles left on the remaining fuel
- Mileage completed or left on this trip
- The location of the vehicle at any single time
- How to get from here to there
- Road conditions and suggested detours

Some of this information is displayed continuously or at the touch of a button (Figure 12-8). The operator can select the desired information and even select USC or metric measurements. Like operating data, this information can be displayed by analog or digital method, but digital is the most common. Many options offered on the 1999 high line models will become standard features on most vehicles before 2010.

One option that is becoming more common came from the military and space exploration. The **Global Position Satellite (GPS)** is actually a network of satellites deployed in orbit around the Earth. A vehicle-mounted receiver translates data from several satellites and calculates the exact location of the vehicle within a few yards. The resulting display provides the vehicle's position on the Earth in an easy-to-read and easy-to-understand format. Using this information and an on-board mapping system, the operator can plan a route to an unfamiliar destination, around an unexpected detour, or to find out how to get back on route.

The future implications of GPS are not easy to define, but a vehicle with an automatic electronic "take me home, James" chauffeur feature is not unrealistic. This feature is used by manufacturer-sponsored roadside assistance options. A call on a cell phone provides GPS data that can have police or assistance vehicles on hand in a short period of time. One manufacturer has a system that will unlock car doors! A call to a toll-free number and, if the proper password and code is presented, a signal is bounced off a satellite to the vehicle commanding the anti-theft control module to unlock the doors.

The satellite network known as the *Global Position Satellite (GPS)* consists of hundreds of satellites that communicate with each other and transmitters/receivers on earth.

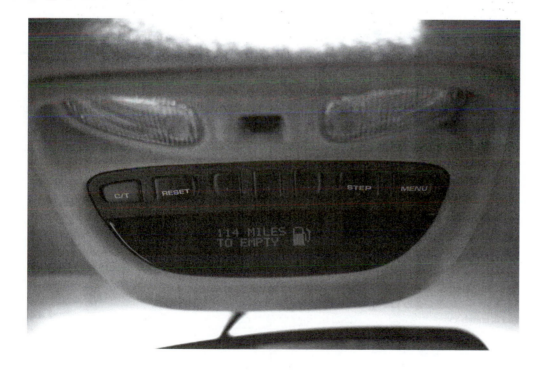

Figure 12-8 This computer can compute current and historical fuel mileage, time and mileage left on a trip, inside and outside temperatures, and other data that may assist the driver. (Courtesy of Chevrolet Motor Division, General Motors Corporation)

One manufacturer of large trucks has a satellite transmitter/receiver. If the truck breaks down, the driver calls the dispatcher and gives the code for that truck. The maintenance department communicates via satellite with the truck's PCM. The PCM transmits operating data and **snapshot** data back via the satellite. The technician selects the most probable repair parts based on that data, proceeds to the downed vehicle, and makes the necessary repairs. Vehicle down time is greatly reduced.

Shop Manual
pages 337–343

A *snapshot* is stored data that can be retrieved and studied at a later time. It is usually data of a particular happening at a particular time such as a misfiring cylinder under heavy load. It may be referred to as a "movie."

Instrument Panel Operation

The instrument panel houses most of the gauges and warning lights. Most informational data is displayed in an overhead console or near the mid-point of the dash (Figure 12-9). This keeps the small space in the panel clear for essential operating data. A **printed circuit board** is located behind the instrument panel. This board supplies the necessary circuits between the gauges and warning lights to their individual sensors or computers. A printed circuit board is used to reduce or eliminate the number of wires needed to connect all the different components. A printed circuit board is basically a very early, crude model of an integral circuit. It is being replaced by much better versions that are smaller and capable of many more circuits.

The panel's power and all other electrical connections to the remainder of the vehicle are made using one or two large, multiple-pin connectors. Large, molded, or shell terminals are plugged into the connector on the vehicle side. The panel side of the connector is an integral part of the board. The conductors are downsized and attached to the board's circuits using small soldered connections.

The *printed circuit board* has been replaced by integrated circuits. They are much smaller and have more resistance to damage by fair wear and tear.

Shop Manual
pages 343–351

Lighting Systems

Most of the exterior lighting is required by federal and state regulations. The only mandated interior lighting is the instrument panel. Each of the mandated lighting systems, interior or exterior, must meet requirements for the amount of illumination generated and for positioning. Other lights may be provided for passenger comfort and convenience provided they do not distract from the mandated lights.

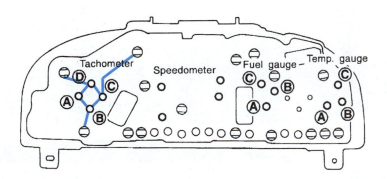

Figure 12-9 Small, flat strips of conductors connect the various gauges and lights in the instrument panel. Note the suggested circuit indicated by the highlight. (Courtesy of Nissan North America)

Lamp Specifications

The vehicle manufacturer specifies the type of lamp. There are many types used on a modern vehicle. Lamps for the marking, turn, brake, and headlight systems must meet illumination capacity specified by federal and state regulations. Other lamps like the courtesy lights are selected by type to provide the level of illumination needed and the mounting space available.

The lamps are connected to the circuits using several types of connections. Most turn and brake lamps use **bayonet** housings to lock the lamp in place (Figure 12-10). A single filament lamp will have locking tabs across from each other. Dual-filament, lamp-locking tabs are offset from each other to prevent improper installation. Headlight lamps use weather-resistant, hard-plastic connectors. Interior lamps may be the bayonet-type or the slide-in-type and are usually single-filament lamps (Figure 12-11).

A *bayonet* connection is so named because of its push in, twist, and lock steps to make the connection. Military bayonets are fitted to a rifle in this manner.

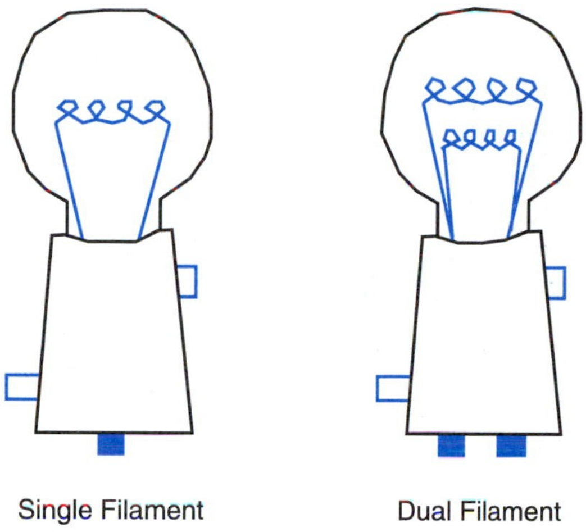

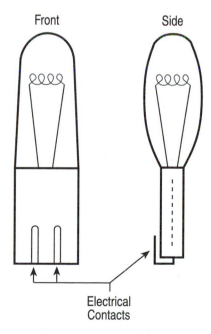

Single Filament   Dual Filament

Figure 12-10 The offset pins ensure that the two contacts at the bottom of the lamp connect to the correct circuit.

Front   Side

Electrical Contacts

Figure 12-11 Slide-in lamp contacts are small wires folded up over the base. The female socket will have two brass strip contacts.

Exterior Lighting

The mandated exterior lights are:

- *Headlights*—one on each front corner of the vehicle, white, must provide high and low levels of illumination.
- *Marking lights*—one on each corner of the vehicle, yellow in front, red in rear.
- *Turn signal lights*—one on each corner of the vehicle, yellow in front, red or yellow in rear.
- *Brake or stop lights*—one on each rear corner, one following the driver's direct line of sight, red.

Headlights

The *aim* of the headlight is set by federal regulations.

Headlight systems may use two or four lamps provided they meet the illumination and position requirements. The two-lamp system uses two filaments in each lamp (Figure 12-12). One filament is for the high beam and the second is for the low beam. The high beam is the strongest and provides illumination over a higher area. This tends to blind on-coming driver's as the vehicles approach each other. The headlight system must be equipped with a switch for the driver to change to low beam without losing visibility of the road. The low beam filament produces less illumination and is **aimed** to the right and low to provide coverage in the area near the vehicle (Figure 12-13).

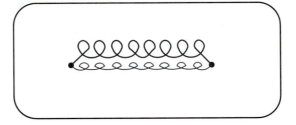

Figure 12-12 A two-lamp headlight system has two filaments in each lamp. Each filament has a different resistance.

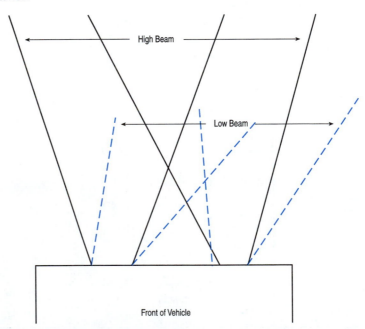

Figure 12-13 As indicated by the highlight, the low beam illumination is to the lower right, and not as bright as the high beam illumination.

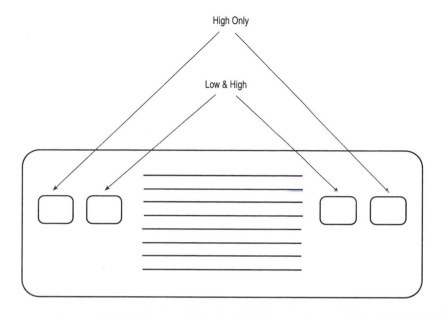

High Only

Low & High

Figure 12-14 The smaller, four-lamp system allows the vehicle's grille and hood to be slightly lower and provides a wider field of illumination on high beam.

The four-lamp system has two headlamps at each front corner of the vehicle (Figure 12-14). Two inside or outside lamps are for high beams while the other pair provides low beam. In this case, the high/low beam switch usually turns the high beam pair on and off. The two low beam lamps are lit in both modes.

The high/low beam selector or **dimmer switch** is most commonly part of the **multi-function lever** mounted to the left of the steering column (Figure 12-15). This lever also controls the turn signal and may include wiper and cruise control switches. A blue light mounted in the top center of the instrument panel indicates that high beam has been selected. On older vehicles and some present-day heavier trucks, the high/low beam switch is mounted in the floorboard and is operated by the driver's foot. Automatic switching was tried on some of the more expensive vehicles in the past with mixed results. One of the problems then was the state of electronics. The sensor used to detect oncoming headlights was not always accurate, requiring the driver to make a panic changeover manually. Today's electronics are more reliable and automatic high/low switching is starting to reappear.

The *dimmer switch* was adopted because the headlights were much dimmer on low beam.

A *multifunction switch* on the steering column was feasible until the use of electronics became widespread. Electronics can do the same work as mechanical switches but are hundreds of times smaller.

Headlamp High/Low-Beam Changer

To change the headlamps from low beam to high or high to low, pull the turn signal lever all the way toward you. Then release it.

When the high-beams headlamps are on, this light on the instrument panel also will be on.

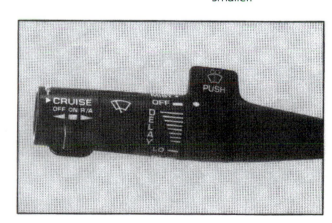

Figure 12-15 The multi-function lever houses several switches beside the hi/lo beam selector. (Courtesy of Chevrolet Motor Division, General Motors Corporation)

Headlamps must be aimed to prevent blinding other drivers and for the best illumination of the area ahead of the vehicle (Figure 12-16). In most cases, the lamps are installed into an adjustable mount. The mount is connected to the frame by springs and adjusting screws. There are usually only two screws per mount, one each for vertical and horizontal adjustment.

Lamps used for the headlight systems can be sealed beam or a single bulb installed into a reflective chamber. The sealed beam lamp has filaments or an iodine vapor bulb installed in a chamber lined by highly reflecting material (Figure 12-17). The chamber is sealed by the permanent installation of the lens. Any damage to the lamp requires replacement of the entire sealed beam lamp. Sealed beam lamps may be found on the two- and four-headlamp systems.

With the manufacturer's desire and need to increase fuel mileage, the front of many vehicles had to be sloped to reduce air resistance. The federally required height-above-ground of the lights and the bulky sealed beam lamps hindered the engineers. The solution called for a redesigned headlight system. The manufacturers removed the filaments from the lens and reflector assembly. The assembly was shaped to fit the lower front contours of the vehicle while maintaining the height-above-ground requirements. A separate lamp was then installed into the rear of the assembly. The lamp is a high-intensity halogen bulb capable of producing comparable illumination as the sealed beam (Figure 12-18). Failure usually requires only the changing of the bulb, but not the lens and reflector. Most vehicles now use this design and a two-lamp headlight system. Some newer vehicles are equipped with low-voltage, high beam headlights. The lights are on any time the engine is operating. This safety feature better highlights the vehicle to other drivers and pedestrians during daylight. They are referred to as *daytime running lights*.

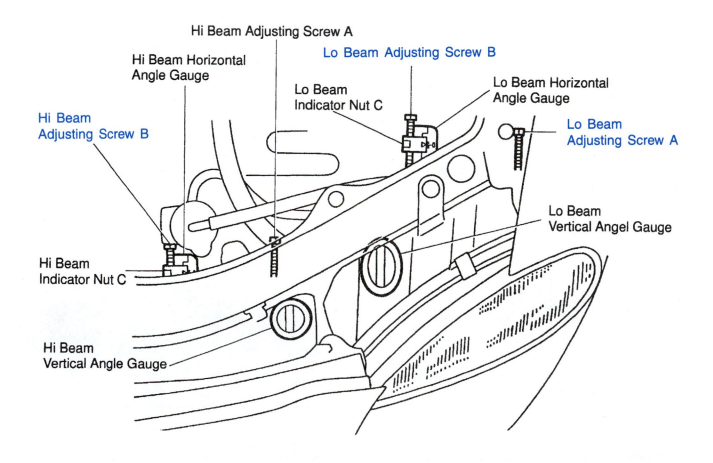

Figure 12-16 The newer headlight system has built-in angle gauges to help aim the lamps. However, the gauges must be properly set in the vehicle. (Reprinted with permission)

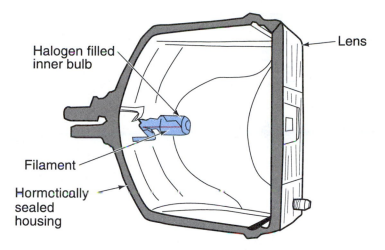

Figure 12-17 The interior of the sealed beam lamp is reflective and has the lamp or filaments mounted in the center of the cavity.

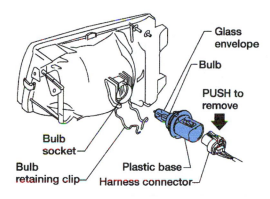

Figure 12-18 This small bulb may have one or two filaments for high and low beams or the vehicle may use two bulbs. (Courtesy of Nissan North America)

Marking Lights

Marking or parking lights are used to identify the vehicle when visibility is low but headlights are not needed. Marking lamps are usually single-filament in front and dual-filament in the rear. The marking lamp is also used in most vehicles as the turn signal lamp. The lights are turned on and off with a separate set of contacts within the headlight switch. Most of the newer vehicles leave the marking lights on during headlight operations.

Turn Signals

The lamps in this system are used to communicate the driver's intentions to other persons on the road. Most turn signal systems use the marking lamps as the signaling devices. The turn signal switch is mounted behind the steering wheel on the column. It is controlled by a lever extending to the left from the column and has three positions: off, left, and right (Figure 12-19).

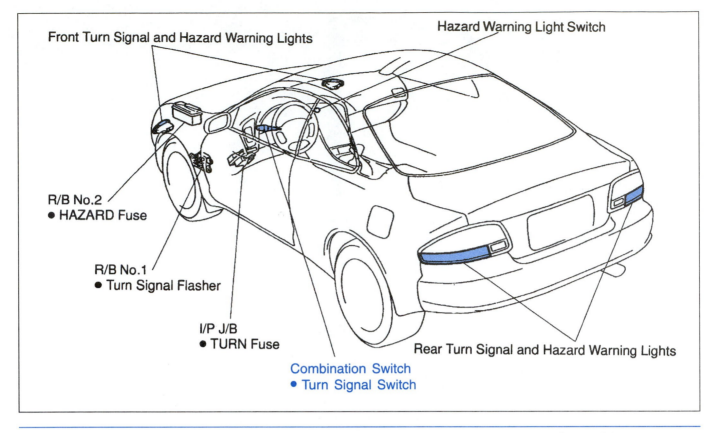

Front Turn Signal and Hazard Warning Lights

Hazard Warning Light Switch

R/B No.2
● HAZARD Fuse

R/B No.1
● Turn Signal Flasher

I/P J/B
● TURN Fuse

Rear Turn Signal and Hazard Warning Lights

Combination Switch
● Turn Signal Switch

Figure 12-19 All standard modern vehicles in the United States have the turn signal lever on the left side of the steering column. (Reprinted with permission)

The *flasher* was so named because its actions make the turn signal lamps flash on and off.

A *bimetallic* strip has two pieces of metal laid side by side. One piece reacts to heat faster than the other one does.

An *emergency flasher* is similar to the turn signal flasher except it should have a longer life and conduct a heavier current.

Moving the end of the lever downward supplies power from the fuse through a **flasher** to the left marking lamps at the front and rear of the vehicle (Figure 12-20). The flasher is a **bimetallic** circuit breaker. As current flows through it, the heat causes the breaker to trip and open the circuit. As it cools, the breaker resets and closes the circuits to the lamps. This process cycles the flashers on and off between .5 second and 1.0 second on the average. The result is a flashing light at each end of the vehicle on the left side. The same process happens for a right selection except for the lamps that are flashing. Adding additional lamps such as those for a trailer can overload the flasher. Installing a heavy-duty flasher is usually required before adding additional lighting.

The rear turn signal lens may be amber or red. If the brake lamp is housed in the same reflective and lens assembly as the signal, the lens will be red. This is usually referred to as the *single lamp system* because the marker, turn, and brake use one lamp with dual filaments. The lens for a *two-lamp system,* which uses a separate lamp for turns, has amber for signaling and red for the marker or stop. Some vehicles are equipped with three lamps, one each for marking, turning, and braking. The single requirement is that the turn lens is amber or red and the marking and brake lights are red.

The four-way flasher or **emergency flasher** works much the same as a turn signal. A separate switch is used to make all four turn signal lamps flash together. The four-way circuit bypasses the turn signal circuit and directly feeds each of the four lamps. Brake applications will override the four-way on the rear lamps to alert following drivers to the situation. Brake applications will not override regular turn signals.

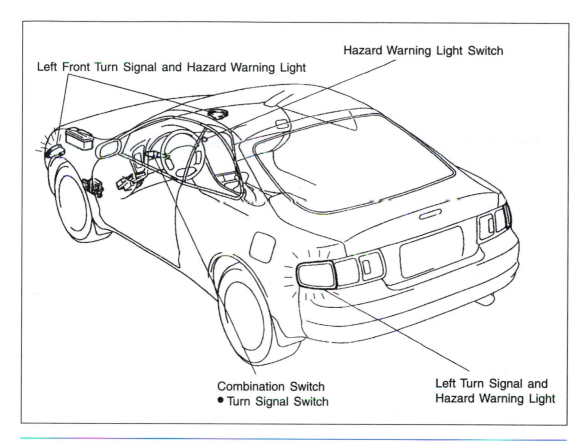

Left Front Turn Signal and Hazard Warning Light

Hazard Warning Light Switch

Combination Switch
● Turn Signal Switch

Left Turn Signal and
Hazard Warning Light

Figure 12-20 The turn signal lever can be moved up for right turns and down for left turns as shown here. (Reprinted with permission)

Brake Lights

Brake lenses must always be red and have much higher illumination than the other lamps near them. A mechanical switch on the brake pedal rod controls the brake lights. Power is fed from a connection in the turn signal switch to the brake switch or directly from the battery (Figure 12-21). When the brake switch is closed, current travels to the rear stop or brake lamps. If a turn signal is on, the flasher interrupts the current to that lamp. When the four-way circuit is operating, the brake light current enters below the flasher and provides a normal steady current to the lamps.

A third brake light was mandated by federal regulations to gain quick the attention of a following driver. The third brake light is officially known as the **Center High-Mounted Stop Light (CHMSL).** It may be located on the lower edge of the rear window or mounted to the trunk lid (Figure 12-22). The third brake light is within the normal line of sight for a following driver. The lamp is wired to the regular brake light circuits in such a way to block out the turn signal circuit. In this manner, the third brake light will only function if one of the two brake lights is on. A newer version of this system has a flasher in series with the third brake light. This will produce a blinking light directly in the following driver's line of sight.

The *Center High-Mounted Stop Light (CHMSL)* is a federally-mandated third brake light, usually located on the lower rear window or on the trunk lid of a vehicle.

Tag Lights

The tag light is switched on any time the marking or headlight is selected. The light is used to illuminate the rear license plate so peace officers can read the tag. Sometimes, two lamps are used. A working tag light is required in all states.

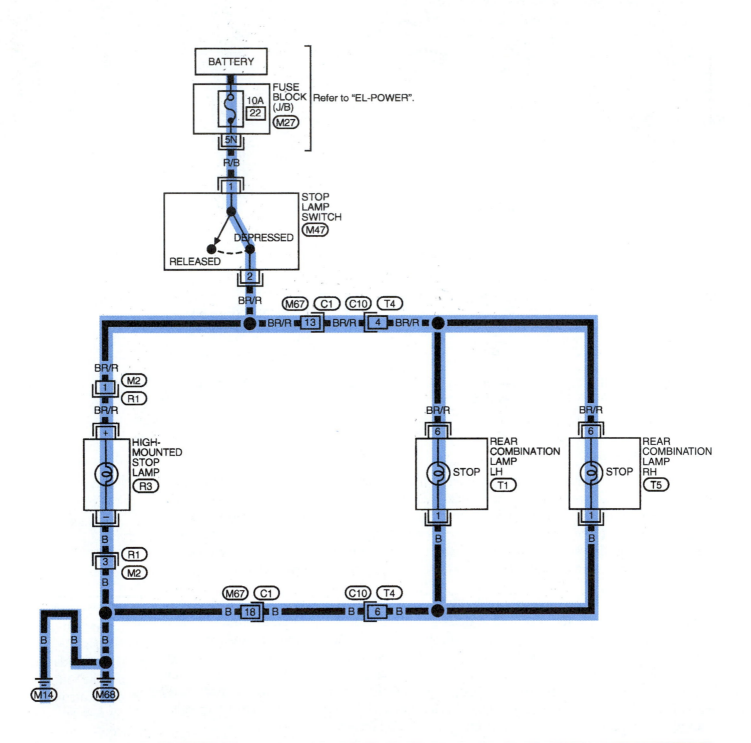

Figure 12-21 The three stop lamps are on parallel circuits, but they share a common ground. (Courtesy of Nissan North America)

Coupe:

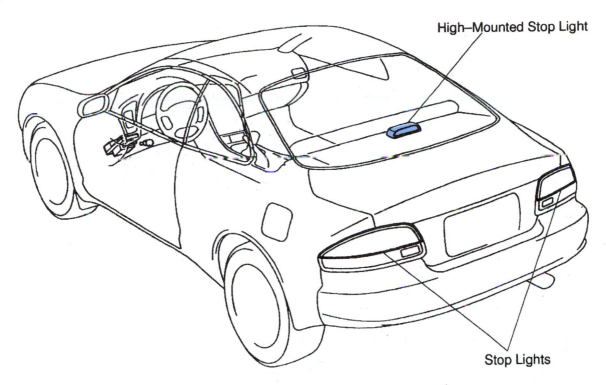

High–Mounted Stop Light

Stop Lights

Liftback:

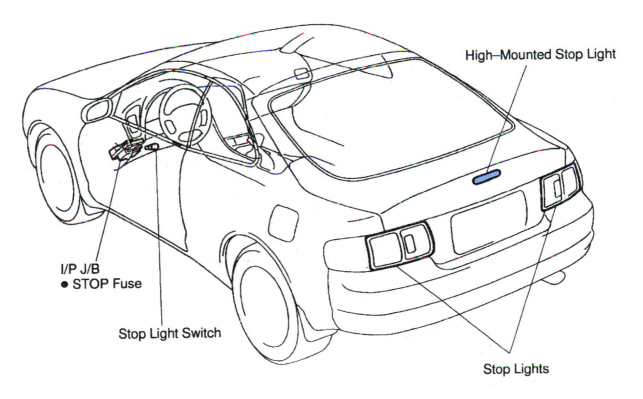

High–Mounted Stop Light

I/P J/B
● STOP Fuse

Stop Light Switch

Stop Lights

Figure 12-22 The CHMSL must be centered on the rear of the vehicle and not more than three inches below the bottom of the rear windshield. (Reprinted with permission)

Other Exterior Lights

There are usually only three other exterior light systems offered. They are the engine compartment light, trunk light, and turning lights. The hood and trunk lights are provided on most vehicles and are simple circuits. Raising the hood or trunk closes a mercury or mechanical switch and the circuit. The two lighting systems are provided as a convenience item. The turning lights are more complicated and are not offered on many vehicles.

The turn or cornering lights have a clear lens lamp mounted on the front sides of the vehicle, normally just behind the turn signal assembly. When a turn signal is selected with the headlights on, the circuit closes a relay that turns on the cornering light for that side. The lamp emits a high-intensity, low-range light to aid the driver in negotiating a corner. This system, like the automatic headlight dimmer, is not very popular and is offered on limited vehicles.

Interior Lighting

Backlit means the light is coming from behind or from the side of the object being viewed.

Interior lighting is provided for entry or exit from the vehicle and illuminates the passenger compartment. The instrument gauges are **backlit** by lamps inserted from the rear of the panel. Twist locked into cavities in the board are the illumination lamps. The lamps are positioned to light the gauges indirectly. The front side of the panel is coated with a non-glare finish. Contacts in the headlight switch turn on the lamps any time the marking or headlights are selected. The intensity of the illumination can be controlled by a rheostat mounted within the headlight switch or on a separate switch on the dash. The lights can be turned off with the rheostat switch without affecting the exterior lights. The headlight switch or the rheostat does not affect the warning lights in the instrument panel.

Included in the group known as the instrument lights are lamps to highlight various controls located on the dash. Most dash panels house the controls for heating and air conditioning, radio, blower, and on-board computer input keys. Lamps placed behind the control panels illuminate the markings and the buttons.

Dome and Courtesy Lights

The dome light and courtesy lights are switched on when a door is opened (Figure 12-23). Most have separate switches to control individual lights when the door is closed (Figure 12-24). The

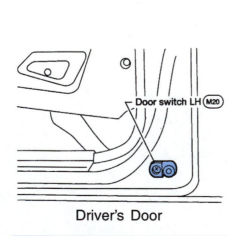

Driver's Door

Figure 12-23 A typical location for a door switch. This switch will turn on the dome lights and the "door ajar" lamp if equipped. (Courtesy of Nissan North America)

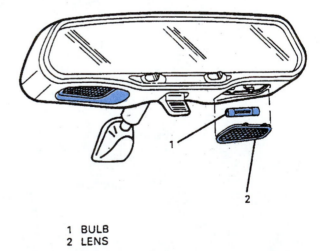

1 BULB
2 LENS

Figure 12-24 This room (dome) lamp has a three-position switch plus a switch in each front door post. This type of light fixture may be known as a "map light." (Courtesy of Chevrolet Motor Division, General Motors Corporation)

dome light is mounted near the center of the roof. Courtesy lights, if equipped, are usually mounted on the kick panels, under the kick panels, or on the doors. Rear courtesy lights are usually mounted to the left and right side of the passengers. In addition to the courtesy lights listed, reading or map lights may be positioned to shine toward each seat. This allows a passenger to read or search for an item without hindering the driver's front or rear visibility. Overhead dome lights tend to block the driver's vision through the inside rearview mirror.

Glove Box and Key Lights

The glove box light works similar to the hood and trunk lights. Opening the glove box door closes a mechanical switch. A small lamp is lit to illuminate the interior of the glove box. This is a common problem when tracing a drain on the battery. Items placed in the box may prevent the door from opening the switch and the light slowly drains the battery overnight.

The slot for the key on the outside of the door and the ignition switch may have a small lamp, possibly an LED, for illumination. This helps the driver insert the key in the door quickly at night and helps prevent marring the door's finish. The ignition switch is in an awkward position for the typical interior light. Providing a small lamp near the switch helps the driver insert the key at night.

Accessories

Shop Manual
pages 352, 358

Accessories include systems like power door locks and windows, custom seat covers and carpets, trailer hitches, and many other much-liked but not-always-necessary items. In this section, we will discuss power door locks and windows, anti-theft, and power seats as examples of system operation. The next section will cover a major and expensive accessory: climate controls commonly known as heat and air conditioning systems.

Power systems like door locks, windows, and seats were designed to help the driver control security and provide comfort without distracting from driving conditions. The door locks have recently been incorporated into anti-theft systems and remote control.

Door Locks

The electrical/mechanical mechanism for the locks is basically a reversible polarity solenoid moving the manual lock mechanism. There is a solenoid for each door controlled by parallel-wired switches in the front doors (Figure 12-25). Operating either switch locks or unlocks all doors. A switch built into the key lock on the outside of the driver's door will also perform the same function. On some vehicles, one twist of the door key will unlock the driver's door and a second twist operates the other door locks.

The solenoid is positioned to move the locking mechanism when activated. The solenoid can either push or pull the locking mechanism. The operation depends on the type of lock and the space available to install the solenoid. Most vehicles have manually-operated locks to back up the electric system.

With the availability of dependable electronics, the door locks can be activated with a small remote control device. The body computer receives the remote signal and operates the solenoids. Anti-theft security systems are connected to the body computer so the alarm system is turned on or off whenever the locks are activated (Figure 12-26). Many vehicles have automatic systems that lock the doors when the vehicle reaches about 20 miles per hour. The locks must be unlocked with the key, switch, or the remote control unit. A common fuse protects all of the lock circuits.

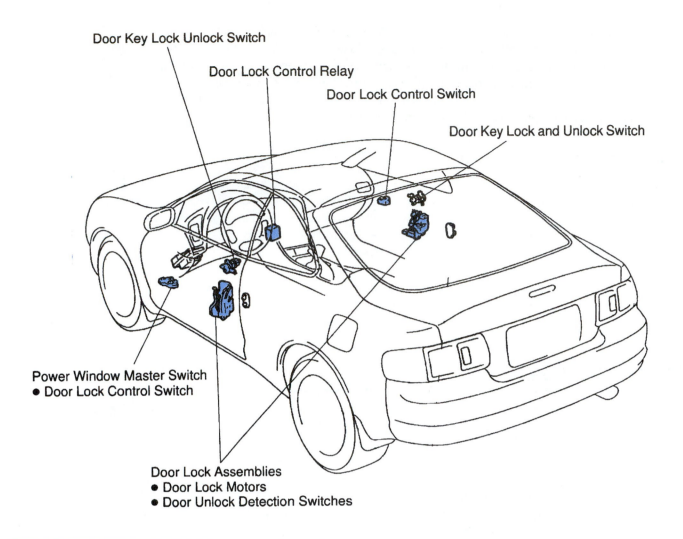

Door Key Lock Unlock Switch

Door Lock Control Relay

Door Lock Control Switch

Door Key Lock and Unlock Switch

Power Window Master Switch
● Door Lock Control Switch

Door Lock Assemblies
● Door Lock Motors
● Door Unlock Detection Switches

Figure 12-25 Note that the door lock switch is mounted within the master switch cluster. (Courtesy of Nissan North America)

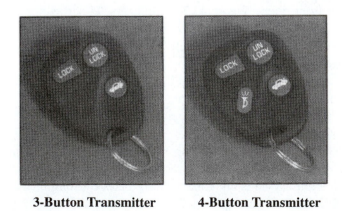

3-Button Transmitter   **4-Button Transmitter**

Figure 12-26 The horn button on the right remote is used to sound the alarm in a panic situation and will arm or disarm the anti-theft systems. (Courtesy of Chevrolet Motor Division, General Motors Corporation)

Anti-Theft Systems

Anti-theft systems range from a toggle switch placed in the ignition circuit to ignition keys that must identify themselves to the PCM before the vehicle can be powered up. A toggle switch can easily be found and flipped on. This may prevent some thefts but a professional thief will be gone with the vehicle in two or three minutes.

Most anti-theft systems in use today rely on noise and flashing lights to alert anyone nearby that the vehicle is being stolen. However, once the thief is inside and the engine is running, the anti-theft system will usually disarm and the vehicle can be driven safety.

A later system uses an ignition key with an embedded resistor pellet. When the key is inserted into the ignition switch, a small current is sent through the key and pellet. If the resistor is present and correct, the current flows to the anti-theft module to power up the vehicle.

In 1999 a new system was installed on some General Motors vehicles. Ignition keys containing small **transponders** used with specific security codes. The system's control module was placed around the steering column under the column covers. A round antenna ran from the module and encircled the ignition key slot. The anti-theft module and the PCM were programmed to recognize the key's security code and save it. This was done when the vehicle started for the first time at the assembly plant. It was also done at the dealership for replacement of additional keys. The key's code was cross-matched against other selected components so the key could not be used in another car. When a driver inserted the key into the ignition switch, the control module asked for identification from the key. The key transmitted its code and the anti-theft control module compared it to the saved codes for the key and the selected components. If all codes were correct, the module signaled the PCM to switch on the electrical systems. Presently, up to eight key codes can be stored in the module and PCM. If a key, control module or PCM is replaced, the system has to be reprogrammed.

A transponder will transmit its code by radio or radar when actuated by a specific signal from a radio or radar transmitter (interrogator).

Windows

Power windows use reversible electric motors to raise or lower the windows through gears or a plastic strip punched to fit over gear teeth. There is a window switch at each door with a **master switch** on the driver's side. The master switch may also have a lock switch that deactivates the other switches. The non-driver window motors are parallel to the master switch.

The master switch consists of a switch for each window motor and the window lock-out switch, if so equipped (Figure 12-27). The other switches placed on the side panel will only operate that window. A common fuse or circuit breaker protects all of the window circuits.

The output gear moves the mechanical device that actually moves the window on the motor. In a straight gear system, the output gear is meshed with a large, flat sector gear. The sector is attached to the arms that push up or pull down on the glass frame. Plastic strap (ribbon) or steel stranded cable mechanisms use the motor's output gears to pull the cable or

The master switch houses a switch for each window and can lock out all other individual window switches.

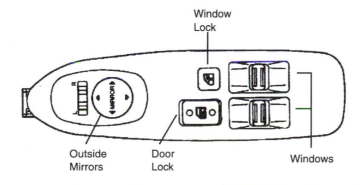

Figure 12-27 Note the three different systems controlled from this master switch cluster. (Reprinted with permission)

ribbon which are attached to the glass frame (Figure 12-28). The ribbon is punched at even intervals and fits over the output gear. Two or three pulleys and a metal guide keep the ribbon or cable in contact with the gears and pulleys to prevent slippage.

Some vehicles are equipped with automatically-lowering driver's side windows. When the switch is operated under specified conditions, the window, once started, will lower automatically to its lowest position. This allows the driver to direct attention to the environment. The window will not automatically rise because of the possible harm to any body part that is resting on or extended out the window.

Seats

Power seats are usually classified as four-way or six-way. Four-way seats can move electrically forward or backward and they can tilt the backrest. Six-way types have the same movements plus tilting of the seat bench. The systems may be on each of the front seats or just on the driver's.

The controls are usually mounted at the outside end of the seat bench, but may be mounted in the door or console. There will be two or three switches depending on the system. One will control the forward and rearward movement of the seat while another controls the tilt of the backrest. A third switch is used to control the motors to tilt the bench.

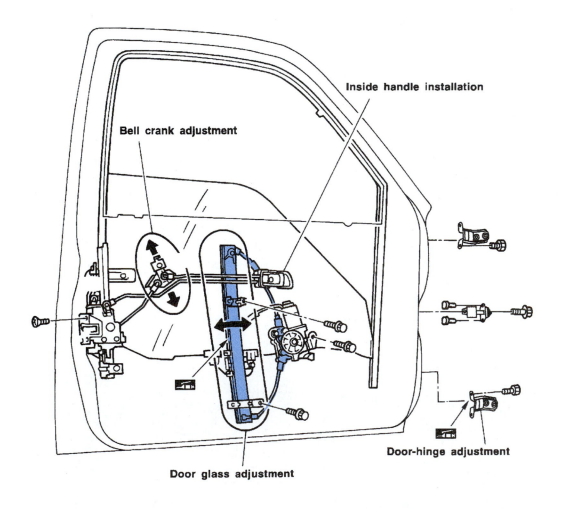

Figure 12-28 This power window unit uses a cable fed through guides to raise and lower the window. (Courtesy of Nissan North America)

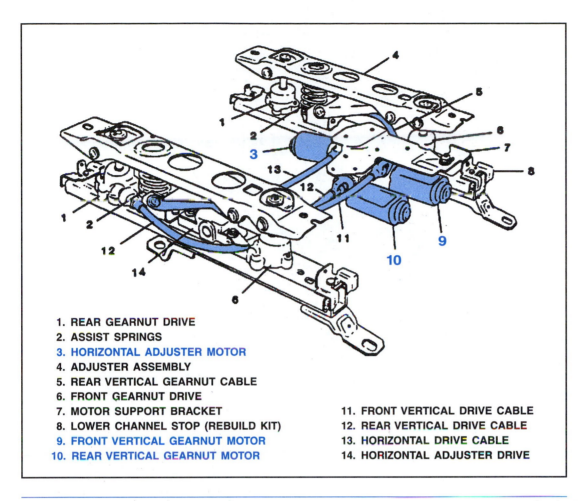

1. REAR GEARNUT DRIVE
2. ASSIST SPRINGS
3. HORIZONTAL ADJUSTER MOTOR
4. ADJUSTER ASSEMBLY
5. REAR VERTICAL GEARNUT CABLE
6. FRONT GEARNUT DRIVE
7. MOTOR SUPPORT BRACKET
8. LOWER CHANNEL STOP (REBUILD KIT)
9. FRONT VERTICAL GEARNUT MOTOR
10. REAR VERTICAL GEARNUT MOTOR

11. FRONT VERTICAL DRIVE CABLE
12. REAR VERTICAL DRIVE CABLE
13. HORIZONTAL DRIVE CABLE
14. HORIZONTAL ADJUSTER DRIVE

Figure 12-29 This six-way power seat uses three motors to move the seat. (Courtesy of Chevrolet Motor Division, General Motors Corporation)

The actual movement of the seat components is done with reversible electric motors and gears, cables, or levers (Figure 12-29). Most systems use an output gear meshed with a *toothed rail*. The rail is attached to the seat component. As the gear turns, the rail and component are moved. The motors provide a great deal of torque for their size and energy usage and can cause damage to anything left under the seat.

There are power seat systems on the market that are wired to a body computer. Sensors on the undercarriage of the seat provide position data to the computer. The computer can be programmed to remember the data and relate it to a particular operator. Selecting the proper code activates the seat and returns it to the position recorded in the computer's memory.

Climate Control Systems

Shop Manual
pages 352–357

Climate control system is a new term for the heating and air conditioning systems. The term is used to designate either a straight heating system or a heater/air conditioning combined system. The heater system is used to warm the interior of the vehicle during cold weather. Naturally, the air conditioning is used to do the opposite.

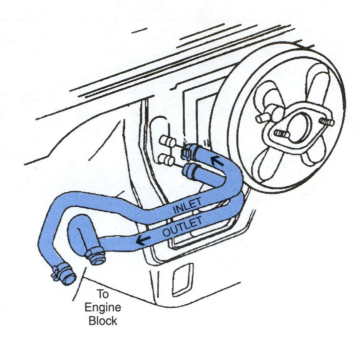

Figure 12-30 Heater hoses run from near the front of this GM 3800 engine. (Courtesy of Chevrolet Motor Division, General Motors Corporation)

Heater Systems

The heating system uses the engine's coolant to heat air that is being forced into the passenger compartment. Hoses run from the engine block to a small heater core and back to the engine block (Figure 12-30). The core is located in front of the front passenger seat, either within the dash or on the engine side of the firewall. A mechanical or vacuum-operated valve controls the flow of hot coolant into the core. A variable-speed motor drives a blower and forces air around the heater core fins and through ducts into the passenger compartment. The hot coolant in the heater core heats the passing air. The core is similar to a radiator in appearance and operation.

The control panel mounted in the dash houses buttons or levers to direct the air, change blower speed, and temperature. The operator can select outlets that direct warm air to the windshield, dash panel, or to the floor. The air lever moves blend doors within the ductwork mechanically with cables or it controls vacuum to small vacuum motors. The blower switch has three or four speed settings. The temperature selector is used to adjust the amount of hot coolant allowed into the heater core or to block the flow completely. The heater system depends on the engine and its thermostat for operation. A poorly performing engine or a stuck thermostat may result in poor heater operation.

Air Conditioning

Before discussing the air conditioning (A/C) system, a quick review of heat and pressure is needed. When a liquid is heated enough, it changes to a vapor. Adding pressure to the liquid raises the boiling point and reduces vaporization. By the same standard, removing heat from a vapor causes the element to cool and liquify. Pressure and heat are the two theories that make air conditioning systems work.

The A/C is usually split into two sections or sides: high and low (Figure 12-31). The high side extends from the **compressor**, through the **condenser,** and to the metering valve. From the compressor to the condenser, the refrigerant is in the form of a high-pressure vapor before

The *compressor* is a high-pressure pump built specifically for air conditioning systems. Other types of compressors are used to pressurize air or another medium.

A *condenser* cools down a vapor and assists in changing the liquid from a vapor state to a liquid state (condensation).

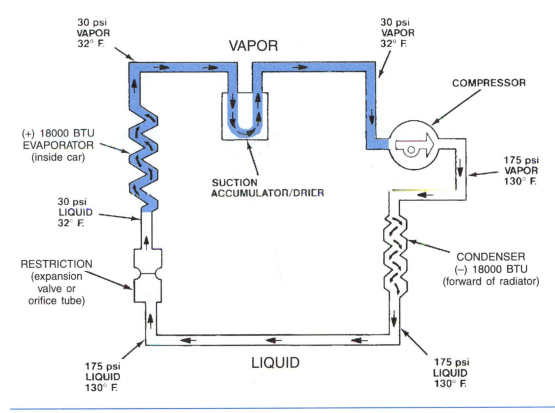

Figure 12-31 The refrigerant is a vapor from the metering valve to the suction (intake) side of the compressor. This is the low side. (Courtesy of Ford Motor Company)

changing into a high-pressure liquid within the condenser. From the metering device, through the evaporator, and to the compressor is the low side. The amount of liquid entering the **evaporator** is controlled, which reduces the pressure. The lower pressure and heat from the passing air vaporizes the refrigerant and a low-pressure vapor exits to the compressor. The condenser and evaporator are similar in form to the radiator and heater core, respectively.

An A/C system consists of several more components than the heater system. The cooling agent in an A/C unit is a **refrigerant** circulating under pressure through the system. The refrigerant has properties that may affect the environment and human health. The next section covers the refrigerant and its hazards.

A belt-driven compressor at the front of the engine pressurizes the refrigerant (Figure 12-32). Pressure or temperature switches protect the compressor by switching it on and off as needed. A condenser mounted in front of the radiator cools the refrigerant while a dash-mounted evaporator core allows the cool refrigerant to absorb heat from the air being directed into the passenger compartment. Depending upon the design, the A/C system may have either an expansion valve or an orifice tube to control refrigerant flow into the evaporator.

In addition, some systems will have an **accumulator** or **receiver/dryer** to filter and clean the refrigerant (Figure 12-33). An orifice tube is used with accumulator systems and expansion valves are used with receiver/dryers. The lines are usually made of aluminum with a high-pressure, flexible-hose section. High-side pressures may reach 300 psi or more on a hot day. Low-side pressures are proportional to the high-side pressures and usually range from 30 psi to 50 psi. The operating pressure rises as the outside temperature increases.

The *evaporator* helps a liquid heat enough so it will change from liquid to vapor (evaporation).

A *refrigerant* is a chemical that can absorb, transport, and release a great deal of heat in a fairly small space under easy-to-create conditions.

An *accumulator* separates vapor from liquid, but does not capture moisture (water). A *receiver/dryer* also separates liquid and vapor, but it will capture moisture (water) present in the refrigerant.

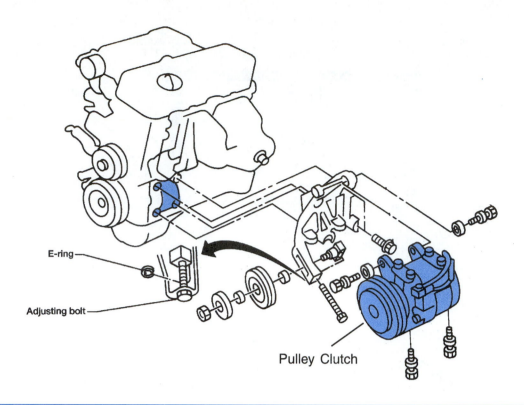

E-ring

Adjusting bolt

Pulley Clutch

Figure 12-32 This compressor is mounted at the very bottom of the engine block. (Courtesy of Nissan North America)

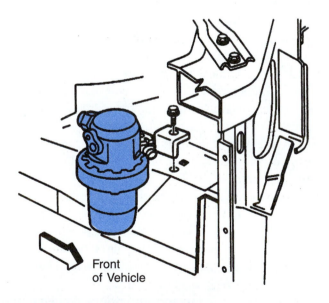

Front
of Vehicle

Figure 12-33 This accumulator is mounted to the left front side of the engine compartment. (Courtesy of Chevrolet Motor Division, General Motors Corporation)

When the A/C switch is closed and the engine is running, current flows through low or high-pressure switches to the magnetic coil and clutch. The clutch locks the compressor shaft to the pulley driven by the belt. The compressor draws in low-pressure vapor refrigerant from the evaporator, pressurizes it, and pumps it to the condenser. Air flowing over the condenser fins cools and liquifies the refrigerant. The incoming high-pressure vapor at the top of the condenser forces the refrigerant down through and out of the condenser and back to the metering device. Before entering the evaporator, the refrigerant passes through an expansion valve or orifice tube. Too much refrigerant forces the evaporator pressure upward and heat cannot be transferred. Too little refrigerant and the evaporator freeze over. The warm air passing over the evaporator is cooled by transferring its heat to the liquid refrigerant. The refrigerant is heated, vaporized, and loses pressure quickly. The low-pressure vapor is forced through a receiver/drier and back to the compressor. The receiver/drier has a cleaning filter and a moisture-absorbing material, **desiccant,** to remove any water in the refrigerant.

The expansion valve is a pressure- or temperature-sensitive metering device (Figure 12-34). If the pressure or temperature within the evaporator or low side is too high, less refrigerant is

Desiccant is a white, powered compound that can absorb many times its weight in water or moisture.

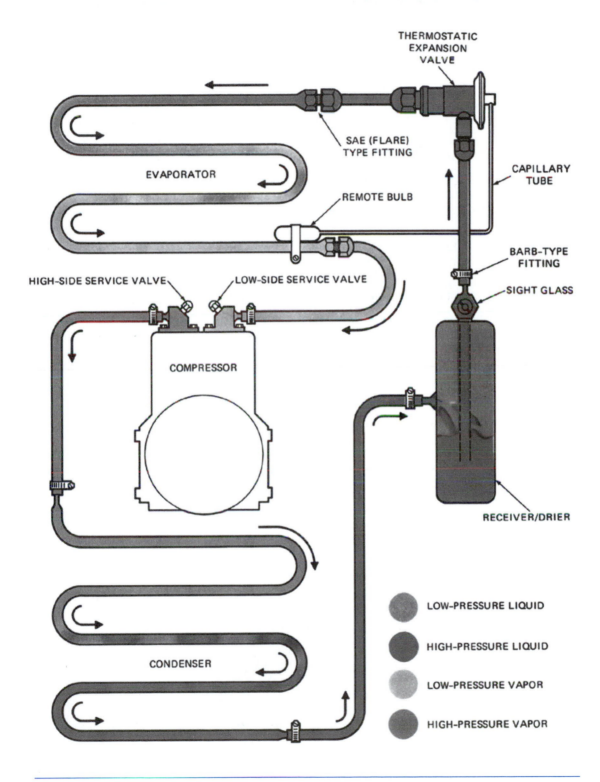

Figure 12-34 Many expansion valves indirectly measure evaporator pressure by directly measuring its external temperature. (Reprinted with permission)

allowed to enter the evaporator. The valve may be mounted on the evaporator's intake line or within the evaporator housing.

The orifice tube is basically a series of small screens (Figure 12-35). The very small holes in the screens restrict the refrigerant, thereby controlling the volume of refrigerant entering the evaporator. Most orifice tubes are about $2^1/_2$ inches to 3 inches long and roughly the diameter of a pencil. A tube is inserted into the low-side line at the last connection before the evaporator. Most orifice tubes can be replaced by disconnecting the line and pulling the tube out with a special wrench. The new tube is pushed into place and the lines reconnected. Some Ford vehicles have the tube inserted deep into the line. The line must be replaced to replace the tube.

The high-pressure switch is mounted in the high side, either on the line or on the compressor. The low-pressure switch is on low side and usually mounted on or near the receiver/drier, but may be mounted on the compressor. Some units use a dual pressure switch to sense both high and low pressures (Figure 12-36) The switches are used to protect the system when the pressures do not meet specifications. It should be noted that many systems do not have high-pressure switches. Instead, they rely on the expansion valve to control high-side pressure. Almost all systems have a low-pressure switch to protect against low refrigerant charge and compressor damage. The compressor and other working components are lubricated in part with lubricant that is circulating through the system with the refrigerant.

Special oil is added to the system to lubricate the compressor. Most of the lubricant is retained within the compressor, but a limited amount is circulated with the refrigerant to lube the hoses, switches, and the valves of the compressor. The older R12 systems use different oil than the newer R134a units. The R134a units use PAG or Ester oils. The new and old oils cannot be mixed (Figure 12-37).

Since the air conditioning system is installed in vehicles with a heater, the blower and ductwork are shared (Figure 12-38). In most cases, the air conditioning controls are on the same panel as the heater controls. The most common A/C controls are buttons labeled "MAX" and "NORMAL" or something similar. The **MAX** switch turns on the A/C and opens the door that draws air from inside the passenger compartment. This allows cool air from the interior to be circulated back through the system. The passenger compartment is cooled quicker using this setting. The NORMAL setting turns on the A/C, but draws in outside air. Normally MAX is used to quickly cool a vehicle upon startup and NORMAL is used after the initial cool down to keep the vehicle comfortable.

When set on "MAX," the air conditioning unit sounds much louder. This is caused by the sound of the blower being heard through the inside air intake door instead of the firewall.

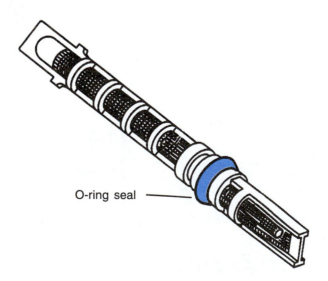

O-ring seal

Figure 12-35 A typical orifice tube. (Courtesy of Chevrolet Motor Division, General Motors Corporation)

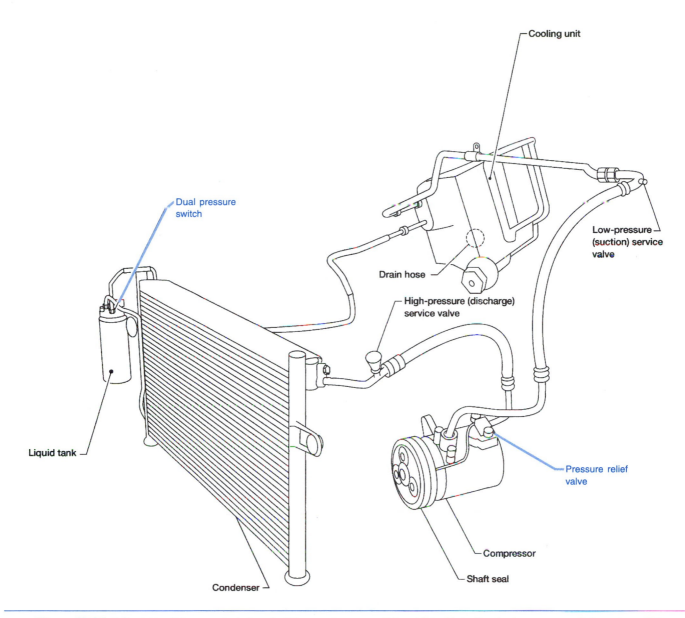

Cooling unit

Dual pressure switch

Low-pressure (suction) service valve

Drain hose

High-pressure (discharge) service valve

Liquid tank

Pressure relief valve

Compressor

Shaft seal

Condenser

Figure 12-36 A liquid tank is mounted ahead of the condenser on this vehicle. Note the dual pressure switch on top of the tank and the pressure relief valve on the compressor. (Courtesy of Nissan North America)

Figure 12-37 The three main types of air conditioning lubricants.

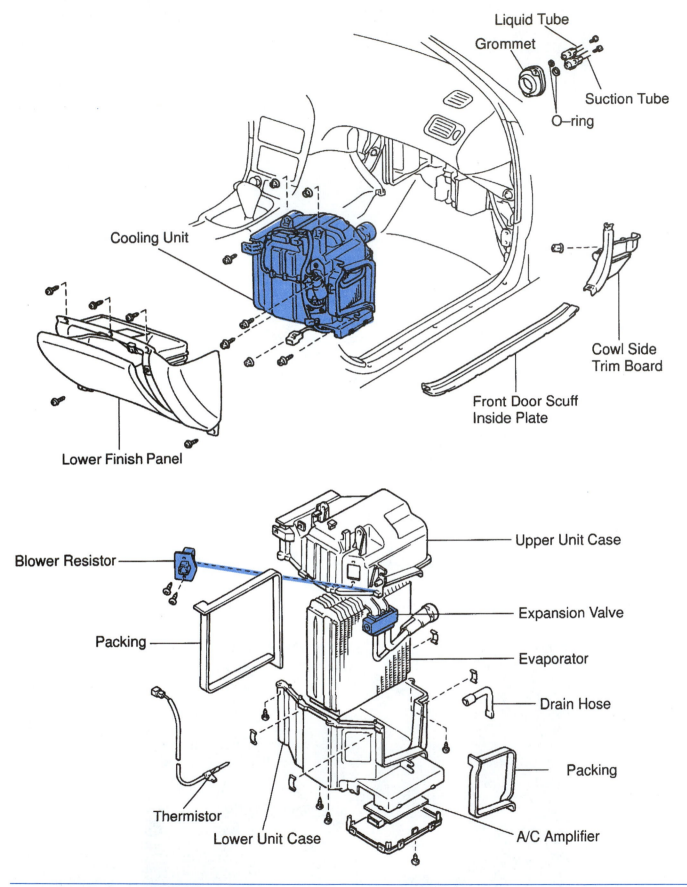

Liquid Tube
Grommet
Suction Tube
O-ring

Cooling Unit

Cowl Side
Trim Board

Front Door Scuff
Inside Plate

Lower Finish Panel

Blower Resistor

Upper Unit Case

Packing

Expansion Valve

Evaporator

Drain Hose

Packing

Thermistor

Lower Unit Case

A/C Amplifier

Figure 12-38 This package contains only the air conditioning evaporator. Note the mounting of the blower resistor and expansion valve. (Reprinted with permission)

Some vehicles are equipped with automatic temperature control (ATC) systems. The operator selects the desired in-car temperature and the ATC control module will select and operate the climate control system to fulfill the operator's request. The system may operate components of the heater and the air conditioning system at the same time or use just one of them to keep the temperature at the desired level. The module can also control blower speed operation.

The newest ATC systems allow passengers in different seats to select a temperature for their area. This individual choice provides individual comfort to each passenger instead of just the driver.

EPA and Air Conditioning Refrigerant

Shop Manual
page 352

The Environmental Protection Agency (EPA) is charged with protecting the environment of the United States. One of its responsibilities is clean air. Gas- or diesel-powered vehicles are prime causes of dirty air. The Clean Air Act of 1990, as amended, is the latest law authorizing the EPA to take action concerning air quality. One of the issues addressed is A/C refrigerant.

The **ozone layer** helps protect the Earth and its inhabitants from harmful **ultraviolet** rays and some of the heat from the sun. The older air conditioning systems, in homes and automobiles, used **carbon tetrachloride** or R12 as the refrigerant (Figure 12-39). If refrigerant is released into the atmosphere, it breaks apart and the chloride rises to the level of the ozone layer. There, the chloride atom acts to break down the ozone molecule. The result is direct ultraviolet radiation reaching the Earth's surface.

The *ozone layer* is located about 9 to 18 miles (15 to 39 km) above the earth's surface. It protects the earth from the sun's ultraviolet rays.

Ultraviolet is the term given to electromagnetic radiation with wavelengths shorter than violet. It is not visible to the human eye, but is dangerous to humans.

When the *carbon tetrachloride* molecule is broken by sunlight or age, the chloride atoms rise to the ozone layer and destroy the ozone atoms.

Figure 12-39 A typical container of R12 refrigerant. It can be purchased in 30-pound and 50-pound containers also. The containers are white or grayish-white.

The EPA and the Clean Air Act specifically prohibit the release of any refrigerant into the atmosphere, imposing very heavy fines for violators. Technicians working on A/C systems are required to pass an EPA-approved test before they can legally handle refrigerant. The test involves the control of refrigerant, the effect of refrigerant on the ozone layer, and some of the legal requirements of the Clean Air Act. The test does not include technical information on mechanical diagnosing and repairing an A/C system. The test is available from several approved sources including ASE and MACS. Some parts stores have the test booklet on hand for anyone interested in taking the test. The test is basically open-book, with twenty-five multiple-choice questions. Complete the test and mail the answer sheet with the noted fee to the address given. The results, with an identification card (if passed), are sent back. It should be noted that similar requirements are applied to fix A/C units like the ones used in homes and businesses and refrigerators, private and commercial. Any person desiring to purchase any refrigerant must show proof that he/she passed the EPA test by presenting the identification card.

The EPA requires very specific actions to occur during **recovery, evacuating,** and **recharging** an air conditioning unit. All of the requirements are based on controlling refrigerant leaks into the atmosphere. The requirement allows each shop to make individual decisions on repairing a system or just adding refrigerant to a leaking system. The customer can request that a leaking system be topped off with refrigerant, but the shop may legally refused to so.

In addition to testing and control procedures, the EPA directed research and development of an ozone-friendly refrigerant. Research in this area has been ongoing since the mid-1970s, but the growing evidence of ozone depletion increased the amount of funding and the number of laboratories.

A new refrigerant, R134a, was developed and is being used by most vehicles produced since 1994 (Figure 12-40). R134a consists of hydrogen-based molecules that do not affect the ozone layer. However, regulations on releasing refrigerant still apply and the EPA test on refrigerant handling must be completed before buying or using R134a. Systems using R134a nor-

Recovery is the process of capturing refrigerant in an A/C system. *Evacuating* is removing all air and moisture from the system after recovery. *Recharging* is putting refrigerant back into the system after repairs.

Figure 12-40 R134a containers are light blue and can be bought in 30- and 50-pound containers. The color is required by the EPA as an identification of the chemical inside.

mally run slightly higher operating pressures and there is a small loss in system efficiency. There are other ozone-friendly refrigerants on the market, but R134a is the most popular.

In order to encourage the public to use the new refrigerant, federal, state, and local governments have imposed a progressively increasing tax on R12. In ten years, the price of R12 has gone from 79 cents for a 14-ounce can to 18 dollars or more at the parts store. Most repair centers charge 30 dollars or more for a pound (16 ounces) of R12. Most R12 units can be *retrofitted* to R134a. In most cases, the system is emptied of R12 and most of the old lubricant, seals, and gaskets are replaced. The R134a refrigerant and the correct oil is then charged into the systems. Many shops recommend that a damaged R12 system be retrofitted during routine repairs. The savings on the refrigerant will pay for the kit to make the change. The price of the old and new lubricants is about the same.

Restraint Systems

Restraint systems hold passengers in their seats and away from vehicle parts during an accident. This not only protects passengers from further injury, but helps the driver maintain control of the vehicle.

Seat Belts

Lap seat belts have been around for years. One end of the lap belt is anchored to a small axle assembly, anchored to the vehicle frame, and snapped locked into a second anchor at the other end of the belt. The belt, if properly used, reduces the chance of a passenger slamming forward into the steering wheel, dash, or other interior component. But more important, it would keep the person *inside* the vehicle. Many people died in survivable crashes because they were thrown from their vehicle and then run over by the same vehicle. Later, lap belts were used with shoulder belts and the term *seat belt* came to mean a restraint with both lap and shoulder belts.

The shoulder belt functions similar to the lap belt, but it is anchored at the top of the doorpost behind the driver's or passenger's shoulder. It should extend over, down, and across the shoulder and connect to the lap belt at the traveler's opposite hip. It was, and still is, designed to work in conjunction with the seat belt. The shoulder belt can work well alone in certain types of crashes on certain size people. But the lap and shoulder belts are most effective when used together. When used properly, the latest design has both belts permanently connected together near the person's hip.

The belts are wrapped around a small axle at each main anchor. The axle has a ratchet or lock on one end. One type of ratchet has a small weight that is sensitive to the rapid deceleration that occurs during an accident. During sudden deceleration, the weight swings forward and locks the axle, thereby stopping the movement of the belt. In some cases, a quick jerk on the belt will lock it. Releasing the force on the belt allows the axle to unlock.

There are two major operational problems with seat belts. The critical point to seat belt operation is that vehicle occupants must hand-connect their belts in almost every belt system. If belts are not connected, they do nothing to protect drivers and passengers. A second problem involves pregnant women. A seat belt must be positioned low on the abdomen (Figure 12-41) or injury to the unborn infant might occur. The belt will still work adequately even though it is set low. Seat belts will leave uncomfortable bruises and marks on passengers or drivers during an accident, but they will heal in a few days.

CAUTION: Never work in or around the instrument panel and steering column until the air bag has been disarmed. Serious injury or death could result from a deploying air bag.

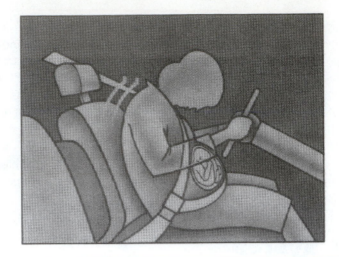

Figure 12-41 A pregnant women should wear the seat belt low to protect the child. (Courtesy of Chevrolet Motor Division, General Motors Corporation)

Air Bags

Supplement restraint systems (SRS) are restraint systems designed to improve the seat belt system. SRS is not designed to protect any person who does not use the seat belts.

Supplement restraint systems (SRS) are passive restraints requiring no action by the passengers or driver. They are generally referred to as *air bag systems* and are designed to keep anyone from hitting the front of the dash and the steering wheel. The original system included a single air bag mounted in the steering wheel. Later systems added a passenger air bag mounted in the right side of the dash. Unfortunately, air bags have caused the death of some motorists. A shorter person might sit closer to the steering wheel and, as a result, is too close to the deploying air bag. When a child is buckled into a child seat and placed in the front passenger position, the deploying passenger-side air bag could injure or kill the child. A public outcry directed at the National Transportation Safety Board prompted further research into a safer air bag system. Regulations and standards were adopted to allow vehicle manufacturers to reduce the force of air bag deployment. In addition, a switch was installed to turn off the passenger-side bag (Figure 12-42). The new regulations also allow the owner to request that the driver's air bag be deactivated under certain, specific conditions, such as when the regular driver is of smaller build and has to sit closer to the steering wheel. The ideal solution is to reduce the force of deployment to meet individual needs. However, that option is not feasible now.

The latest air bag system includes the two bags mentioned previously plus air bags mounted in the doors. The side bags are designed to reduce injuries during a crash to the side of the vehicle.

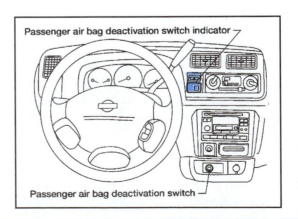

Figure 12-42 A passenger air bag deactivation switch. (Courtesy of Nissan North America)

Air Bag Operation

The air bag system relies on reliable electronics to function properly. It draws power directly from the battery and is turned on by switching the ignition to run. Impact or arming switches are located behind the front bumper and inside the front doors if equipped with side bags.

A crash that is strong enough to trip (close) an arming switch signals the air bag control module to deploy the air bags. A frontal crash will fire both driver and passenger air bags.

The folded air bag is placed over a chemical that will produce a large volume of gas when ignited. When the control module ignites the chemical, it expands and rapidly inflates the air bag. The system requires a backup power source in case the battery is destroyed before the air bag can be deployed. This can cause serious injury or death if the system is not deactivated during repair work.

A **capacitor** is used as the reserve or backup power. The capacitor is charged each time the ignition is switch on and will store an electrical current rated at about 36 volts. If the battery is destroyed and the capacitor loses power, the capacitor will discharge. This provides enough power to operate the module and deploy the air bags. During normal operation, the capacitor slowly loses its charge after the ignition is switched off. During the drain down, the air bag can still be fired if the capacitor senses a sudden loss of power or receives a surge of electrical current.

Before beginning work inside the vehicle, the capacitor must be drained of electrical current. The acceptable method is to remove the negative cable from the battery and wait awhile. The amount of time is set by the air bag and vehicle manufacturer and is shown in the service manual.

To further protect the technician, it is suggested that the air bags be disconnected after the capacitor has discharged. There are two bright yellow plugs located near the bottom of the steering column under the dash. Once the yellow plugs are disconnected, the battery can be reconnected to supply power for testing of the failed components. When all of the repairs are completed and covers are installed, disconnect the negative cable again. The yellow plugs can now be connected without any danger of electric surge.

> A *capacitor* is an electric current storage device.

Summary

- ❏ Passenger comfort devices are designed to make the driver and passenger comfortable during vehicle operations.
- ❏ Passenger conveniences are those items that make it easier for the operator or passenger to control vehicle components or provide some type of service device to the passengers.
- ❏ Warning systems indicate to the driver that a dangerous or damaging condition in the vehicle may occur.
- ❏ Warning systems have bright red or amber lights and may have audible signals to alert the driver to some condition.
- ❏ Information systems provide data pertaining to vehicle operations.
- ❏ Instrumentation usually refers to the lights, gauges, and other devices that present information to the driver.
- ❏ Information may be conveyed in digital, audible, analog, text, or color displays.
- ❏ Most exterior lighting is required to meet federal and state regulations concerning illumination and positioning.
- ❏ Headlights may have either two or four lamps with high and low beams.
- ❏ Except for the turn signals, all of the rear mounted lights must be red.
- ❏ Brake lights override the rear four-way flashers but not the regular turn signals.
- ❏ A third brake light is mounted high to attract the attention of following drivers.

Terms to Know

Accumulator

Analog

Backlit

Bayonet

Capacitor

Carbon tetrachloride

Center High-Mounted Stop Light (CHMSL)

Communication buss

Compressor

Condenser

Cornering lights

Desiccant

Digital

❑ The only interior light mandated by regulations is the lighting of the instrument panel.

❑ The instrument panel is finished with a non-glare coating.

❑ Most instrument panels use circuit boards.

❑ Each power lock circuit is parallel wired to the other lock circuits.

❑ One fuse protects all of the locking circuits.

❑ Most power door lock systems are also part of the vehicle's anti-theft system.

❑ Each of the power window circuits is parallel to each other and the master switch.

❑ The master switch can control all of the windows plus lock out the other windows switches.

❑ Power seats can have four- or six-way positions.

❑ The ability to tilt the seat's bench is the difference between four- and six-way seat systems.

❑ Some seat systems can remember a designated position.

❑ Climate control includes heating and air conditioning systems.

❑ Most A/C units operate at high pressure.

❑ The condenser cools the refrigerant.

❑ The evaporator cools the air by removing heat from it.

❑ The compressor pumps and pressurizes the refrigerant.

❑ Expansion valves or orifices regulate the amount of refrigerant entering the evaporator.

❑ The EPA is responsible for directing research and implementing procedures to protect the environment.

❑ The Clean Air Act of 1990, as amended, charges the EPA to clean vehicle emission discharges.

❑ Chloride in R12 destroys ozone molecules.

❑ R134a is an ozone-friendly refrigerant.

❑ A person must be certified by an EPA-approved test before buying refrigerant or servicing A/C systems.

❑ Seat belts most be connected to protect the motorist.

❑ Supplemental restraint systems are known as air bags.

❑ A capacitor is used as backup power for the air bag system.

Review Questions

Short Answer Essays

1. List and label the color of the lights or lenses required on mandated exterior lighting systems. Include options, if any.

2. Describe the major differences between four-way and six-way power seats.

3. Explain how non-remote-controlled power door locks are operated from outside the vehicle.

4. List the major components on the high side of an A/C system.

5. Describe how a high-pressure liquid becomes a low-pressure vapor in an A/C system.

6. Explain the process that makes the turn signal lamps blink.

7. Describe the purpose of the CHMSL.

8. Explain the purpose of the printed circuit board used in the instrument panel.

9. List and describe the purpose of the most common dash mounted warning lights.

10. Explain the difference and operation of two-lamp and four-lamp headlight systems.

Fill-in-the-Blanks

1. The left, rear power window switch is _____ _____ _____ _____ with the master switch.

2. The power door locks are protected with a(n) _____ _____ _____ .

3. The ignition key slot may be illuminated with a(n) _____ _____ .

4. The rear power door locks may be unlocked by turning the door key _____ _____ _____ .

5. The instrument panel is _____ and finished with a _____ coating.

6. The colored band type of gauge displays information similar to a(n) _____ gauge.

7. HUD is used to project information onto the _____ in the _____ line of _____ .

8. Body computers can be programmed to _____ _____ _____ for a designated _____ .

9. The dimmer switch can be mounted on the floor or within the _____ _____ .

10. R12 and R134a purchase and use are regulated by the _____ _____ _____ _____ as required by the _____ _____ of _____ .

ASE Style Review Questions

1. Interior lighting systems are being discussed.
   Technician A says the only mandated interior lighting is the instrument panel.
   Technician B says the courtesy and dome lights may be wired to the same door switch. Who is correct?
   A. A only
   B. B only
   C. Both A and B
   D. Neither A nor B

2. *Technician A* says the lamp or bulb can be replaced in the sealed beam head light.
   Technician B says the need for better aerodynamics required the headlight lamps to be redesigned. Who is correct?
   A. A only
   B. B only
   C. Both A and B
   D. Neither A nor B

3. The rear lights on a vehicle are being discussed.
   Technician A says the turn signals may be amber or red.
   Technician B says three-lamp systems uses dual-filament bulbs. Who is correct?
   A. A only
   B. B only
   C. Both A and B
   D. Neither A nor B

4. Rear dual-filament lamps are being discussed.
   Technician A says one-filament is used for turn and marking lights.
   Technician B says one-filament is used only for brakes. Who is correct?
   A. A only
   B. B only
   C. Both A and B
   D. Neither A nor B

5. Climate control systems are being discussed. *Technician A* says the heater uses engine coolant to heat the passenger compartment. *Technician B* says the condenser in mounted in the dash. Who is correct?
   A. A only
   B. B only
   C. Both A and B
   D. Neither A nor B

6. *Technician A* says R134a molecules are chloride based. *Technician B* says R134a purchase and use are regulated by the EPA. Who is correct?
   A. A only
   B. B only
   C. Both A and B
   D. Neither A nor B

7. Air conditioning systems are being discussed. *Technician A* says the refrigerant changes from a liquid to a vapor in the condenser. *Technician B* says heat from the passing airflow vaporizes the refrigerant. Who is correct?
   A. A only
   B. B only
   C. Both A and B
   D. Neither A nor B

8. The depletion of the ozone layer is being discussed. *Technician A* says R12 released into the atmosphere destroys ozone molecules. *Technician B* says a depleted ozone layer allows more ultraviolet radiation to reach the Earth. Who is correct?
   A. A only
   B. B only
   C. Both A and B
   D. Neither A nor B

9. *Technician A* says information systems uses data from the PCM to compute some information for display. *Technician B* says the information may be computed and displayed by a computer module separate from the PCM. Who is correct?
   A. A only
   B. B only
   C. Both A and B
   D. Neither A nor B

10. *Technician A* says the expansion valve regulates high-side pressure on an A/C system. *Technician B* says an orifice may be used to control refrigerant flow into the evaporator. Who is correct?
    A. A only
    B. B only
    C. Both A and B
    D. Neither A nor B

GLOSSARY

Adjustable wrench: A wrench that adjusts to fit bolts of different sizes.

Llave ajustable: Llave que se puede ajustar para utilizarse con pernos de diferentes tamaños.

Aerodynamics: The science and study of objects moving through air.

Aerodinámica: La ciencia e estudio de los objetos moviendose por el aire.

Air gauge: A gauge to check air pressure, commonly to check tire inflation.

Medidor de aire: Un medidor para revisar la presión del aire, suele usarse para revisar la inflación de los pneumáticos.

Air-operated ratchet driver: A ratchet handle powered by air for fast removal or installation of bolts or nuts.

Trinquete accionado por aire: Palanca para trinquetes accionada por aire y utilizada para remover o instalar rápidamente pernos o tuercas.

Air spring: A suspension spring that uses an air bladder and piston connected to the suspension components to provide the spring action.

Muelle de aire: Muelle de suspensión que utiliza un saco de aire y un pistón conectados a los componentes de la suspensión para proveer un movimiento amortiguador.

Allen wrench: A wrench used to tighten or loosen Allen-head screws.

Llave Allen: Llave utilizada para apretar o aflojar tornillos Allen.

Ammeter: An electrical meter used to measure amperage or current flow.

Amperímetro: Instrumento eléctrico utilizado para medir la intensidad eléctrica o el flujo de corriente.

Amperage: The movement of electrical current through an electrical circuit.

Amperaje: Movimiento de corriente eléctrica a través de un circuito eléctrico.

Analog meter: A meter that uses a pointer and scale for readings.

Instrumento analógico: Instrumento provisto de un indicador y una escala para las lecturas.

Antilock brake system: A system that uses a computer to determine if a wheel is locking up during braking and a control unit to pulse the brakes on the wheel that is in danger of locking up.

Sistema de frenos antibloqueo: Sistema de frenos que utiliza una computadora para determinar si se produce un bloqueo en alguna de las ruedas durante el frenado y una unidad de mando para pulsar los frenos de la rueda en la cual existe el riesgo de que se produzca un bloqueo.

ARTA: The Automatic Transmission Rebuilders Association offers technical certification for automatic transmission technicians.

ARTA: La asociación de renovadores de las transmisiones automáticas que provee la certificación técnica para los técnicos de las transmisiones automáticas.

Backlit: Placing light behind or at an angle to an object being lit. This reduces glare on the object.

Trasiluminado: Poniendo la luz detrás o en un ángulo al objcto que se alumbra. Esto disminuye el deslumbre en el objeto.

Ball bearing: A bearing made of precision-machined balls trapped in a cage. Commonly used for wheels and axles.

Cojinete balero: Un cojinete hecho de las bolas de maquinado preciso atrapados en una jaula. Suelen usarse para las rudas y los ejes.

Ball joints: The suspension components used to attach the control arm to the steering knuckles and that allow wheel movement.

Juntas esféricas: Componentes de la suspensión utilizados para fijar el brazo de mando a los muñones de dirección y que permiten el movimiento de la rueda.

Ball-peen hammer: A type of hammer with a head with one round face and one square face.

Martillo de bola: Tipo de martillo que tiene una cabeza con una cara redonda y la otra cuadrada.

Battery: One of the main parts of the automotive electrical system that uses two dissimilar metals and an acid to develop electrical energy.

Batería: Uno de los componentes principales del sistema eléctrico del automóvil que utiliza dos metales disímiles y un ácido para producir energía eléctrica.

Bayonet: A type of locking connection used primarily on automotive lamps; the method of locking a bayonet onto a rifle.

Bayoneta: Un tipo de conexión de bloque que se usa primariamente en las lámparas automotrices; el metodo de fijar una bayoneta en un rifle.

Bearing journal: The highly machined part of a component where the bearing is fitted.

Muñon del cojinete: La parte altamente acabado a máquina del componente en donde se coloca el cojinete.

Bearing load: The amount of load a bearing can support without damage to the bearing or load. Loads may be axial or radial.

Carga del cojinete: La cantidad de la carga que puede soportar un cojinete sin sostener daños al cojinete o a la carga. Las cargas pueden ser axial o radial.

Bearing races: The highly machined surfaces of a bearing where the balls or rollers operate. *See* ball bearing.

Pistas de los cojinetes: Las superficies altamente acabadas a máquina de un cojinete en donde operan las bolas o los rodillos. *Ver tambien* cojinete balero (ball bearing).

Bias: A tire that is constructed with the plies arranged on a bias crossing each other.

Bias: Llanta fabricada con las estrias arregladas para que se crucen entre sí.

Block: The part of the engine assembly that houses the internal engine components and provides mounting surfaces for external engine components.

Bloque: La parte de la ensambladura del motor que contiene los componentes internos del motor y provee las superficies para instalar los componentes externos del motor.

Bolt: A threaded fastener used with a nut to hold automotive parts together.

Perno: Asegurador fileteado utilizado con una tuerca para sujetar piezas automotrices.

Box-end wrench: A wrench designed to fit all the way around a bolt or nut.

Llave de cubo estriado: Llave diseñada para cubrir completamente un perno o una tuerca.

Brake pads: The frictional parts used to contact the rotor during braking to stop the rotor and stop the car.

Cojines de freno: Piezas de fricción que hacen contacto con el rotor durante el frenado y ejercen presión sobre éste para detener la marcha del vehículo.

Brake shoes: The friction covered parts used in the drum brake system to stop the brake drum from rotating.

Zapatas de freno: Piezas cubiertas de fricción utilizadas en el sistema de frenos de tambor con el fin de impedir la rotación del tambor de freno.

Brake vacuum booster: Brakes that reduce driver braking effort using engine intake manifold vacuum acting on a diaphragm connected to the master cylinder push rod.

Frenos hidráulicos auxiliares de vacío: Frenos que disminuyen los esfuerzos del conductor utilizando el vacio del difusor de entrada del motor que acciona un diafragma conectado a la varilla de empuje del cilindro maestro.

Brake warning light: A red light in the dash that alerts the driver to low brake fluid or when the parking brake is on.

Luz indicadora del freno: Una luz roja en el tablero de instrumentos que informa al conductor de un nivel bajo de fluido de freno o de que el freno esta en uso.

Brass hammers: Tools with heads made from soft brass that are used to hammer on parts without damaging them.

Martillos de latón: Herramientas con cabezas hechas de latón suave que se utilizan para martillar piezas sin deteriorarlas.

Bushing: A thin-walled, pipe-type component used to support a shaft in the housing, thereby reducing shaft wobble and wear.

Buje: un componente tubular de muro delgado que sirve para soportar el eje en la caja asi disminuyendo la oscilación y el desgaste del eje.

Business ethics: The practice of conducting business by being fair and honest to customers, employees, and the owner.

ética de negocio: La práctica de dirigir el negocio en una manera justa y honesta a los clientes, los empleados y el dueño.

Butt connector: An electrical connector used to connect two pieces of wire together.

Extremo de conectador: Conectador eléctrico utilizado para conectar dos alambres.

Caliper: The hydraulic component of disc brakes that activates the brake pads that stop the rotor during braking.

Calibre: Componente hidráulico de los frenos de disco que acciona los cojines de freno que detienen el rotor durante el frenado.

Camshaft bearing: Bearings formed in the shape of a small, highly machined pipe section similar in shape to a bushing, but much stronger and with a better finish.

Cojinete del árbol de levas: Los cojinetes formados en la forma de una sección de tubo altamente acabado parecido en forma a un buje pero mucho más fuerte y mejor acabado.

Capacitor: An electrical storage device that discharges its current on an electrical signal. Commonly used as backup power for the air bag system.

Capacitador: Un dispositivo de almacenaje eléctrico que descarga su corriente con una señal eléctrica. Suele usarse para respaldo de un sistema de bolsa de aire.

Carbon tetrachloride: A nonflammable, colorless, poisonous gas used in refrigerant and spray propellant.

Carbon tetracloruro: Un gas no inflamable, sin color, tóxico que se usa en el propelente de atomización y refrigeración.

Catalyst: An element that can cause a chemical or physical change in other elements without changing itself. Commonly used in catalytic converters to change reduce harmful elements.

Catalizador: Un elemento que puede causar un cambio físico o químico en otros elementos sin cambiarse. Suelen usarse en los convertidores catalíticos para cambiar los elementos nocivos.

Chisel: A bar of hardened steel with a cutting edge ground on one end; driven with a hammer to cut metal.

Cincel: Barra de acero templado dotada de un filo cortante en uno de los extremos y que es accionada por un martillo para cortar metal.

Carcass: The remainder of a tire after most of the tread has been removed. A good carcass can be treaded again in a process called "recapping."

Carcasa: Lo que resta de un pneumático después de que la mayoría de la banda de rodadura se haya quitado. Un alma buena puede recibir un rodamiento nuevo en un proceso que se llama "recauchutado."

Catalytic converter: A mechanical device placed in the vehicle's exhaust system to reduce harmful emissions of which the most common are hydrocarbons, carbon monoxide, and nitrogen oxides. *See* catalyst.

Convertidor catalítico: Un dispositivo mecánico colocado en el sistema de escape del vehículo para disminuir los emisiones nocivos los más comunes que son los hidrocarburos, el monóxido de carbón, y los óxidos nítricos. *Ver tambien* catalizador (catalyst).

Cavitation: The process where a liquid pump draws in air instead of liquid; can cause serious damage to the pump and other components.

Cavitación: El proceso por el cual una bomba de líquidos aspira el aire en vez de un líquido; puede causar los daños serios a la bomba y otros componentes.

Center High-Mounted Stoplight: A brake (stop) light placed high on the rear of the vehicle in the direct line of sight of following drivers. Intended to reduce rear end collisions by attracting following drivers' attention quicker.

Lámpara de parada montada en alto en centro: Un luz del freno (de parada) colocado en alto en la parte trasera del vehículo en línea directa de vista de los conductores que siguen. Su intención es atraer la atención de los conductores más rapidamente y asi disminuir las colisiones traseras.

Center of gravity: The point at which an object, theoretically, can be balanced. It is the point where the shape, size, and weight of the object is equal, but not necessarily the physical center.

Centro de gravedad: La punta en la cual, teóricamente, ese objeto puede ser balanceado. Es la punta en donde la forma, el tamaño, y el peso del objeto es iguala, pero que no necesariamente es el centro físico.

Circuit: The path followed by electrical current or hydraulic fluid. Usually includes a power source, conductors, controls, and a load.

Circuito: El rumbo que transcorre un corriente eléctrico o un fluido hidráulico. Suele incluir un fuente de potencia, los conductores, los controles, y una carga.

Circuit breaker: A circuit protection device that uses a bimetal arm, which moves away from a contact and opens the circuit during overload.

Interruptor para circuito: Este dispositivo de protección del circuito utiliza un brazo bimetal que se aleja de un contacto y abre el circuito en caso de una sobrecarga.

Clutch switch: A safety switch that is closed when the clutch pedal on a manual transmission is pressed down. Prevents the engine from cranking when the clutch pedal up. *See* park/neutral switch.

Interruptor del embrague: Un interruptor de seguridad que se cierre cuando el pedal del embraque de una transmisión manual se oprime. Previene que se arranca el motor cuando el pedal del embrague se levanta. *Ver tambien* interruptor de park/neutral (park/neutral switch).

Coil pack: A set of two or more ignition coils in one unit.

Paquete de bobina: Un conjunto de dos o más bobinas de ignición en una unedad.

Coil spring: A suspension spring made from a round bar of spring steel wound in the shape of a coil.

Muelle helicoidal: Muelle de suspensión hecho de una barra redonda de acero templado devanado en forma espiral.

Combination wrench: A wrench with one box end and one open end.

Llave de combinación: Llave con un cubo estriado y el otro abierto.

Combination pliers: General-purpose pliers with a slip joint for two jaw openings.

Alicates de combinación: Alicates para uso general con una junta deslizante para que las tenacillas se puedan ajustar en dos posiciones.

Combination switch: A component of the brake system that contains a warning light switch, metering valve, and proportioning valve.

Conmutador de combinación: Componente del sistema de frenos que contiene un conmutador para la luz de aviso, una válvula de medida, y una válvula dosificadora.

Computerized service information system: System in which a personal computer is used to access service information stored on compact discs.

Sistemas computarizados de información de servicio: Sistema en el cual se utiliza una computadora personal para introducir la información de servicio almacenada en discos compactos.

Continuity: A complete path for current flow in an electrical circuit.

Continuidad: Trayectoria completa para el flujo de corriente en un circuito eléctrico.

Control arms: The "A"-shaped suspension components that pivot and allow suspension movement.

Brazos de mando: Componentes de la suspensión en forma de "A" que giran y permiten el movimiento de la suspensión.

Conversion factor: A mathematical factor used to convert English measurements to metric measurements or the reverse.

Factor de conversión: Un factor matemático que se usa para convertir las medidas inglesas a la medidas métricas o de reverso.

Coolant: A liquid circulated around hot engine parts to remove the heat and prevent damage.

Refrigerante: Líquido que circula por las piezas calientes del motor para remover el calor y prevenir averías.

Cords: The layers of rubber and other materials used to support the tire's tread.

Cuerdas: Las capas de caucho o de otras materias que se usan para soportar el la banda de rodadura del neumático.

Cotter pin puller: A tool used to hook and remove a cotter pin.

Sacapasador: Herramienta utilizada para prender y remover un pasador de chaveta.

Crumple zones: The sections of a vehicle designed to absorb impact forces during a crash. Used to reduce the amount of impact delivered to the passenger compartment.

Zonas de aplastamiento: Las secciones de un vehículo diseñadas para absorbar las fuerzas del impacto durante una colisión. Se usan para disminuir la cantidad del impacto entregado al compartimento del pasajero.

Current: The flow of electricity measured in ampere or the number of electrons passing a single point in one second. One ampere is about 168 billion billion electrons.

Corriente: El flujo de la electricidad medida en amperios o la cantidad de los electrones pasando una sóla punta en un segundo. Un amperio es aproximadamente 168 billones de electrones.

Dealership: An automotive business that sells new vehicles and has a shop for repairs. Usually sells and services only certain brands of vehicles.

Distribuidor: Un negocio automotríz que vende los vehículos nuevos y que tiene un taller para las reparaciones. Suele vender y ofrecer los servicios para ciertas marcas de vehículos.

Decimal: A method used to separate a whole number and part of the number. An example would be 1.1 or one and one-tenths.

Decimal: Un metodo que se usa para separar un número integro de una parte de un numero. Un ejemplo sería 1.1 o uno y un decimo.

Diagonal cutting pliers: Pliers with cutting edges on the jaw for cutting cotter pins or wire.

Alicates de corte diagonal: Alicates dotados de filos cortantes en las tenacillas para cortar pasadores de chaveta o alambres.

Dial caliper: A precision measuring device that can be used to measure the inside, outside, or depth of an automotive part.

Calibre de cuadrante: Dispositivo para medidas precisas con el cual puede medirse el interior, el exterior, o la profundidad de una pieza automotriz.

Dial indicator: A precision measuring tool that registers readings on the face of a dial.

Indicador del cuadrante: Herramienta para medidas precisas que registra las lecturas en el cuadrante.

Diaphragm: Flat, flexible devices placed between two chambers that can be used perform or increase the performance of certain actions. Commonly used within the brake vacuum booster to increase the driver's force applied to the brake pedal.

Diafragma: Los dispositivos planos y flexibles que se colocan entre dos cámaras que pueden usarse para funciones o para aumentar las funciones de ciertas acciones. Se usan comunmente con el freno de vacío con asistencia para aumentar la fuerza que coloca el conductor al pedal de freno.

Die: A tool used to cut internal threads.

Troquel: Herramienta utilizada para cortar filetes internos.

Digital: Term normally used to express information in numerical form.

Digital: Un término que suele usarse para expresar una información en forma numérica.

Digital storage oscilloscope: An expensive engine analyzer that can store and display a vehicle's operating information in digit form.

Osiloscópio de almacenaje: Un analizador de motor muy caro que puede almacenar y mostrar la información de operación del vehículo en forma digital.

Dimmer switch: A switch used to change between high and low beam headlights.

Interruptor de resistencia regulable: Un interruptor para cambiar entre los luces de carretera y los de la ciudad.

Double-post lift: A lift that has two lift cylinders or lift mechanisms. It may be mounted in the floor or above the floor.

Ascensor de doble poste: Un izador que tiene dos cilindros o mecanismos de izar. Puede montarse en el piso o arriba del piso.

Drum: The braking device that is attached to and turns with the wheel. Brake shoes inside the drum are used to slow the drum.

Tambor: El dispositivo de frenado que se conecta y gira con la rueda. Las zapatas adentro del tambor se usan para frenar el tambor.

Dual overhead cam: An engine that has two or more camshafts mounted in the cylinder head.

Doble árbol de levas encima de cabeza: Un motor que tiene dos o más árbol de levas en la cabeza del cilindro.

Dynamic balance: A method of electronically dividing a wheel assembly into four equal parts and then balancing each part against each of the other three parts.

Equilibrio dinámico: Un metodo de dividir electronicamente una ensambladura de rueda en cuatro partes iguales y luego equilibrar cada parte contra las otras tres partes.

Electrolyte: A mixture of sulfuric acid and water in a car battery that is extremely corrosive and dangerous.

Electrolito: Mezcla sumamente corrosiva y peligrosa de ácido sulfúrico y agua en la batería de un vehículo.

Electromotive force: The force that moves electrons out of their orbit and is measured in volts. May be referred to as "voltage."

Fuerza electromotor: La fuerza que mueva los electrones fuera de sus orbites y que se mide en voltios. Puede referirse como "voltaje."

Emergency flasher: The circuit breaker used to make the front and rear turn signals flash together. The correct name is hazard light flasher.

Luz de emergencia: El disyuntor de circuito que permite funcionar las señales indicadoras delanteras y traseras a la misma vez. Su nombre correcto es lámpara destello de emergencia.

End play: The amount of movement a shaft can make lengthwise within its mountings.

Juego final del cigüeñal: La cantidad del movimiento longitudinal que puede hacer un eje en su montadura.

Evaporative: The process of a liquid changing into a vapor.

Evaporativo: El proceso de un líquido cambiando al vapor.

Exhaust gas recirculating: The process and devices used to route spent exhaust gases back through the engine to reduce combustion chamber temperature that prevents the formation of nitrogen oxides.

Recirculación del gas de escape: El proceso y dispositivo que se usa para desviar el gas de escape por el motor para disminuir la temperatura de la cámara de combustión que previene la formación de los óxidos nítricos.

Exhaust pipe: The device that routes the exhaust gases from the exhaust manifold on the cylinder head into the exhaust systems.

Tubo de escape: El dispositivo que desvía los gases desde el escape del multiple de escape hacia el sistema de escape.

Exhaust stroke: The stroke of a piston engine when the exhaust valve is opened, the intake is closed, and the upward movement of the piston forces exhaust gases into the exhaust manifold.

Carrera de escape: La carrera de un pistón del motor cuando esta abierta la válvula del escape, la admisión esta abierta, y el movimiento hacia arriba del pistón admite los gases del escape dentro del multiple de escape.

Expansion valve: A valve that opens and closes based on either heat or pressure. Commonly used to meter refrigerant in a climate control system.

Válvula de expansión: Una válvula que se abre o se cierre según o el calor o la presión. Suele usarse para medir al refrigerante en un sistema de climatizaje controllado.

Fasteners: Small threaded and nonthreaded parts that hold automotive components together.

Aaguradorcs. Pequeñas piezas fileteadas y no fileteadas que sujetan componentes automotrices.

Feeler gauge: A measuring tool using precise thickness blades to measure the space between parts.

Calibrador de espesores: Lámina metálica o cuchilla acabada con precisión de acuerdo al espesor utilizada para medir el espacio entre dos piezas.

File: A hardened steel tool with rows of cutting edges used to remove metal for polishing, smoothing, or shaping.

Lima: Herramienta de acero templado compuesta de una barra de filos cortantes que se utiliza para desgastar, pulir o alisar el metal.

Fixed caliper: A brake caliper that is bolted stationarily and uses two or more pistons to apply the brake pads against the rotor.

Calibre fijo: Un calibre de freno estacionario empernado que usa dos o más pistones para aplicar las balatas del freno en el rotor.

Flasher: A repeating-action circuit breaker that opens and closes to make the turn signal lamps flash on and off.

Luz de destello: Un disyuntor de circuito de acción repetitiva que abre y cierre para hacer alumbrarse y apagarse las lámparas indicadoras de viraje.

Foot-pounds: A measurement of twisting force being used to accomplish a task.

Pies libras: Una medida de la fuerza giratoria que se usa para cumplir una tarea.

Fractions: The expression used to describe parts of a whole number. One-half of a whole is expressed as $1/2$.

Fracciones: La expresión que se usa para describir partes de un número total. La mitad de una totalidad se expresa como $1/2$.

Franchise: An independent business operating under the regulations and policies of a larger business. A dealership is a franchise-type operation.

Sucursal: Un negocio independiente que opera bajo las regulaciones y las policias de un negocio más grande. Un distribuidor es una operación de tipo sucursal.

Fuse: A circuit protection device that has a metal strip that melts and opens the circuit during an overload.

Fusible: Dispositivo de protección del circuito con una banda de metal que se funde y abre el circuito en caso de una sobrecarga.

General manager: The person exercising control over the day-to-day operation of a business. May or may not be the owner.

Administrador general: La persona que se encarga de controlar la operación cuotidiana del negocio. Puede ser el dueño o no.

Global Positioning Satellite: One satellite in a system of satellites that are used to determine the position of an object or a point on earth. Used extensively by aircraft, ships, and military.

Satélite de posición mundial: Un satélite en un sistema de satélites que se usan para determinar la posición de un objeto o de un punto en la tierra. Se usa extensivamente por los aviones, los buques, y el militar.

Grease: Term commonly applied to a thick lubricant.

Grasa: Un término que se aplica comunmente a un lubricante espeso.

Grounding switch: A switch placed on the negative side of a circuit. Used to connect the circuit to ground, thereby completing the circuit.

Interruptor a tierra: Un interruptor colocado en el lado negativo de un circuito. Se usa para conectar el circuito a la terra asi completando el circuito.

Hacksaw: A type of saw made to cut metal.

Sierra para metales: Tipo de sierra fabricada especialmente para cortar metal.

Hazardous waste: Any waste that is directly hazardous to health or environment. Most household waste is not hazardous while the majority of industrial waste is.

Desechos tóxicos: Cualquier desecho que es directamente tóxico al salud o al medio ambiente. La mayoría de los desechos caseros no son tóxicos mientras que la mayoría de los desechos industriales lo son.

Heads-up-display: A method of projecting information onto the windshield in the driver's line of sight.

Presentación proyectado en pantalla: Un metodo de proyectar la información en la parabrisas en la línea de vista del conductor.

Hydraulics: The theory, study, and application of using a liquid to transfer or create force.

Hidráulica: La teoría, el estudio, y la aplicación de usar un líquido para transferir o crear la fuerza.

Hydro-boost: A method using the power-steering pump to provide brake boost to the master cylinder.

Asistencia hidráulica: Un metodo que utilisa la bomba de dirección para proveer una asistencia del frenado en el cilindro maestro.

Impact wrench: An air-operated wrench that impacts as well as drives a bolt or nut.

Llave de impacto: Llave accionada por aire que golpea además de accionar un perno o una tuerca.

Independent suspension: A suspension where each wheel can act and react to road conditions without affecting the other wheels.

Suspensión independiente: Una suspensión en la cual cada rueda pueda actuar y reaccionar a las condiciones del camino sin afectar a las otras ruedas.

Index line: A line on a measuring device where all measures are read.

Línea graduada: Una línea en un dispositivo de medición en la cual todas las medidas se leen.

Intake stroke: The downward movement of the piston when the exhaust valve is closed and the intake valve is closed.

Carrera de admisión: El movimiento hacia abajo del piston cuando la válvula de escape esta cerrada y la vávula de admisión esta cerrada.

Interference: The actions of one part that conflicts or interferes with the movement of another part. A type of fit between two components commonly referred to as "press fitted."

Interferencia: Las acciones de una parte que es en conflicto o que interfiere con el movimiento de otra parte. Un tipo de ajuste entre dos componentes que re refiere comunmente como "ajuste de presión."

Investment: The amount of time, material, human resources, and funding needed to establish and maintain a business.

Inversión: La cantidad del tiempo, material, recursos humanos, y los fondos requeridos para establecer y mantener un negocio.

Isolation/dump solenoid: A common term applied to the pressure controller in rear-wheel, anti-lock brake systems.

Solenoide isolación/descarga: Un término común dado al controlador de presion en los sistemas de frenado anti-bloqueo de las ruedas traseras.

Jack stands: Strong supports placed under a car that has been lifted by a floor jack.

Soportes de gato: Soportes fuertes colocados debajo de un vehículo que ha sido levantado con el gato de pie.

Key: A small, hardened piece of metal used with a gear or pulley to lock it to a shaft.

Chaveta: Trozo pequeño de metal templado utilizado con un engranaje o con una polea para fijarlo a un árbol.

Kinetic: Refers to energy being applied.

Kinético: Refiere a la energía que se aplica.

Leaf spring: A type of suspension spring that uses layers of spring steel stacked on each other for spring action.

Muelle de lámina: Tipo de muelle de suspensión que utiliza capas de acero templado montadas la una sobre la otra para lograr un movimiento amortiguador.

Linear: A straight line or a measurement of a straight line or space.

Linear: Una línea recta o una medida de una línea o un espacio recto.

Load-sensing proportioning valve: A brake valve that can change the amount of braking done on the rear wheel based on the load being carried.

Válvula proporcionador sensor de cargas: Una válvula de frenos que puede cambiar la cantidad del frenado por la rueda trasera basado en la carga que se le aplica.

MacPherson strut suspension: A suspension arrangement that uses a strut assembly in place of a shock absorber and top control arm.

Suspensión de montantes MacPherson: Distribución de la suspensión que utiliza un conjunto de montantes en vez de un amortiguador y un brazo de mando superior.

MACS: The Mobile Air Conditioning Society offers technical certification for climate control technicians.

MACS: La Sociedad de Acondicionamiento de Aire Móvil ofrece la certificación técnica para los técnicos de control climático.

Magnetic field: The magnetic force created by electrical current moving through a conductor.

Campo magnético: La fuerza magnética creado por un corriente eléctrico moviendose por un conductor.

Malfunction Indicator Light: A dash mounted light that alerts the driver to a malfunction in the electronic controls for the engine and transmission.

Lámpara indicador de fallos: Una lámpara montada en el tablero de instrumentos que informa al conductor que existe un fallo en los controles eléctricos del motor y la transmisión.

Master cylinder: A device that operates when the driver pushes on the brake pedal to force fluid under pressure to activate wheel brake units.

Cilindro maestro: Dispositivo que funciona cuando el conductor pisa el pedal de freno para hacer que el fluido bajo presión accione las unidades del freno de la rueda.

Master switch: A switch that can control several circuits singularly or together.

Interruptor maestro: Un interruptor que puede controlar a varios circuitos individualmente o juntos.

Maxi-boost: A brake booster using an independent pump and motor to create the boost pressure.

Asistencia máximo: Una asistencia de freno que utiliza una bomba independiente y un motor para crear la presión de asistencia.

Mechanic creeper: A flat, low-riding platform on which a technician can lay and work under a vehicle.

Tabla rodante del mecánico: Una plataforma móvil y bajo en la cual un tecnico puede tenderse para trabajar debajos de un vehículo.

Mechanical advantage: The increase in force between the input and output of a hydraulic circuit.

Ventaja mecánica: El incremento en fuerza entre la entrada y la salida de un circuito hidráulico.

Meshed: The term applied to two or more gears whose teeth work against and with each other.

Endentados: El término aplicado a dos o más engranajes cuyos dientes trabajan contra el uno y otro o juntos.

Meter: A gauge that displays measurements such as a water or gas metered in the home. The process of regulating liquids or vapor.

Medidor: Un medidor que muestra las medidas tal como uno de agua o gas en la casa. El proceso de regular los líquidos o vapores.

Metering valve: A valve used to delay the activation of front disc brakes until rear drum brakes are activated.

Válvula de medida: Válvula utilizada para demorar el accionamiento de los frenos de disco delanteros hasta que se accionen los frenos de tambor traseros.

Muffler: A term applied to a device that reduces noise. A component of the exhaust system.

Amortiguador: Un término aplicado al dispositivo que disminuya el ruido. Un componente del sistema de escape.

Multifunction lever: A lever mounted on the steering column that has switches for more than one electrical system.

Palanca multifunción: Una palanca montada en la columna de dirección que tiene interruptores que controlan más que un sistema eléctrico.

NASCAR: A race series sanctioning organization.

NASCAR: Una serie de carreras sancionada por la organización.

NATEF: The National Automotive Technician Education Foundation certifies automotive training program. A branch of ASE.

NATEF: La Fundación de Educación de Técnicos Automotríz certifica una programa de entrenamiento automotríz. Una parte de ASE.

NATEF key: A copyrighted symbol for NATEF.

Llave NATEF: El símbolo del NATEF protegido por derecho de autor.

Needle bearing: A bearing made of small rollers used in drive shafts.

Cojinete de agujas: Un cojinete hecho con pequeñas rodillas que se usa en los ejes propulsores.

Needle nose pliers: Pliers with long, slim jaws for gripping small objects.

Alicates de puntas de aguja: Alicates que tienen tenacillas largas y delgadas para sujetar objetos pequeños.

Neoprene: A synthetic rubber produced by the polymerization of chloroprene that is highly resistant to oil, heat, light, and oxidation.

Neoprene: Un caucho sintético producido por la polimerización del cloroprene que es altamente resistente al aceite, al calor, a la luz, y a la oxidación.

Nitrogen oxides: A harmful emission chemical produced when the combustion chamber temperature exceeds 2500 degrees Fahrenheit.

Óxidos nítricos: Una emisión nociva química que se produce cuando la cámara de combustión supera 2500 grados Fahrenheit.

Non-independent suspension: A suspension where the suspension and wheels are mounted to a solid axle. The actions of one wheel affect the other wheel.

Suspensión no independiente: Una suspensió en la cual la suspensión y las ruedas se montan en un eje sólido. Las acciones de una rueda afectan a la otra rueda.

Non-inference: A term applied to two or more working, mated components that do not interfere with the movements or actions of each other. An engine that has sufficient clearance in the combustion chamber so the piston will not hit the valve.

Non interferencia: Un término aplicado a dos o más componentes que no interferan con los movimientos o los acciones de cada uno. Un motor que tiene bastante juego en la cámara de combustión para que el pistón no pegará a la válvula.

Nuts: Fasteners with an internal thread that are used with bolts and studs.

Tuercas: Aseguradores con un orificio en que hay labrada una hélice, hecha para que ajuste en ella la de un perno o un espárrago.

Offset screwdriver: A screwdriver with blades arranged at an offset to allow driving screws in small spaces.

Destornillador angular: Destornillador con las cuchillas arregladas en ángulo para poder darles vuelta a los tornillos en espacios muy limitados.

Open-end wrench: A wrench with an opening at the end that can slip onto the bolt or nut.

Llave española: Llave con una apertura en el extremo que puede introducirse en el perno o la tuerca.

Outside micrometer: A measuring tool designed to make very precise measurements of the outside of a part in either U.S. (English) or metric units.

Micrómetro exterior: Herramienta para medidas diseñada con el fin de llevar a cabo medidas sumamente precisas del exterior de una pieza en el Sistema Imperial Británico (EEUU) o el métrico.

Overhead cam: An engine that has one camshaft mounted in each head. A type of engine valve train.

Árbol de levas sobre cabeza: Un motor que tiene un árbol de levas montado en cada cabeza. Un tipo de tren de válvulas del motor.

Overhead valve: A type of engine valve train where the camshaft is in the block and the valves are in the cylinder head.

Válvula sobre cabeza: Un tipo de tren de válvulas de motor cuyo árbol de levas esta en el bloque y las válvulas estan en la cabeza del cilindro.

Ozone layer: An atmospheric layer about 7 to 13 miles above Earth. Provides protection against the sun's heat and radiation.

Capa de ozono: Una capa atmosférica de una distancia aproximadamente 7 a 13 millas sobre la tierra. Provee protección contra el calor y la radiación del sol.

Palladium: A catalyst element used in catalytic converters. A very expensive metal.

Paladio: Un elemento catalítico usado en los convertidores catalíticos. Un metal muy caro.

Parallel circuit: An electrical current having more than one path to ground.

Circuito paralelo: Un corriente elétrico que tiene más que un rumbo a tierra.

Parallelogram: A leaning, rectangular figure. A type of automotive steering system.

Paralelogramo: Una figura rectangular inclinada. Un tipo de sistema de dirección automotivo.

Payload: The amount of weight or load a vehicle can transport within its limits.

Carga máxima: La cantidad de peso o de carga que puede transportar un vehículo en sus limites.

Pin: A small, round length of metal that is driven into two parts to hold them together.

Pasador: Pequeña pieza metálica y redonda que se inserta de manera forzada en dos piezas para sujetarlas.

Pin punch: A tool used with a hammer to drive out pins from automotive parts.

Punzón de pasadores: Herramienta utilizada con un martillo para remover pasadores de piezas automotrices.

Pitch gauge: A tool with toothed blades used to match with threads for identification.

Calibrador de paso: Herramienta con cuchillas dentadas que se aparean con filetes a fin de identificar el tamaño de los mismos.

Planetary gear set: A set of two gears mounted within a third gear. The gears are ring, sun, and planetary.

Juego de engranajes planetarios: Un juego de dos engranajes montados dentro de un tercer engrenaje. Los engranajes son de anillo, sol, y planetarios.

Platinum: A catalyst in a catalytic converter.

Platino: Una catalista en un convertidor catalítico.

Pneumatic: Driven or operated by air.

Neumático: Impulsado o operado por aire.

Polarity: The type of electrical charge either positive or negative.

Polaridad: El tipo de carga eléctrica sea positiva o negativa.

Positive crankcase ventilation: A system used to vent the pressures in the crankcase during engine operation. An emission control system. Commonly referred to as PCV.

Ventilación positiva del cárter: Un sistema que se usa para ventilar las presiones dentro del cárter durante la operación del motor. Un sistem de control de emisiones. Suele referirse como PCV.

Pound-foot: The amount of force needed to move a one-pound mass one foot.

Libra pie: La cantidad de fuerza requerida para mover una masa de una libra a un pie.

Power piston: A piston that produces or increases power or force. Commonly refers to the boost piston in brake boosters.

Piston de potencia: Un piston que produce o incrementa la potencia o la fuerza. Suele referirse como el piston de asistencia en los frenos con asistencia.

Power stroke: The stroke of an engine that delivers power to the crankshaft.

Carrera de potencia: La carrera de un motor que provee la fuerza al cigüeñal.

PowerMaster: A type of brake booster system.

Powermaster: Un tipo de sistema de frenos con asistencia.

Pressure: The force per square inch on a surface. Commonly used in hydraulic systems to determine the amount of force a pressurized liquid exerts against the all of the interior walls of a sealed hydraulic system.

Presión: La fuerza por pulgada cuadrada en una superficie. Comunmente usado en los sistemas hidráulicos para determinar la cantidad de fuerza con que oprime un líquido contra todos los muros interiores de un sistema hidráulico sellado.

Pressure dressing: A first-aid bandage used to apply force to a wound to slow the bleeding.

Vendaje con presión: Un vendaje de primera cura que se usa para aplicar la fuerza en una herida para controlar las hemorragias..

Preventive maintenance: A system of maintenance services that includes routine services and preventive actions to reduce vehicle break down.

Mantenimiento preventativo: Un sistema de servicios de mantenimiento que incluye los servicios rutinas y las acciones preventivas para disminuir las averías de los vehículos.

Primary piston: The piston in the brake's master cylinder that is closest to the driver; the master cylinder's rear piston.

Piston primaria: El piston en el cilindro maestro del freno que queda más cerca al conductor; el piston trasero del cilindro maestro.

Printed circuit board: An older type electronic board that has been replaced with integrated circuits.

Placa de circuito impreso: Un tipo de tabla electrónica anticuado que se ha reemplazado con los circuitos íntegros.

Proportioning valve: A valve used to lower the hydraulic pressure going to the rear drum brakes to ensure balanced braking.

Válvula dosificadora: Válvula utilizada para disminuir la presión hidráulica que se transmite a los frenos de tambor traseros con el propósito de asegurar un frenado equilibrado.

Rack and pinion steering gear: A type of steering gear that uses a toothed rack driven by a pinion gear.

Engranaje de dirección por piñón y cremallera: Tipo de engranaje de dirección que utiliza una cremallera dentada accionada por un piñón.

Radial ply tire: A tire constructed with plies that run at right angles to the circumference of the tire.

Llanta con la estria radial: Llanta fabricada con las estrias arregladas en ángulos rectos con respecto de la circunferencia de la llanta.

Ratchet handle reamer: A reamer fitted with a ratchet handle.

Escariador de llave de trinquete: Un escariador que tiene una palanca de trinquete.

Rear wheel antilock: An antilock brake system first required on 1988 model light truck and vans. It was the first mass-produced, antilock system sold on U.S. domestic vehicle.

Antibloqueo de la rueda trasera: Un sistema de frenos de antibloqueo requerido primero en los modelos de camiones ligeros y vans de 1988. Fue el primer sistema de antibloqueo producido en masa y vendido en un vehículo domestico de los EU.

Receiver/dryer: A climate-control component used to filter and clean the refrigerant.

Recibidor/ desecador: Un componente de control climatizaje que se usa para filtrar y limpiar el refrigerante.

Rectifier: A component in the alternator used to change alternating current to direct current.

Rectificador: Un componente en el alternador que se usa para cambiar el corriente alterno al corriente directo.

Reservoir: A storage tank for liquid with automatic exits for feeding the liquid into a hydraulic device.

Depósito: Un tanque de almacen para un líquido con las salidas automáticas para suministrar el líquido a un dispositivo hidráulico.

Resistance: The opposition to current flow in an electrical circuit.

Resistencia: Oposición que presenta un conductor al paso de la corriente eléctrica en un circuito eléctrico.

Resonator: A sound-reducing device usually placed near the exit end of the exhaust system.

Resonador: Un dispositivo para disminuir el ruido que suele ser colocado cerca de la extremidad de salida del sistema del escape.

Rhodium: An element used as a catalyst in a catalytic converter.

Rodio: Un elemento usado como catalizador en un convertidor catalítico.

Right-to-Know: A series of laws and regulations that require the employer to notified employees of safety hazards in the work place.

Derecho a saber: Una serie de leyes y regulaciones que requieren que el patrón notifica los empleados de los peligros a la seguridad en el establecimiento de trabajo.

Rivet: A soft, metal pin with a head at one end that is used to hold two parts together.

Remache: Chaveta blanda de metal con una cabeza a un extremo utilizada para sujetar dos piezas.

Rod bearing: The bearing between the piston's connecting rod and the crankshaft. An insert-type bearing.

Cojinete de la biela: El cojinete entre la biela del piston y el cigüeñal. Un cojinete de tipo inserción.

Roller bearing: Slender, machined rollers trapped between cages or separators. Used on shafts and wheel assemblies. *See* ball bearing.

Cojinete de rodillos: Los rodillos acabado por máquina atrapados en jaulas o en separadores. Se usan en los ejes y las ensambladuras de ruedas. *Ver tambien* cojinete de bolas (ball bearing).

Room temperature vulcanizing: A chemical sealant that cures at room temperature. Commonly known as RTV.

Vulcanización a temperatura ambiente: Un sellante químico que cura en temperaturas de ambiente. Se conoce comunmente como RTV.

Rotary: A component that functions in a rotary motion. An engine that uses rotary actions instead of pistons to produce and transmit power. Mazda's rotary engine.

Rotario: Un componente que funciona en un movimiento rotativo. Un motor que usa las acciones rotarias en vez de los pistones para producir y transferir la potencia. El motor rotario de Mazda.

Rotor: The rotating component of a disc brake that is attached to the wheel hub and is stopped to stop the car.

Rotor: Componente giratorio de un freno de disco que se conecta al cubo y que es retenido para detener la marcha del vehículo.

Runout: The amount of wobble in a disc, wheel, or shaft.

Corrimiento: La cantidad de oscilación en un disco, una rueda, o un eje.

Scan tool: An electronic tool used to communicate with the vehicle's computer to retrieve DTCs and operating data.

Herramienta exploradora: Una herramienta electrónica que se usa para comunicar con la computadora del vehículo para recobrar los DTC y los datos de operación.

Scissor lift: An automotive lift that functions similar to the actions of a pair of scissors.

Izador tijera: Un izador automotríz que funciona parecido al acción de un par de tijeras.

Screw: A fastener that fits into a threaded hole.

Tornillo: Asegurador que se inserta dentro de un agujero filetado.

Screwdriver: A tool used to turn screws.

Destornillador: Una herramienta que se usa para dar vuelta a los tornillos.

Secondary Air Injection: An older emission control system that injected fresh air into the exhaust system to promote catalyst in the converter.

Inyección de aire secundario: Un sistema de control de emisiones anticuado que inyectaba aire fresca al sistema de escape para promover un catalizador en el convertidor.

Secondary circuit: The electrical circuit between the ignition coil and the spark plug.

Circuito secundario: El circuito eléctrico entre la bobina del encendido y la bujía.

Secondary piston: The front piston in the master cylinder. *See* primary piston.

Piston secundario: El piston delantero en el cilindro maestro. *Ver tambien* piston primario (primary piston).

Series circuit: An electrical circuit with only one path to for current flow.

Circuito en serie: Un circuito eléctrico con un sólo rumbo para el flujo del corriente.

Series-parallel: A series circuit that controls or regulates a parallel circuit.

Serie en paralelo: Un circuito de serie que controla o regula un circuito paralelo.

Service manager: The person responsible for controlling and managing the service department of an automotive business.

Gerente de servicio: La person que se encarga del control y la administración del departamento de servicio de un negocio automotríz.

Service manual: A paper book or computerized database with technical information and specifications for vehicle repairs.

Manual de servicio: Un libro de papel o un archivo computerizado con la información técnica y las especificaciones para las reparaciones en los vehículos.

Service writer: A person who meets the customer and initiates the repair order. Reports to the service manager.

Representante de servicio: Una persona que recibe a la clientela y inicia el orden de reparación. Da información al gerente de servicio.

Setback: The angle formed between two wheels on the same axle when the wheels are not aligned with each other.

Retroceso: El ángulo formado entre dos ruedas en el mismo eje cuando las ruedas no son alineadas la una con la otra.

Shank: The part of a bolt between the end of the threaded end and the bottom of the bolt head or the non-cutting portion of a drill bit.

Vástago: La parte de un perno entre la extremidad de la parte enroscada y la extremidad de la cabeza del perno o la parte de una broca de taladro que no corta.

Shoe web: The metal frame of a brake shoe where the friction material is bonded or riveted.

Alma de la zapata: El armazón metal de un zapato (balata) de freno en donde esta pegada o remachada la material de fricción.

Single-post lift: An automotive lift with one lift cylinder, which is usually mounted in the floor.

Izador de un poste: Un izador con un cilíndro de izar, que típicamente esta montado en el piso.

Sliding caliper: A disc brake caliper that has all of the applied pistons on one side (usually inside the rotor) causing the whole caliper to slide away from the brake rotor which applies the outer pad.

Calibre deslizante: Un calibre de freno disco que tiene todos los pistones aplicados en un lado (típicamente dentro del rotor) causando que todo el rotor desliza del rotor de frenos que aplica la balata exterior.

Small hole gauge: A measuring tool used with an outside micrometer to measure the inside of small holes.

Calibrador de agujero pequeño: Herramienta para medidas utilizada con un micrómetro exterior para medir el interior de agujeros pequeños.

Snap ring: An internal or external expanding ring that fits in a groove and works to hold parts together.

Anillo de muelle: Anillo interior o exterior expansible que se inserta dentro de una ranura para sujetar piezas.

Snap ring pliers: Pliers that fit into internal- or external-type snap rings and allow them to be removed safely.

Alicates de anillo de muelle: Alicates que se insertan dentro de anillos de muelle interiores o exteriores con el propósito de permitir una remoción segura de los anillos.

Snapshot: Short electronic data records (memory) stored in a scan tool and collected during a road test for playback later. May be called a "movie."

Instantáneo: Unos datos cortos eléctronicos (memoria) almacenado en un detector y colectados durante un ensayo sobre carretera para recobrarse despues. Pude llamarse una "pelicula."

Socket drivers, handles, and attachments: Devices that fit into the square hole in the socket wrench and are used to drive socket wrenches.

Empujadores, palancas, y equipo adicional: Dispositivos insertados dentro del agujero cuadrado de la llave de cubo que se utiliza para activar la misma.

Socket wrenches: Wrenches that fit all the way around a bolt or nut and can be detached from a handle.

Llaves de cubo: Llaves que cubren completamente un perno o una tuerca y que pueden removerse de la palanca.

Solder: A tin-lead or copper-zinc metal alloy that melts easily and is used for making electrical connections.

Soldadura: Una aleación metálica de estaño-plomo o cobre-zinc que se derrite fácilmente y se usa para formar las coneccions eléctricas.

Speed sensor: An electronic sensor that signals the PCM of wheel or vehicle speed. Used with antilock brake systems and engine/transmission controls.

Sensor de velocidad: Un sensor electrónico que señala al PCM de la rueda de la velocidad del vehículo. Se usa con los sistemas de frenos antibloqueo y los controles de motor/transmisiones.

Spindle: A short, machined, high-strength shaft for supporting a wheel assembly. Used on non-driving wheels.

Husillo: Un eje, corto, maquinado de acero de alta tensión para soportar una ensambladura de rueda. Se usa en las ruedas que no impulsan.

Splines: Internal or external teeth cut in a part used to hold it in place in another part.

Lengüetas: Dientes internos o externos labrados en una pieza para que ésta se mantenga en su lugar dentro de otra pieza.

Split ball: A type of small hole gauge used with a micrometer to measure small bores.

Bola hendida: Un tipo de medidor de hoyos pequeños usado con un mircrómetro para medir los pequeños huecos.

Square-cut O-ring: A specially-cut O-ring commonly used as a seal for disc brake caliper and caliper piston.

Sello en O cortado cuadrado: Un sello en O cortado especialmente que se usa comunmente como un sello para un calibre de frenos de disco y piston de calibre.

Start: The act of the engine beginning to run under its own power. Takes place during and after cranking.

Marcha: El acto del motor en funcionar independentemente. Ocurre despues del arranque.

Starter punch: A punch with a taper on the end that allows the starting of pin removal.

Punzón iniciador: Punzón dotado de un extremo afilado para facilitar la remoción de un pasador.

Static balance: Balancing a tire using only two planes of the wheel assembly instead of four quarters. *See* dynamic balancing.

Equilibrio estático: Equilibrar un neumático usado dos planos de la ensambladura de la rueda en vez de los cuatro cuadros. *Ver tambien* equilibrio dinámico (dynamic balancing).

Stepped feeler gauge: A type of feeler gauge where the end is .002 inch thinner than the remainder of the blade. May be referred to as a "go/nogo" gauge.

Galga palpadora de pasos: Un tipo de medidor de hoja en el cual la extremidad mide .002 de una pulgada más delgada del resto de la hoja. Puede referirse como un medidor "va/no va."

Stud: A fastener that has threads at both ends and is used with a threaded hole and a nut.

Espárrago: Asegurador provisto de filetes a ambos extremos que se utiliza con un agujero fileteado y una tuerca.

Supplemental restraint system: The official designation for the air bag system.

Sistema restricción suplementaria: La designación oficial del sistema de bolsa de aire.

Supplier: A maker and/or seller of automotive parts. May be a local parts store or national manufacturer. *See* vendor.

Abastecedor: El fabricador y/o vendedor de las partes automotivos. Puede ser una tienda de refacciones local o un fabricante nacional. *Ver tambien* vendedor (vendor).

Tail pipe: The section of the exhaust extending from the rear of the muffler to the rear of the vehicle.

Tubo de escape: La sección del escape que se extiende de la parte trasera del amortiguador hacia la parte trasera del vehículo.

Tap: A tool used to cut external threads.

Aterraje: Herramienta utilizada para cortar filetes externos.

Tapered roller bearing: A roller bearing where the rollers are laid at an angle to the centerline of the bearing, thereby creating a bearing with the hole on one side larger than the hole on the opposite side.

Rodamiento de cilindros cónico: Un cojinete de rodillos en el cual los rodillos se colocan a un ángulo al línea central del cojinete creando un cojinete con el hoyo de un lado más grande que el hoyo del lado opuesto.

Taps: A set of taps are used to make or clean internal threads in a fastener.

Terrajas: Un conjunto de terrajas que se usan para hacer o limpiar las roscas interiores en un asegurador.

Telescopic gauge: A tool used along with a micrometer to measure the inside of holes or bores.

Calibrador telescópico: Herramienta utilizada con un micrómetro para medir el interior de agujeros o calibres.

Terminal connector: An electrical connector installed on the end of a wire to connect it to an electrical component.

Conectador de borne: Conectador eléctrico instalado en el extremo de un alambre para conectarlo a un componente eléctrico.

Thermostatic air cleaner: An air cleaner that uses a temperature sensor and vacuum to direct warm air into the filter.

Limpiador de aire termostático: Un limpiador de aire que usa un sensor de temperatura y un vacío para dirigir el aire cálido hacia el filtro.

Thimble: The rotating end of a micrometer used to move the spindle to and from the work piece.

Manguito: La extremidad rotativa de un micrómetro usado para mudar el husillo hacia o del pieza de trabajo.

Threaded fasteners: Fasteners that use spirals called "threads" to hold automotive parts together.

Aseguradores fileteados: Aseguradores que utilizan espirales llamados "filetes" para sujetar piezas automotrices.

Thread restoring file: A special file used to repair or straighten threads with minor damage. *See* thread chasers.

Lija restaurador de roscas: Una lija especial que se usa para reparar o enderezar las roscas con daños mínimos. *Ver tambien* peine de roscar (thread chasers).

Tire bead: The part of the tire that fits and seals against the wheel.

Costura del neumático: La parte de un neumático que queda y sella en la rueda.

Toe-in: When the distance between the leading edges of the tires on the same axle is less than the trailing edges of the same tires. *See* toe-out.

Convergencia: Cuando la distancia entre los bordes delanteros de los neumáticos del mismo eje es menos que los bordes traseros de los mismos neumáticos. *Ver tambien* divergencia (toe-out).

Toe-out: When the distance between the leading edges of the tires on the same axle is more than the trailing edges of the same tires. *See* toe-in.

Divergencia: Cuando la distancia entre los bordes delanteros de los neumáticos del mismo eje es más que los bordes traseros de los mismos neumáticos. *Ver tambien* convergencia (toe-in).

Toe-out-on-turn: The difference between the leading edges of the steering tires and the trailing edges of the same tires. Is usually greater than toe. Cannot be corrected using alignment methods.

Divergencia en vuelta: La diferencia entre los bordes delanteros de las ruedas de dirección y los bordes traseros de los mismos neumáticos. Suele ser mayor que convergencia/divergencia. No puede ser corregido usando los metodos de alineamiento.

Tolerances: The amount of allowable error from desired specifications.

Tolerancia: La cantidad de error que se permite en los especificaciones deseados.

Tone ring: A toothed ring placed around a shaft or inside a rotor that causes a sensor to pulse a signal each time a tooth passes the sensor.

Anillo de tono: Un anillo con dientes que se coloca alrededor de un eje o dentro de un rotor que causa que un sensor impulsa una señal cada vez que ese diente pasa por el sensor.

Torque stick: A limited torque tool used with an impact wrench to tighten a fastener to a specified torque.

Pala de torsión: Una herramienta de torsión limitada que se usa con una llave de impacto para apretar un asegurador a una torsión especifica.

Torsion bar: A type of suspension spring that uses the twisting of a bar attached to the frame and suspension to control suspension movement.

Barra de torsión: Tipo de muelle de suspensión que utiliza el movimiento de torsión de una barra fijada al armazón y a la suspensión para controlar el movimiento de ésta.

Torque wrench: A wrench designed to tighten bolts or nuts to a certain tightness or torque.

Llave dinamométrica: Llave diseñada para apretar pernos o tuercas a una tensión o torsión específicas.

Traction control: An electronic system that controls power to a spinning wheel. Shares some components with the antilock brake systems.

Control de tracción: Un sistema electrónico que controla el poder a una rueda giratoria. Comparte algunos componentes con los sistemas de frenos antibloqueantes.

Transponder: A small, electronic device that will transmit its identification code to a receiver when commanded to so by radar or radio signal.

Transpondedor: Un dispositivo electrónico pequeño que transmite su código de identificación a un recibidor bajo la comanda de un señal de radar o de radio.

Tread: The part of the tire that contacts the road and provides friction for vehicle operation.

Banda de rodadura: La parte del neumático que toca el camino y provee la fricción para la operación del vehículo.

Trigger: A specific electronic or mechanical command that will cause a specific action to take place.

Gatillo: Una comanda específica electrónica o mecánica que causará ocurir una acción especifica.

Twist drill: A drill bit with cutting edges and flukes running in a twisting pattern up the length of the bit's shank.

Barrenadora de columna: Una brocade taladro cuyos bordes cortantes y las colas estan puestos para que dan la vuelta alredeor de la longitud del váatago del taladro.

Ultraviolet: Radiation rays whose wavelengths are outside of a human's vision range.

Ultra violeta: Los rayos de radiation cuyos ondas son afuera del rango de la vista del ser humano.

Unibody: A vehicle design in which the body doubles as most of the vehicle's frame. Uses sub-frames to support the engine and driveline components.

Monocasco: Un diseño del vehículo en el cual la carrocería tambien funciona como el armazón del vehículo. Usa los bastidores auxiliares para soportar los componentes del motor y del eje propulsor.

Uniform tire quality grading: A system designating the quality of a tire's traction, tread wear, and temperature resistance.

Evaluación uniforme de la calidad de una llanta: Sistema que indica la calidad de la tracción de una llanta, el desgaste de la huella, y la resistencia a la temperatura.

Unibody: A body design that uses sheet metal as the support structure and not a separate frame.

Unibody: Diseño de carrocería cuya estructura de soporte es una hoja de lata y no un armazón separado.

Unified system: A system of thread classification based on the U.S. (English) system of measurement.

Sistema unificado: Sistema de clasificación de filetes basado en el Sistema Imperial Británico (EEUU) de medida.

Valve train: A group of engine components that open and close the engine valves during the four-stroke cycle.

Tren de válvulas: Conjunto de componentes del motor que abre y cierre las válvulas del motor durante el ciclo de cuatro tiempos.

Vendors: A direct seller to the public or businesses or both. Normally not a manufacturer. *See* supplier.

Vendedores: Un vendedor directo al público o a los negocios o los dos. No suele ser fabricante. *Ver tambien* abastecedor (supplier).

Vernier scale: The 1/10,000ths markings on a micrometer.

Gama Vernier: Las calibraciones de 1/10,000 en un micrómetro.

Vise grip pliers: Pliers that can be locked on a part to hold it tightly like a hand-held vise.

Tenazas de sujeción: Alicates que pueden enclavarse a una pieza para sujetarla con firmeza como un tornillo operado manualmente.

Volatile: The flammability of an element.

Volátil: La inflamabilidad de un elemento.

Voltage: The amount of electrical pressure pushing current through an electrical circuit.

Tensión: Cantidad de presión eléctrica que empuja la corriente a través de un circuito eléctrico.

Washers: Fasteners used with bolts, screws, studs, and nuts to protect machined surfaces and prevent nut loosening.

Arandelas: Aseguradores utilizados con pernos, tornillos, espárragos, y tuercas para proteger superficies maquinadas y prevenir que las tuercas se aflojen.

Weather-tight: An electrical connector with seals to prevent moisture or dirt from entering the connector.

A prueba de agua: Un conector eléctrico con los sellos para prevenir que entra en el conector la humedad o el polvo.

Wheel cylinder: The hydraulic part on the drum brake that activates the brake shoes when the driver pushes the brake pedal.

Cilindro de rueda: Pieza hidráulica en el freno de tambor que acciona las zapatas de freno cuando el conductor pisa el pedal de freno.

Worm gear: A spiral gear on a straight shaft that drives another gear. Usually found in recirculating ball and nut steering gears.

Engranaje sinfín: Un engrenaje espiral que propulsa otro engrenaje. Se encuentran comunmente en los engrenajes de dirección tipo bola y tuerca.

Yoke: The part of the drive shaft where one side of the u-joint fits.

Brida: La parte de un eje propulsor en donde queda un lado de la junta.

INDEX